MATHEMATICAL
MODELING

No. 6

Edited by
William F. Lucas, Claremont Graduate School
Maynard Thompson, Indiana University

Thomas L. Vincent Alistair I. Mees
Leslie S. Jennings
Editors

Dynamics of Complex Interconnected Biological Systems

With 91 Illustrations

Birkhäuser
Boston · Basel · Berlin

Thomas L. Vincent
Department of Aerospace and
Mechanical Engineering
University of Arizona
Tucson, AZ 85721
USA

Alistair I. Mees
Mathematics Department
University of Western Australia
Nedlands 6009
Western Australia
Australia

Leslie S. Jennings
Mathematics Department
University of Western Australia
Nedlands 6009
Western Australia
Australia

Library of Congress Cataloging-in-Publication Data
Dynamics of complex interconnected biological systems / Thomas L. Vincent, Alistair I. Mees, Leslie S. Jennings, editors.
p. cm. — (Mathematical modeling ; no. 6)
Proceedings of a workshop held in Albany, Western Australia, Jan. 1–5, 1989 and sponsored by the Dept. of Industry, Technology and Commerce (Australia) and the National Science Foundation (USA).

1. Biological systems—Mathematical models—Congresses. 2. Game theory—Congresses. 3. Chaotic behavior in systems—Mathematical models—Congresses. I. Vincent, Thomas L. II. Mees, A.I. III. Jennings, Leslie S. (Leslie Stephen) IV. Australia. Dept. of Industry, Technology and Commerce. V. National Science Foundation (U.S.) VI. Series: Mathematical modeling (Boston, Mass.) ; no. 6.
QH323.5.D96 1990
574.01'1—dc20 90-581

ISBN-13: 978-1-4684-6786-4 e-ISBN-13: 978-1-4684-6784-0
DOI: 10.1007/978-1-4684-6784-0

Softcover reprint of the hardcover 1st edition1990

Camera-ready copy prepared from the editors' TEX file.

9 8 7 6 5 4 3 2 1

PREFACE

This volume contains the proceedings of the U.S. Australia workshop on Complex Interconnected Biological Systems held in Albany, Western Australia January 1-5, 1989. The workshop was jointly sponsored by the Department of Industry, Trade and Commerce (Australia), and the National Science Foundation (USA) under the US–Australia agreement.

Biological systems are typically hard to study mathematically. This is particularly so in the case of systems with strong interconnections, such as ecosystems or networks of neurons. In the past few years there have been substantial improvements in the mathematical tools available for studying complexity. Theoretical advances include substantially improved understanding of the features of nonlinear systems that lead to important behaviour patterns such as chaos. Practical advances include improved modelling techniques, and deeper understanding of complexity indicators such as fractal dimension.

Game theory is now playing an increasingly important role in understanding and describing evolutionary processes in interconnected systems. The strategies of individuals which affect each other's fitness may be incorporated into models as parameters. Strategies which have the property of evolutionary stabilty result from particular parameter values which may be determined using game theoretic methods. Since the main feature of living systems is that they evolve, it seems appropriate that any model used to describe such systems should have this feature as well. Evolutionary game theory should lead the way in the development of such methods.

The workshop brought together researchers in Australia and the USA who had worked on these problems or on methodologies which would be suitable for solving them. The participants included applied mathematicians, control theorists, mathematical biologists, and biologists. Each participant was invited to give an informal presentation in his or her field of expertise as related to the overall theme. The formal papers (contained in this volume) were written after the workshop so that the authors could take into account the workshop discussions, and relate their work to that of other participants. To further encourage this exchange, each paper contained in this volume was reviewed by two other participants who then wrote formal comments. These comments, with the author's reply in some cases, are appended to each paper. We feel that these comments and replies form a very valuable part of this volume in that they give the reader a share in the workshop experience. They also provide a thread which helps join many of the papers together. Indeed, the reader may wish to peruse the comments first.

The papers are grouped according to the the classifications Modelling, Tools, and Games. Part I, Modelling, includes papers dealing with general problems of modelling; 3 papers on models of specific biosystems; and 2 papers on less detailed but broader models which attempt to capture features of classes of biological phenomena. Part II, Tools, presents results and methods useful in analysis of the apparently very diverse range of biological systems which have been modelled. There are 4 papers on complex behaviour of simple systems, including the question of whether simple models can be built from complex data; and 3 papers on other tools and methods. Part III, Games, has 2 papers on evolutionary games, and 2 papers on game-theoretic aspects of human intervention in biological systems.

We would like to acknowledge the financial support of DITC and NSF which made the workshop possible. Ms Cathy Nicotina provided invaluable assistance in organising the conference. Cynthia Belonogoff supervised the typesetting using the TEX system.

Thomas L. Vincent
Alistair I. Mees
Leslie S. Jennings

Tucson, Arizona
Perth, Western Australia
August 1989

PARTICIPANTS

The name, postal address and hence affiliation, telephone number, facsimile number and E–mail address if available are included for completeness. Note that E–mail addresses sometimes have to be rearranged for local computer systems.

Participants from the United States of America
Professor Tom Vincent
Department of Aerospace and Mechanical Engineering, University of Arizona, Tucson, AZ 85721, USA.
Telephone: (602) 621–2325 Fax: (602) 621–8191
E–mail: Bitnet% "vincent1tl@arizrvax"

Professor Joel Brown
Department of Biological Science, University of Illinios at Chicago, Chicago, IL 60680, USA.
Telephone: (312) 996–4289 Fax: (312) 996–9239
E–mail: Bitnet% "u30119@uicvm"

Professor Yosef Cohen
Department of Fisheries and Wildlife, University of Minnesota, St. Paul, MN 55108, USA.
Telephone: (612) 625–2255 Fax: (612) 625–5299
E–mail: Bitnet% "sbpq@uminn1"

Professor Walt Grantham
Department of Mechanical and Materials Engineering, Washington State University, Pullman WA 99164–2920, USA.
Telephone: (509) 335–3224 Fax: (509) 335–7632
E–mail: Bitnet% "grantham@wsuvm1"

Professor Bob McKelvey
Department of Mathematical Sciences, University of Montana, Missoula, MT 59812, USA.
Telephone: (406) 243–2622 Fax: (406) 243–2327

Professor Paul Rapp
Department of Physiology and Biochemistry, The Medical College of Pennsylvania, 3300 Henry Avenue, Philadelphia, PA 19129, USA.
Telephone: (215) 842–7051 Fax: (215) 849–1380
E–mail: Bitnet% "P_RAPP@BRYNMAWR

Australian Participants

Professor Alistair Mees
Department of Mathematics, The University of Western Australia, Nedlands, WA 6009, Australia.
Telephone: (9) 380–3375 Fax: (9) 381–6427
E–mail: alistair@madvax.uwa.oz.au

Dr Nick Caputi
Fisheries Department, Western Australia Marine Research Laboratory, P.O. Box 20, North Beach, WA 6020, Australia.
Telephone: (9) 447–1366 Fax: (9) 447–3062

Dr Micheal Deakin
Department of Mathematics, Monash University, Melbourne, VIC 3217, Australia.
Telephone: (3) 565–4454 Fax: (3) 565–4403

Dr Phil Diamond
Mathematics Department, University of Queensland, St. Lucia, QLD 4067, Australia.
Telephone: Fax: (7) 371–5896
E–mail: pmd@axiom.maths.uq.oz.au

Dr Norm Hall
Fisheries Department, Western Australia Marine Research Laboratory, P.O. Box 20, North Beach, WA 6020, Australia.
Telephone: (9) 447–1366 Fax: (9) 447–3062

Dr Sean McElwain
Department of Mathematics, University of Newcastle, Newcastle, NSW 2308, Australia.
Telephone: (4) 968–5606 Fax: (4) 960–1661
E–mail: mcelwain@nucs.nu.oz.au

Dr Mary Myerscough
School of Chemistry, Macquarie University, North Ryde, NSW 2113, Australia.
Telephone: (2) 805–8297 Fax: (2) 887–4752
E–mail: CM_GRAY@MQCCVAXD.MQ.OZ

Dr Mike Fisher
Department of Mathematics, The University of Western Australia, Nedlands, WA 6009, Australia.
Telephone: (9) 380–3367 Fax: (9) 381–6427
E–mail: fisher@madvax.uwa.oz.au

Mr James Glover
Department of Mathematics, The University of Western Australia, Nedlands, WA 6009, Australia.
Telephone: (9) 380–3338 Fax: (9) 381–6427

Dr Les Jennings
Department of Mathematics, The University of Western Australia, Nedlands, WA 6009, Australia.
Telephone: (9) 380–3361 Fax: (9) 381–6427
E–mail: les@madvax.uwa.oz.au

Mr Kevin Judd
Department of Mathematics, The University of Western Australia, Nedlands, WA 6009, Australia.
Telephone: (9) 380–3338 Fax: (9) 381–6427
E–mail: kevin@madvax.uwa.oz.au

Associate Professor Tony Pakes
Department of Mathematics, The University of Western Australia, Nedlands, WA 6009, Australia.
Telephone: (9) 380–3368 Fax: (9) 381–6427

Associate Professor Kok Lay Teo
Department of Mathematics, The University of Western Australia, Nedlands, WA 6009, Australia.
Telephone: (9) 380–3363 Fax: (9) 381–6427
E–mail: teo@madvax.uwa.oz.au

CONTENTS

PREFACE v
PARTICIPANTS vii

PART I MODELLING

MODELLING BIOLOGICAL SYSTEMS
Michael A.B. Deakin 2

A LENGTH–STRUCTURED MODEL OF THE WESTERN ROCK LOBSTER FISHERY OF WESTERN AUSTRALIA
N.G. Hall, R.S. Brown, N. Caputi 17

LEGUMES AT LOGGERHEADS: MODELLING COMPETITION BETWEEN TWO STRAINS OF SUB-CLOVER
Anthony G. Pakes 40

TWO DIMENSIONAL PATTERN FORMATION IN A CHEMOTACTIC SYSTEM
M.R. Meyerscough, P.K. Maini, J.D. Murray, K.H. Winters 65

MATHEMATICAL MODELLING OF THE CONTROL OF BLOOD GLUCOSE LEVELS IN DIABETICS BY INSULIN INFUSION
M.E. Fisher 84

PART II TOOLS

MODELLING COMPLEX SYSTEMS
Alistair I. Mees 104

DETECTING FOLDS IN CHAOTIC PROCESSES BY MAPPING THE CONVEX HULL
J. Glover 125

CHAOS IN COMPLEX SYSTEMS
Kevin Judd 139

A CHAOTIC SYSTEM: DISCRETIZATION AND CONTROL
Walter J. Grantham, Amit M. Athalye 155

IMPULSIVE EVOLUTION EQUATIONS AND POPULATION MODELS
Phil Diamond ... 175

SCALING AS A TOOL FOR THE ANALYSIS OF BIOLOGICAL MODELS
D.L.S. McElwain ... 204

A NUMERICAL ALGORITHM FOR CONSTRAINED OPTIMAL CONTROL PROBLEMS WITH APPLICATIONS TO HARVESTING
L.S. Jennings, K.L. Teo ... 218

PART III GAMES

STRATEGY DYNAMICS AND THE ESS
Thomas L. Vincent ... 236

COMMUNITY ORGANIZATION UNDER PREDATOR–PREY CO-EVOLUTION
Joel S. Brown ... 263

THE EXPLOITERS CONSERVATIONISTS GAME: HOW TO BE AN EFFECTIVE CONSERVATIONIST
Yosef Cohen ... 289

ANALYSING THE HARVESTING GAME OR WHY ARE THERE SO MANY KINDS OF FISHING VESSELS IN THE FLEET?
Robert McKelvey ... 306

Part I

Modelling

Modelling Biological Systems

MICHAEL A. B. DEAKIN

Abstract

Biological systems are complex systems, almost by definition. Thus models of such systems necessarily simplify. That even simple models can lead to very complicated behaviour leads us to re-examine the goal of mathematical modelling. We may distinguish two loose categories of model: the "pure" model whose end is the understanding of phenomena, and the "applied" model directed to their control. Central to both is the notion of prediction, and it is this that is seen to be likely to be possible only to a limited degree. These matters are further compounded by the corresponding difficulty in solving the inverse problem of inferring the model from the phenomena. This leads us to query the way in which we might seek to understand the system (in the pure case) and to control it (in the applied one).

With Galileo and Newton we see the fruition of the view that mathematics is the natural language of science, and Newton's *Principia*, for example, represents a major achievement in mathematical physics: the mathematical argument advancing the physical understanding in deep and complex ways. The work of these pioneers was greatly extended by later generations of mathematicians; we see Euler and Laplace, for example, tackling and in large measure solving extremely complicated problems of physics or astronomy, and doing so by the use of mathematics (in many cases developed for the purpose).

So successful had that endeavour become that we find Laplace (1814) writing:

> "An intelligence which, at one given instant, knew all the forces by which the natural world is moved and the

position of each of its component parts, if as well it had the capacity to submit all these data to Mathematical analysis, would encompass in the same formula the movements of the largest bodies in the universe and those of the lightest atom; nothing would be uncertain for it, and the future, as also the past, would be present to its eyes. The human mind affords, in the perfection which it has been able to give to Astronomy, a faint glimpse of such an intelligence."

And the passage continues in this vein.

This quotation, a famous one, is noteworthy for two reasons. Laplace, by 1814, is right to boast of "the perfection [of] Astronomy". The dynamics of the solar system (as indeed he goes on to say) were very largely understood, through the efforts of Laplace himself and of others, by the year 1814. They may be thought of as forming a paradigm for science — a mathematical paradigm at that, and one allowing completely accurate prediction of the future and completely accurate reconstruction of the past.

Furthermore, this paradigm was seen as extending beyond the realm of astronomy — to give a completely deterministic "clockwork universe" in which everything was foreordained in that in principle it could be known by such an intelligence as Laplace envisaged — which contemporary philosophers refer to as the "Laplace demon".

Indeed this paradigm came to be seen as conflicting with the freedom of the human will and there are sentences in Laplace's account that may be read as endorsing the view that our freedom of action is illusory. Maxwell and Boussinesq, in particular, found this conclusion unpalatable and in their efforts to refute it went some way towards our modern theories of catastrophe and chaos. (For more detail, see Deakin, 1988.)

Now astronomy can provide such a paradigm, I would argue, because its objects and basic laws are inherently *simple.* To a good approximation, it suffices for many purposes to consider a planet (say) as a moving point endowed with a single scalar characteristic — its mass.

Biology, by contrast, tends to deal with objects that are not simple. While it is true of course that for some simple purposes, such as determining the response of a biological object to a gravitational field, these complexities may be ignored, this is usually not relevant. For the biologist wishes to study precisely those *specifically biological* properties of living organisms that also tend to make them complex.

Thus biological science proceeded by entirely different paths from the physical sciences and it is only this century that specifically *mathematical* theories of biological phenomena have come into prominence.

True there were earlier attempts to discuss biological phenomena in mathematical terms. Thus Euler (1775/1862) is credited with an early version of the *Windkessel* model of blood-flow. But this is not regarded today as a major contribution to its subject. Furthermore, even though our understanding and even basic knowledge of the circulatory system is far from complete, we now have much more such available to us than Euler had. Even so, with its compliant (indeed active) boundaries, its branched and twisted geometry, its non-steady flow, its non-Newtonian working fluid and its visco-elastic walls, the study of the circulation is a very difficult problem in hydrodynamics. We do not now have, and probably should not hope for, a theory of the circulation that matches the paradigm so eloquently described by Laplace.

I would suggest (somewhat tentatively, for there may be studies that I have overlooked) that the first really important contribution to mathematical biology was Malthus' (1798) *Essay on the Principle of Population.* This, and its subsequent revisions may be summarised quite succinctly as "Populations, and in particular the human population, have an inherent tendency to grow exponentially, but the resources available for such growth are such as ultimately to limit it".

This, of course, is a modern restatement, but it does no violence to Malthus and indeed it highlights the essential simplicity of his insight. But it should not in any way be seen as belittling his achievement. The insight still lies behind our concerns over population growth and its consequences, it greatly influenced Darwin and it finds practical application in economic planning.

But compare this with the Laplacean paradigm. We do not say, even in principle, that the Malthusian law can predict, in accurate detail, Australia's population even in one, let along (say) ten, years' time. We do use the law, with curve-fitting to existing data, to estimate population trends and we revise these estimates by taking accurate census and building corrected data into our models. Indeed we construct much more elaborate and detailed models, extending (though still consonant with) the Malthusian idea, but taking into account such other variables as age, sex and geographic distribution.

In the longer term, of course, we need to look at the limits to growth, and one model of this is the Pearl-Verhulst logistic law. Now this *has* found application, particularly in the case of natural populations. For example, it has led to the sometimes useful distinction between r- and K-selection. But again we are looking not at a perfect *fit*, but at an approximate fit at best, or, in many cases, no fit at all in the strict sense, but a convenient theoretical description.

Furthermore, we might, either on theoretical grounds (the self-similarity property, say) or because it fits the existing data better, prefer the Gompertz curve to the logistic. And notice here that it isn't a case of the one's being right and the other's being wrong - though there will be cases in which it will clearly be preferable to use the one rather than the other.

However, we should be clear about one thing. Mere concordance with data is not enough to ensure a good model. In the area of demography, for example, von Foerster *et al* . (1960) modified the Malthusian differential equation from

$$\dot{N} = rN \tag{1}$$

to

$$\dot{N} = rN^{1+\alpha} \tag{2}$$

and fitted the solution curve to existing data. This yielded a *positive* value of α, so that the resulting value of N became infinite in finite time. Doomsday, the predicted date for this catastrophe, was Friday, 13 November, 2026.

Despite the disbelief, not to say derision, with which this study was greeted (Correspondence, 1961), its authors continued to defend it, and indeed its predictions on population levels in 1970, 1975 (see Serrin, 1975) were quite accurate, erring in fact on the side of underestimation.

This study may serve the purpose of startling people into a realisation of the magnitude and urgency of the world population problem — and thus essentially justifying the exercise as a piece of propaganda. But it should be noted that no real *modelling* is involved here. There is no *theory* to justify Equation (2) over Equation (1). The extra α is a fudge factor — no more, no less.

Let us now look at extensions of Equation (1) into 2-species contexts. The most influential and most discussed of these is the Lotka-Volterra model of predator-prey interaction. The equations are well-known and it is provable that all trajectories in phase-space are closed curves. This means that the model itself is structurally unstable. Nowadays we would see this as a consequence of general theory. In the specific case, this was in fact first proved by Lotka himself (1920). (For the record, Lotka's proof constitutes an independent discovery of some of Liapunov's results — then unknown in the West.)

This case should occupy our attention for a number of reasons. In the first instance it should be noted that, like the Pearl-Verhulst equation, and despite its very obvious shortcomings, it has become influential in the ecological wisdom.

But, beyond this, we should look at the different ways in which this, now standard, model has been viewed.

In the first instance, it can hardly be said to have a great record of success. Huffaker (1957) attempted a laboratory realisation of the equations using two species of mite, but he was able to achieve only a qualified agreement. Leigh (1968) analysed the well-known lynx-hare data and concluded that it did not exemplify a Volterra oscillator.

Nonetheless it is a reasonable model to consider, for the assumptions underlying it are the simplest and most natural ones available. It does predict oscillations in the number of both predator and prey and also suggests ways to control these by moving the population toward equilibrium, if this is desirable. This is probably as far as the model should realistically be taken. (Though it should be noted that a 1-predator, 2-prey model predicts the extinction of one of the prey — very much an effect that worries (e.g.) whalers.)

Volterra himself (1937, etc.) however considered the generalisation to n species given by

$$\dot{N}_i = \epsilon_i N_i + \sum_{j=1}^{n} a_{ij} N_i N_j, \qquad 1 \leq i \leq n. \tag{3}$$

In the case where

$$a_{ij} = -a_{ji}, \tag{4}$$

most notably, he was able to produce analogues of many of the results of classical dynamics. It will be noted that Equations (4) preclude self-limitation of the Pearl-Verhulst type. It is further a ready corollary that only equilibria involving an *even* number of species can occur, a result which Volterra himself regarded as significant, though few today would agree with him.

Indeed, Volterra's analysis of Equations (3) and (4) may be appreciated now more as a contribution to the theoretical aspects of dynamical systems theory than as saying anything of biological significance. As far back as 1941, Whittaker almost conceded the point — though he gave more weight to the analogy with physical science than we would today.

Volterra saw Equations (4) as exemplifying conservative (as opposed to dissipative) systems, and he envisaged that, just as in mechanics, where, strictly speaking, systems are dissipative, but the study of conservative systems can tell us much, so in biological studies the same methodology could be followed.

However, in the case of physical systems, conservative models quite frequently give good approximations to reality, and furthermore some dissipative systems can be modelled in relatively simple ways. This would seem to be the result of the comparative simplicity of basic physical objects. Thus, it often suffices in mechanics to characterise an object by a single parameter: its mass.

Nor, in many cases, are we surprised that these few characteristics stand in relatively simple relations to one another. All electrons have the same charge and mass, for example. How different from the biological case, where in ways that tend to be important in many contexts, every elephant is different from every other elephant.

This indeed may be taken almost as a basic principle of biological modelling. Hardin (1960) has dubbed it *The Axiom of Inequality.* If an effect depends on the precise satisfaction of an equality it will not be observed. The same principle has emerged with Thom's (1972) programme for applying catastrophe theory in speculative ways in the social and biological sciences: examine what happens *generically* and base all theories and predictions on that.

There can be exceptions. For example, there seem to be good biochemical reasons for regarding some alleles as obligatory dominants, so that the AA, Aa genotypes are assigned the same fitnesses. And of course even if we knew that dominance was *not* complete (as with, say, sickle-cell anemia), we might for some purpose (e.g. analysing this condition in the USA) assign the same fitness to the different genotypes and be confident of a good approximation.

But by and large the Axiom of Inequality holds sway. In particular, therefore, we may be very critical of Equations (4) as describing the real world. Biomathematics, as Whittaker rightly intuited, has not taken the road explored by Volterra — except for the basic theory of the two-species (and some three species) interactions, where the virtues of simplicity and immediacy referred to above override the objection of structural instability. No version of the Lotka-Volterra theory, however, has remotely matched the Laplacean paradigm.

In fact that paradigm may be seen as increasingly less relevant to the needs of biological models. While it is a useful thing to have (say) the Pearl-Verhulst logistic law, it is also useful to have the Gompertz law — although the details of the two curves differ. Similarly there are other models: differential equations with delay, discrete time models, integro-differential equation models, for example. Each has its own properties and in some cases produces plausible approximations to field or laboratory data. This can be useful for general understanding, for the formulation of hypotheses as to the causes of observed patterns, or even, to a point, for prediction and control. But we do not expect the agreement that the Laplace demon is seen as embodying.

All modelling, mapping as it does the real world prototype onto an abstract mathematical structure, involves some imprecision or loss of detail. And because biological prototypes are, in the main, complex, their mapping onto simple abstract structures necessarily involves crude approximations indeed.

The ready corollary of this rather obvious remark is that calculations based on the simplified model cannot be trusted very far — apart from the initial simplification, we have the propagated effects, which, in the main, will magnify.

Compare this again with the situation in mathematical physics. Newtonian mechanics, to give an example, within its domain of validity, describes certain objects very well indeed — say the motions of the various objects making up the solar system. Thus we were able, to very great accuracy, to say when to begin looking for the return of Halley's comet and to say also where to look.

Even in the case, by contrast, of a very well-founded and well validated biophysical model, the Hodgkin-Huxley equations (say), we do not expect this degree of precision. These equations have successfully described an underlying ionic *mechanism* and have successfully predicted qualitative and even, in places, quantitative results — but their ability to give accurate results in all cases is limited, as the continued proliferation of modifications and reformulations attests. (See, for two recent examples, Strandberg, 1986 and Drouhard, 1987.)

Closely allied to this aspect of matters is the question of the sustaining of long chains of reasoning. Mathematics is quintessentially the science of ensuring the validity of long sequences of deductions, and where its premisses fit the data well, so then do the conclusions of those deductions. This aspect is not evident in biological applications of mathematics — Volterra's equations for ecosystem behaviour have already been noted in this connection.

There are, however, cases in which we do need to predict, as accurately as possible, some specific behaviour. E.g. we may need to legislate to ensure the continued viability of a fishery. In such an instance, it makes sense to resort to *simulation*, which may be thought of as an attempt at very detailed modelling, but of a *specific* situation.

Consider, for definiteness, the modelling of an ecosystem by means of Equations (3) (but without Equations (4)). We will need to produce values of the n quantities ϵ_i, the n^2 coefficients a_{ij}, the n initial values $N_i(0)$ and indeed of n itself. Even in the case of this last, we will probably deliberately simplify, employing a lower, even a *much* lower value, than the true one.

Our estimates of many of these quantities, even the $N_i(0)$, which in principle are accessible to direct observation, are likely to be crude in the extreme. The corollary of this is that such models require to be run *interactively*: they only have value if they are subject to continual testing in the field, and that field data, as they are acquired, are fed back into the model, which is thus subject to the possibility of continuous updating and revision.

This attempt at detailed and specific simulation is one end of the range as far as biological modelling is concerned. The other is the discussion of (essentially) whole *classes* of models in the search for robust qualitative results. Consider, as an example, Rescigno and Richardson's (1967) discussion of two-species systems. Specifically these authors consider the use of a pair of coupled ordinary and autonomous differential equations:

$$\begin{aligned} \dot{N}_1 &= N_1 K_1(N_1, N_2) \\ \dot{N}_2 &= N_2 K_2(N_1, N_2) \end{aligned} \tag{5}$$

as the model. K_1, K_2 are supposed to possess continuous first derivatives and are subject to various further constraints according as the interaction is competitive, exploitative or mutualistic. The analysis is rather general and little is really deduced, but it is seen for example that the predator-prey case may, but need not, lead to oscillatory behaviour. Oscillations do not occur in the case of competition, but Rescigno (1968) showed that competition between *three* species can lead to oscillations.

There is in fact quite a lot of work along lines such as these — essentially dynamical systems theory built on a biological motivation. It is not quite clear yet quite how biologically significant this will turn out to be.

These then are the various types of model that can be used in biology. It will be clear that the Laplacean paradigm applies to none of them. Rather, what is sought is either (a) a generalised understanding of underlying phenomena, or (b) specific predictions in specific cases (or, of course, (c) some hybrid of these two). In neither case can we hope to sustain accurate predictions over long periods of time nor can we with any confidence rely on lengthy chains of deductive inference. The models are too imprecise, the approximations too crude.

It is against this background that we should view the impact of chaos and strange attractors. Interestingly enough, the psychological impact of chaos is not so much that complex systems can generate complicated behaviour: *that* conclusion would hardly be surprising.

The real force of chaos as a mathematical phenomenon and its importance for the theory of modelling is that even simple systems can exhibit very complicated dynamical *behaviour*. Indeed the realisation that simple systems of three autonomous differential equations, and even simpler systems consisting of a single non-linear difference equation, could lead to chaos reopened old questions.

First and foremost, clearly no chaotic system can obey the Laplacean paradigm. This is the point that Maxwell and, to a lesser degree, St. Venant, anticipated in their rejection of that paradigm as being universally applicable in science (see Deakin, 1988). Of course, they did not couch matters in these terms.

What these authors were concerned to stress, what is still relevant today, and what our modern understanding of chaotic phenomena has served to highlight is the concept of stability. If the long-term behaviour of the system is an unstable function of the initial conditions, or of the parameters involved in the model, or of the functional form of the model, then our ability to predict in the long term is negligible.

Furthermore, and to add emphasis to the point, as our own knowledge is necessarily imperfect at least to *some* extent, the scientific process in the presence of chaos is thus constrained to abandon the Laplacean ideal and to settle for something rather more modest.

At issue here is the repeatability of experiments. This can be, and probably has been, interpreted to mean that small perturbations in the conditions of the experiment should produce *permanent* small perturbations in the outcome. In other words, the condition is one of uniform stability.

Such a draconian requirement is not, however, a mandamus. Experiments take finite time and can, within limits, be repeatable, up to experimental error, within a time that may shorten as we restrict the extent of the error we agree to accept.

This, as I see it, is the way we should understand the impact of chaos on the "simulation" end of the biomathematical spectrum. Indeed it's a rather minimal impact, for we are discussing complex (rather than simple) systems and these may well be seen already as likely to result in complex observed behaviours.

Our need to do the best we can, in the short, realistic, term; to predict results of specific actions in specific situations; to consider different scenarios; to attempt to control perceived adverse effects while maximising beneficial ones — all these needs remain, will be met to a point, and are little modified (if, in practice, at all) by our awareness of chaotic phenomena.

In other words, insofar as the strictly *applied* end of the biomathematical spectrum is concerned, recent theory may have made people more aware of the limitations of their methodology, but its actual impact on that methodology has not been very great, for it is of itself a methodology whose limitations are all too obvious.

Consider the very comparable case of weather prediction. The system is a physical one, but one of great complexity — a complexity rivalling that of biological phenomena. Experimentally, we have improved and continue to improve both the accuracy and the duration of validity of short-term forecasts. The goal of long-term forecasting remains, however, as elusive as ever. Lorenz's (1964) discovery of chaos in a very simple climatic model may have added a certain degree of confirmation to the experience, but by itself (without that experience) it would mean rather little to the practical forecaster.

When, however, we come to the other, theoretical, end of the spectrum, matters are rather more complicated. If we go back to Equations (5) and generalise them to n species, we get

$$\dot{N}_i = N_i K_i(N_1, N_2, \ldots, N_n) \qquad 1 \leq i \leq n. \tag{6}$$

Smale (1976) considered the Rescigno-Richardson programme for this generalisation, in the case where the n species were all in competition. In a very pithy paper, he showed that once we have $n \geq 5$, we can say next to nothing about the behaviour of the system. "In fact these equations are compatible with any dynamical behaviour ..... ."

Now, of course, this is pure dynamical systems theory. The model considered by Smale contains only minimal biology, but in a sense that is the point. The idea of the Rescigno-Richardson programme (and others like it) was to prove robust qualitative results from minimal assumptions.

What this *may* mean is that, in order to achieve meaningful biological results, we need to build more biology into the theoretical models. For, in truth, *no* ecosystem on earth contains as few as two (or even five) species, but neither does any ecosystem on earth comprise only competitors. It may be that if we could build a more realistic biology into the models, more definite results might emerge.

Mathematical models tend to begin by observing the regularities in nature. The classic, and influential, case, as we have seen, was astronomy. Laplace's encomium of Astronomy would not have been possible without Newton's success with his law of universal gravitation. That owed its genesis to Kepler's laws of planetary motion.

But note that the word "law" is used differently in the two cases. While Kepler's laws are *descriptions* of observed phenomena (efficient descriptions, let me hasten to add — curve fitting of a very high order), Newton's law is an *explanation*. It says that if we will assume a force of gravitational attractions varying as the inverse square of the distance, then we can *derive* Kepler's laws as consequences. Not only does it make the description even more efficient, it makes a convincing attempt to tell us "what's really going on". Kepler, however, had observed the underlying regularities in nature.

Thus, what Newton had done was essentially to solve an *inverse* problem. From regularity of data, he inferred a *mechanism*. Now there have been successes of this type in biology and in quantitative biology at that. Mendel's laws are perhaps the best known example, and fittingly enough, they have been the basis of many of the very best studies in the biomathematical field.

But in many areas of biology, ecology being one, we still await reliable and agreed notions of what underlying regularities exist. In such a case, it is reasonable to posit, to some extent *a priori*, a model, to analyse it and to seek correspondence with field or laboratory data. The fact that two or more such models may fit the data is not necessarily of concern — as long as we recall that models are models and not ultimate truth. This is the case even in physics, of course, but it is more necessary to keep it firmly in mind when dealing with the biological area.

This is particularly true if we are comparing chaotic output, for if the output is chaotic we see no underlying regularity and the inverse problem clearly has no unique solution. But we may on some reasonable ground (as e.g., simplicity of formulation) prefer to adopt one model above others — just as in an earlier case we imagined a researcher preferring the Gompertz curve to the logistic.

It really comes down to this: that we will continue to need and to use mathematical models but we need to know their limits.

REFERENCES

[1] Correspondence. 1961. On the topic of "Doomsday". *Science* **133**, pp. 936-946.

[2] Deakin, M.A.B. 1988. Nineteenth century anticipations of modern theory of dynamical systems. *Arch. Hist. Ex. Sci.* **39**, pp. 183-194.

[3] Drouhard, J.P. 1987. Revised formulation of the Hodgkin-Huxley representation of the sodium content in cardiac cells. *Comp. Biom.* **20**, pp. 333-350.

[4] Euler, L. 1775/1862. Principia pro motu sanguinis per arterias determinando. *Op. Post. Math. Phys.* **II**, pp. 814-823.

[5] Hardin, G. 1960. The competitive exclusion principle. *Science* **131**, pp. 1292-1297.

[6] Huffaker, C.B. 1957. Experimental studies on predation: Dispersion factors and predator-prey oscillations. *Hilgardia* **27**, pp. 343-383.

[7] Laplace, P.S. 1814. Essai philosophique sur les probabilités. Reprinted in Laplace's collected works, Vol.7.

[8] Leigh, E.R. 1968. The ecological role of Volterra's equations. *Lect. Math. Life Sci.* **1**, pp. 1-61.

[9] Lorenz, E.N. 1964. The problem of deducing the climate from the governing equations. *Tellus* **16**, pp. 1-11.

[10] Lotka, A.J. 1920. Analytical note on certain rhythmic relations in organic systems. *P.N.A.S.* **6**, pp. 410-425.

[11] Malthus, T. 1798. *An Essay on the Principle of Population, etc. London.* Variously expanded, modified and reprinted.

[12] Rescigno, A. 1968. The struggle for life: II. Three competitors *Bull. Math. Biophys.* **30**, pp. 291-298.

[13] Rescigno, A. and Richardson, I.W. 1967. The struggle for life: I. Two species. *Bull. Math. Biophys.* **29**, pp. 377-388.

[14] Serrin, J. 1975. Is "Doomsday" on target? *Science* **189**, pp. 86-88.

[15] Smale, S. 1976. On the differential equations of species in competition. *J. Math. Biol.* **3**, pp. 5-7.

[16] Strandberg, M.W. 1985. Homomorphism on a physical system of the Hodgkin-Huxley equations for the ion conductance in nerve. *J. Theor. Biol.* **117**, pp. 509-527.

[17] Thom, R. 1972. *Stabilité structurelle et morphogénèse.* Benjamin: Reading, Mass.

[18] Volterra, V. 1937. Principes de biologie mathématique. *Acta Biotheor.* **3**, pp. 1-36.

[19] von Foerster, H., Mora, P.M. and Amiot, L.W. 1960. Doomsday: Friday, 13 November, A.D. 2026. *Science* **132**, pp. 1291-1295.

[20] Whittaker, E.T. 1941. Vito Volterra. *Obit. Not. F.R.S.* **3**, pp. 690-729. Reprinted in amended form as a preface to V. Volterra's *Theory of Functionals and of Integral and Integro-Differential Equations*, Dover Reprint, New York, 1959.

PARTICIPANT'S COMMENTS

Biological systems are complex in the sense that they are high dimensional, containing many variables, but many high dimensional systems have low dimensional attractors; such as in fluid dynamics where an infinite dimensional system has a finite dimensional attractor. This may mean that we have to give up the hope of realistically modelling the entire system in favour of a model which captures the essence of the attractor as implied by collected data. Indeed in prediction and control it is irrelevant which of the many different almost equivalent models is chosen.

Low dimensional models can be chaotic, but it is important to notice that chaos has a rich and regular structure. Processes and mechanisms that can produce this regularity are being investigated and modelled.

The question which needs to be answered is whether biological data possesses this kind of regularity and if it does can models of the chaos producing processes be built.

In a different vein stochastic models have been enormously sucessful for biological systems and in some cases are superior to deterministic models. However, we should still see how much slack can be taken up by simple nonlinear deterministic models, since a stochastic component can always be added to finally tighten a model onto experimental data.

Kevin Judd

This paper outlines some of the history and philosophy of mathematical modelling of biological systems. It provides an excellent perspective from which to view the current work in this field.

Deakin's discussion of the work of Lotka and Volterra is particularly apt. The conference came back to the Lotka–Volterra model again and again. For example, Vincent discussed a version of the Lotka–Volterra equations in his paper on evolution under chaotic dynamics where the focus on an individual is replaced by the focus on phenotypes. A version of the Lotka–Volterra equations was used by Grantham to display chaotic behaviour in a relatively simple system.

Chaos also plays a part in Deakin's history and once again he placed the current work into perspective. However, his observations about predictability may be a little pessimistic. The papers by Rapp and Mees, for example, show that there are now techniques, such as the Embedding Theorem and Takens' Theorem which reveal a structure to chaotic phenomena.

Deakin draws a clear distinction between "pure" and "applied" models with the former used to further understanding of phenomena and the latter directed to their control. The conference provided excellent examples of both types of models with, for example, the papers by Brown and Myerscough aimed at obtaining a deeper insight into "idealized" systems whereas the papers by Hall *et al* and Fisher were directed at control.

Sean McElwain

REPLY

Two matters raised by the reviewers call for further comment from me. The first relates to the predictability of chaotic systems. That chaotic systems may possess inherently simple structures is true - Rapp (1987) gave some wonderful examples. This search for a simple "handle" to a complex problem is very much what science is about. Rapp's techniques of time–series analysis produced some very interesting results, and these came with supporting evidence that the underlying attractors were "real" – i.e. significant. Nonetheless we have no theorems as yet to say that attractors need to be low–dimensional. We are not even able to say a priori when we expect low–dimensional attractors and when not. That we may at times discover these, or perhaps more accurately, succeed in fitting simple structures to the phenomena is fortunate and not altogether surprising. Whether it is something we can expect and come to rely on is not yet clear.

The closely related point was made at the conference that Laplace might not have been able to sing the praises of Astromony as he did were it not for the fact that the sun is much larger than the planets, in their turn much larger than their moons which are (in the main) much larger than comets. Were there not orders of magnitude separating the sizes of these different objects, the solar system would not look so orderly. It is only now appearing that this feature may be the result of deep underlying causes rather than serendipitous happenstance.

My text does avoid the topic of stochastic models. Partly because every stochastic model necessarily elaborates a simpler deterministic one, partly because stochastic models, being harder, require more simplifying assumptions if they are to be tractable. Stochasticity may be seen as a way of dealing with chaotic data or as a fundamental property. The very different philosophies implied by these divergent views need not be reflected in the models themselves.

The popular belief, despite Einstein's "Gott würfelt nicht", is that nature, at the sub-atomic level is stochastic. This, however, need not be invoked for most biological situations. Here stochastic models merely cover for our own ignorance or our unwillingness to include too many details. They simplify usefully in other words.

Rapp, P.E. 1987. Oscillations and chaos in cellular metabolism and physiological systems. pp 179-208 of *Chaos* ed. A.V. Holden, Manchester Univ. Press.

Michael A. B. Deakin

A Length-Structured Model of the Western Rock Lobster Fishery of Western Australia

N. G. HALL[1], R. S. BROWN, AND N. CAPUTI[1]

Abstract

By monitoring the settlement of the 9 to 11 month old first post-larval (puerulus) stage of the rock lobsters on coastal reefs, estimates are made of the level of recruitment expected to enter the fishery four years later. These predictions are further improved by introducing into the predictive models the abundance of 4 year old juvenile rock lobsters caught in rock lobster pots.

A model of the fishery, which uses these predictions, has been developed to examine alternative management strategies. The model is a discrete, deterministic simulation model. The state of the system at each time step is represented by the number of rock lobsters within each of a number of batches. Each batch represents rock lobsters of similar carapace length, within the same depth category and fishing region, and with the same characteristics of moult stage, migratory status, and reproductive state. Growth of rock lobsters, through moulting, has a major impact on catchability and hence the fishery, especially since moulting is relatively synchronous. The resulting changes in catchability and its effect on recruitment to the exploited (legal-sized) population are essential elements of the model. An additional feature of the fishery is the more vulnerable, migratory phase of the life history of the rock lobsters. Subsequent to their migration, the rock

[1] Presented by N. G. Hall and N. Caputi

lobsters have a reduced catchability, and are distributed in deeper off-shore waters.

1. Introduction

1.1 The Western Rock Lobster Fishery

The limited-entry fishery for the western rock lobster, *Panulirus cygnus* George, is one of the major rock (spiny) lobster fisheries in the world and the most valuable single species fishery in Australia, accounting for approximately 35% of the country's gross income for all fisheries products (Morgan, 1980b) and over two thirds of Australia's total rock lobster production. The range of *P. cygnus* extends on the western coast of Australia from approximately 21° S to 34° S and includes the Abrolhos Islands (Figure 1). The majority of the 716 fishing boats (at March, 1989) are concentrated between Mandurah in the south and Kalbarri in the north (Figure 1). The boats, which are mostly between 8 and 15 m in length, currently operate 72 960 pots (traps) during a seven and a half month season (15 November to 30 June) that includes a three and a half month Abrolhos Islands season (15 March to 30 June).

The fishery has developed from being under exploited in the 1930's when approximately 250 t of lobster were caught for the local market, through an infant export industry in the late 1940's, producing 1 000 t in 1948, to a catch which, since the early 1960's, has varied from 6 800 t to 12 900 t (Figure 2). The average annual catch is currently worth in excess of $A190 million to the fishermen. The majority of the catch is processed as frozen tails and exported to the United States. Sales of live rock lobsters to Japan are also very important while small but significant markets exist for whole cooked lobsters in France and other European and Asian countries.

Variations in recruitment of western rock lobster are believed to be associated not only with the spawning biomass of the stock, but also with environmental factors, such as the Leeuwin Current, the strength of which is affected by the Southern Oscillation (Pearce and Phillips, 1988). This current may assist the return of the larval stages of the rock lobster to the coast.

The main management objective has been to contain the expansion in fishing effort (Bowen, 1971; Bowen, 1980; Hancock, 1981; Morgan and Barker, 1982). Since 1963 the fishery has been subject to limited entry

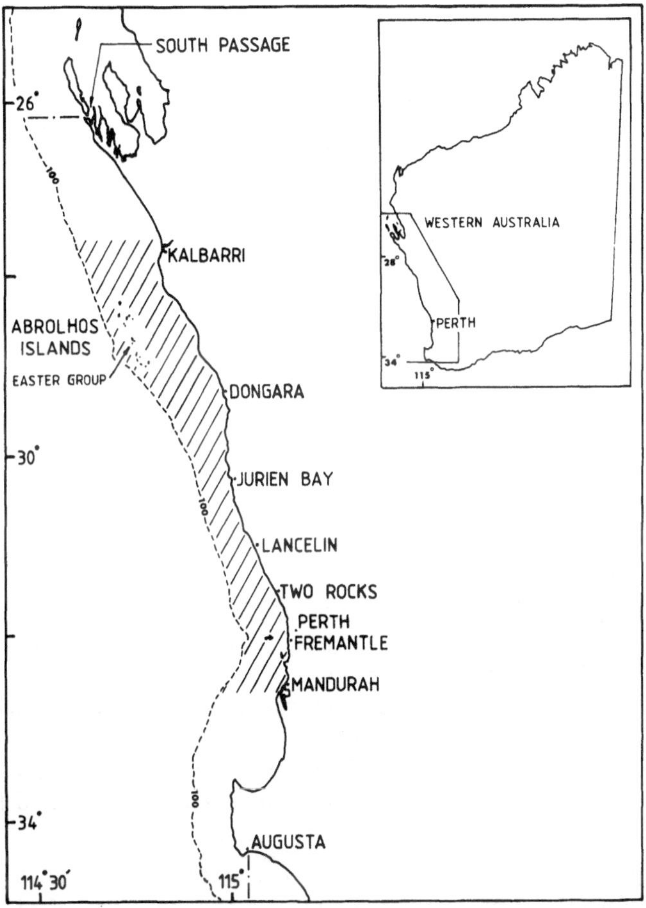

Figure 1. A map of a section of the Western Australian coastline, showing the extent of the western rock lobster fishery and the relative positions of locations mentioned in the text. Fishing activity is concentrated in the hatched area.

with the objectives of optimum utilisation of the resource, reasonable economic return to the fishermen, and orderly exploitation to minimise conflict among professional and recreational fishermen.

Management has been based on research which has led to an understanding of the biology, life history and population dynamics of *P. cygnus* (Hancock, 1981; Bowen and Hancock, 1989). Stock assessment has been undertaken using generalised production models, yield per recruit models, stock/recruitment relationships (Morgan 1980a), and models for predicting recruitment and total catch, based on an index of abundance of puerulus (final larval stage; 9-11 months old) settlement (Morgan *et al.*, 1982; Phillips 1986), and an index of abundance of juveniles (Caputi and Brown, 1986; Caputi *et al.*, 1988).

Predictions based on the puerulus settlement have enabled management options to be examined up to 4 years in advance of the subsequent recruitment of a year-class to the fishery. These predictions have enabled fishermen and their financiers to make decisions with some knowledge of possible trends in the fishery. Information on prediction is regularly sought before major investment decisions are made. The reliability of the predictions of annual catch over the years has given fishermen confidence in the models used for the management of this fishery. This has helped in the acceptance of management decisions. Thus, the utility of predicting catches cannot be over-emphasized.

The managers of the fishery are not only interested in these predictions to know the likely variation in catches, they are also interested in the subsequent variations in the abundance of the spawning stock. When a below average catch was predicted for the 1986/87 season, concern was expressed at the combination of high fishing effort (Figure 2) and a low abundance year-class. In order to avoid a potentially sharp decrease in the spawning stock, a 10% reduction in the number of pots allowed for each fishermen during that fishing season was recommended.

This paper describes a phase of the research/management process, in which information on the various indices of abundance of life history stages, biology, and fisheries parameters (e.g. exploitation rate, etc.) are combined into a model of the fishery to enable consideration of the effects of alternative management regimes. In particular, the model, described in this paper, will be used in the future to examine whether harvesting strategies can be formulated that will enhance the survival of recruits, resulting from predicted poor year-classes of rock lobsters, to the spawning stock, and which will maximise the value to industry of catches resulting from good year-classes.

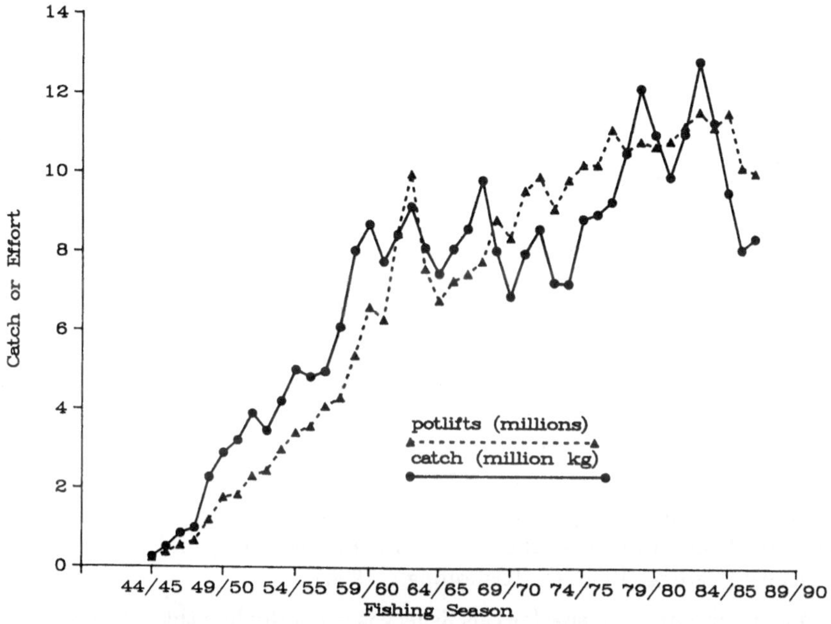

Figure 2. Catch of western rock lobster and nominal fishing effort for the fishing seasons from 1944/45 to 1986/87. Catch statistics to 1970/71 are derived from fishermen's monthly returns, which are processed by the Australian Bureau of Statistics. Values from 1971/72 to 1986/87 are derived from what is considered a more accurate source, the rock lobster processors' monthly returns.

1.2. Life History

The life cycle of the rock lobster is complex and includes a nine to eleven month oceanic larval phase (Phillips, 1981). The puerulus stage completes the oceanic cycle by swimming across the continental shelf and settling in the shallow limestone reef areas along the coast (Phillips, 1981). The peak settlement of puerulus occurs between September and January. The settled puerulus moults into a small juvenile rock lobster, with a carapace length (C.L.) of approximately 8 mm. Based on the time of hatching of the phyllosoma larvae, Morgan *et al.* (1982) adopted January 1, the mid point of the hatching period, as an arbitrary "birthday". Thus the age of newly settled puerulus varies between late 0+ and early 1+ years.

During their first year as benthic juveniles they are known to be solitary and cryptic, living in cracks and crevices in the reefs (P. Jernakoff, Bureau of Rural Resources, Department of Primary Industries and Energy, GPO Box 858, Canberra, Australia 2601, pers. comm.). The development of gregarious behaviour occurs at about 1.5 years of age. During daylight the older juveniles are found under limestone reefs but at night they forage for food in the surrounding seagrass beds (Joll and Phillips, 1984).

When juveniles are between 4 and 6 years old, they migrate as pale coloured "whites" from the shallow reef areas to the outer continental shelf at depths of up to 200 m (George, 1958). This migratory or "whites" phase, as it is known, occurs regularly in November to January each year. The mean size of the "whites" is about 76 mm C.L., the minimum legal size for the fishery (George, 1958; Chittleborough, 1970).

The minimum legal size for this fishery is generally less than the size at first breeding of females though this size varies for different localities along the coast. At the Abrolhos Islands, the majority of females breed below the minimum legal size (C. Chubb, PO Box 20, North Beach, Western Australia 6020, pers. comm.). Juveniles are generally found in shallow waters, 0-40 m, while adult breeding populations are usually at depths from 40 to 80 m; there can be considerable overlap in some locations. Size stratification with depth is not as apparent at the Abrolhos Islands (which rise steeply from the outer continental shelf) and breeding adults are present in both shallow and deep water.

Reproductive maturity of females on the coast generally occurs 12-24 months after the attainment of legal size, i.e. 6-7 years of age (Morgan *et al.*, 1982). Mating takes place from July to January and the female carries the black spermatophore on the sternal plates of the cephalothorax. In October-December, the eggs are extruded, fertilised and become attached to the pleopods. They are carried on the pleopods for 4 to 9 weeks before hatching, depending on water temperature (Chittleborough, 1976; C. Chubb, pers. comm.). The majority of females have two broods of eggs during the breeding season (C. Chubb, pers. comm.).

1.3. Need for Modelling

In 1973, Morgan (1977) estimated that 78% of rock lobsters reaching the age of first capture will be caught in subsequent fishing seasons, which corresponds to an exploitation rate of 48% per year. Using the same estimates of catchability and natural mortality, and with effort applied over a shorter fishing season, the estimate of annual exploitation for 1982/83 was 60%. Rock lobster tagging studies at Two Rocks (Figure 1), using animals of slightly below the legal minimum size, resulted in the recapture from different releases of between 40% and 60% within one year.

With an exploitation rate of this magnitude, and in a fishery which over much of its range exploits rock lobsters before they reach maturity, concern is felt for the maintenance of the spawning stock. This concern is heightened in years when the level of recruitment is low, as the inevitable effect is that additional fishing effort will be directed towards the mature rock lobsters. To date, the industry has been fortunate, in that sustained low levels of recruitment, persisting over several seasons, have not been experienced.

Strategies to modify the level of exploitation are available, ranging from a variation in the number of pots to be fished at different times within the fishing season to changes in the duration of the fishing season. The resulting impact of each strategy on the financial return to the fishing industry and the benefits achieved will differ. For example, an important event in the fishery is the annual migration of the rock lobsters, when large catches occur. The catch of these paler coloured migratory animals is of less value to industry than the catch of dark red non-migratory animals. Nearly 50% of the annual catch is currently taken from November to January, when the annual off-shore migration occurs. By reducing the level of exploitation at this period, greater catches of rock lobsters might subsequently be expected in deeper water. However, the cost of exploiting these animals later in the season, as less catchable, but more valuable, non-migratory rock lobsters, would be greater than the cost in less extensive, shallower waters.

Modelling of the rock lobster fishery is intended to address issues such as these, and to examine the impacts of various fishing strategies on the within season catches in each depth zone, and resulting changes to the spawning stock. An essential aspect of the model must be the realism with which it is seen to describe the biological and fishing processes, and the degree to which it produces estimates of catches which correlate well to actual catches achieved by industry. If fishermen do not believe that the model provides an adequate description of the biology, and do not see that the results from the model agree with recorded catches, they are unlikely to accept alternative management strategies based on estimates produced by the model.

2. Model of the Fishery

In the discussion which follows, familiarity with the literature concerning fisheries modelling is assumed.

Our model of the western rock lobster fishery is a discrete, deterministic simulation model. The state of the stock at each time step is represented by the number of rock lobsters within each of a number of batches of rock lobsters. Each batch represents rock lobsters of similar carapace length, within the same depth zone and fishing region, and with the same characteristics of moult stage and migratory and reproductive status. Each time step represents approximately 1 week.

The fishery is divided into 3 management regions: coastal waters north of 30° S, the Abrolhos Islands region, and waters south of 30° S. Each of these is further divided into depth zones of 0-18, 18-36, 36-54, and over 54 m. For each region and depth zone, monthly catch and effort data are available from fishery and research records. There are no major movements of rock lobsters between regions, but migration to deeper water occurs during the annual "whites"' migration.

Rock lobsters are introduced into the model when their year-class reaches age 3 (Figure 3). An index of the number of rock lobsters in each year-class at this time step (Table 1) is obtained from the geometric mean of the observed catch rates in the period from November to January in 0 to 18 m of water, two years later (Caputi and Brown, 1986). Predictions based on the puerulus and juvenile indices may be used to extend this series some years into the future. An estimate of the total number of rock lobsters in year-class 3 within each depth zone and fishing region is obtained by multiplying the index of year class strength by a scaling factor associated with that depth zone and region; these factors are estimated from the model during the parameter estimation phase (Table 2). Analysis of length samples suggests that the sexes are equally represented in these recruits. The newly introduced males and females are allocated to length classes in accordance with an assumed Gaussian size distribution for rock lobsters of age 3.

Table 1: Data requirements of the model.

Based on Fishery's Data or Assumed
Index of Year-class Strength at Age 3
Male:Female Ratio at Age 3
Initial Length Distribution at Age 3
Growth Schedule for each Sex and Region
Duration of each Moult Stage
Relative Catchability at each Moult Stage
Relationship of Migratory Proportion with Length
Relationship of Maturity to Length for Females
Relationship of Berried Proportion with Time of Year
Relationship of Eggs per Brood to Length
Egg Development Time
Proportion producing Two Broods of Eggs
Knife-edged Fishing Selection at Minimum Legal Length
Minimum Legal Length
Rate of Decline of Enhanced Catchability for Migratory State
Instantaneous Coefficient of Natural Mortality
Enhanced Natural Mortality for Migratory State
Rate of Decline of Enhanced Mortality for Migratory Lobsters
Rate of Decline of Migration Rate after Moult
Relationship of Total Weight to Carapace Length
Relationship of Tail Weight to Carapace Length

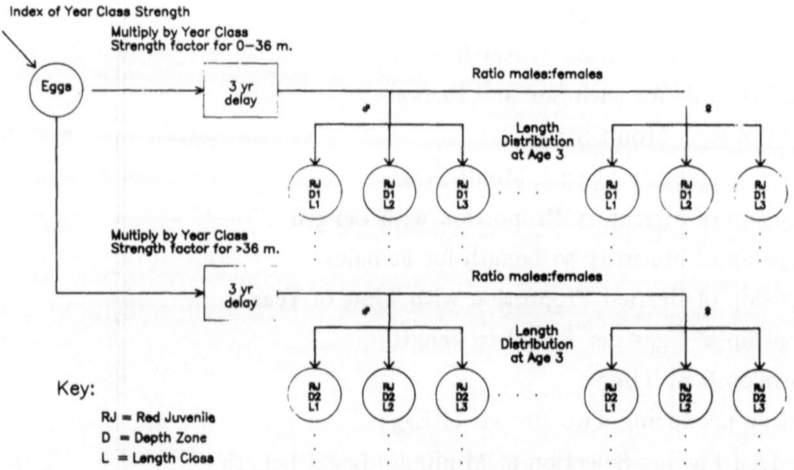

Figure 3. Introduction of a new year-class into the model.

Table 2: Data required for parameter estimation, and estimated parameters

Used during Calibration
Monthly Fishing Intensity by Depth and Region
Actual Monthly Catch by Depth and Region
Estimated from Model during Calibration
Multipliers of Year-class Strength Index by Depth and Region
Relative Catchability associated with Migratory State
Catchability
Instantaneous Migration Rate

Within the stock, a degree of synchrony exists for moulting, with consequent effects on within-season catchability (Morgan, 1977), spawning, recruitment to the exploited stock, and migratory behaviour. Because of these effects, it is essential to model the growth process as a discontinuous growth curve. Growth is assumed to be density independent, but region and sex specific. Growth schedules are provided as input to the computer program. Each schedule contains details, for each of a number of length ranges, of the moult increment, the minimum intermoult period required before moulting might occur, and the calendar months in which rock lobsters of that size range might moult. At each time step, the model examines the growth schedules and updates the moult status of each batch of rock lobsters (Figure 4). The moult status is categorised as post-moult (having a duration of approximately 1 week), early intermoult (approximately 1 week), intermoult (duration assumed to be dependent on size), and pre-moult (approximately 1 week). Females carrying eggs do not moult. In November, when rock lobster moult, a large proportion enter the migratory state; this proportion is related to carapace length.

At the time of moulting, immature females may become mature (Figure 5), where the proportion mature, $p_m(L)$, at carapace length, L, may be represented by a logistic curve:

$$p_m(L) = 1/[1 + \exp\{(a - L)/b\}],$$

where the parameters of the equation are related to the fishing region.

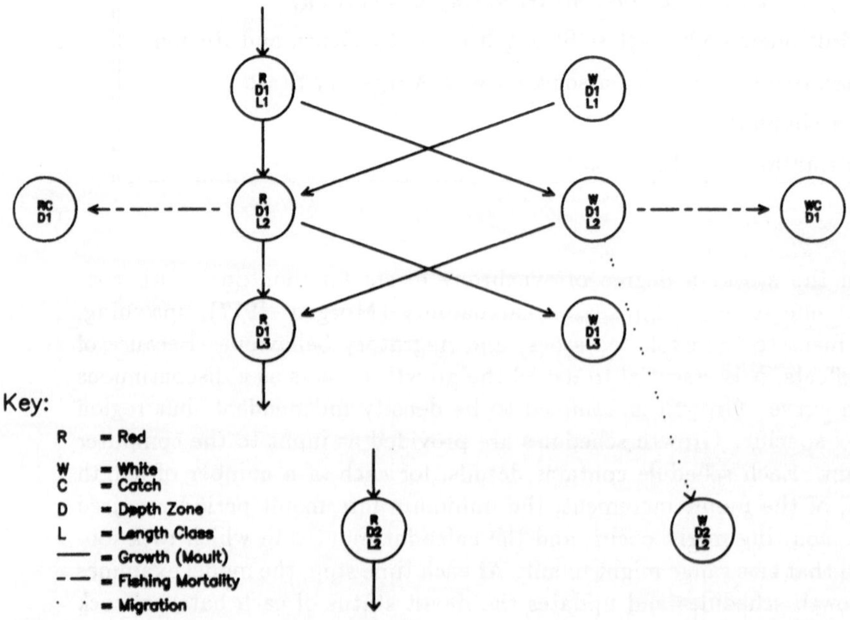

Figure 4. Flow for male rock lobsters of length-class, L2, within the shallowest depth zone, D1, where lobsters moult from length-class L1 to L2, and from L2 to L3. Natural mortality is not included in the diagram. Flows for other length-classes or depth zones are not completely represented.

It follows that, if the carapace length of a batch of immature rock lobsters increases from L_1 to L_2 at a moult, a proportion of the immature lobsters will become mature, where this proportion may be calculated as:

$$\Pr\{\text{maturity at length } L_2 \mid \text{immaturity at length } L_1\} = \frac{p_m(L_2) - p_m(L_1)}{1 - p_m(L_1)}$$

The proportion of mature females which have laid their first brood of eggs increases as the spawning season progresses, and is assumed to have the form of a logistic curve. A batch of rock lobsters which is in the pre-moult stage of the moult cycle, or one which is in a migratory ("whites") state does not produce a brood of eggs. Sufficient time must elapse for the eggs to develop before they hatch, where this period varies with fishing region. Around 70% of the females will produce a second brood of eggs (C. Chubb, pers. comm.). It is assumed that this occurs immediately after the first brood has hatched. A maximum of two broods is assumed.

The number of eggs carried by each female, within each brood, has been estimated by Morgan (1977) to be:

Eggs = 9800 Length - 581850

where the carapace lengths of berried females are measured in mm. The spawning potential of the rock lobster stock is calculated as the total of eggs hatched over all fishing regions during the spawning season.

Knife-edged fishing selection is assumed, as retention of rock lobsters below 76 mm is prohibited, and pots incorporate escape gaps to ensure that capture of undersize rock lobsters is minimised. The retention of berried females is also prohibited.

An estimate of the instantaneous rate of fishing mortality, F, is calculated by multiplying observed fishing intensity (fishing effort per unit of area per unit of time) by a catchability coefficient, q, which is assumed constant over all depth zones and fishing regions. The result is assumed to represent the fishing mortality of non-migratory rock lobsters in the early intermoult stage of the moult cycle; this must then be multiplied by factors to adjust for the effects of other moult stages or alternate migratory status. A problem with the assumption of constant catchability is that the area of habitat utilised by the rock lobster is a small proportion of the total area, comprising the remnants of ancient submerged coastal reefs. These areas are targeted by fishermen when setting pots, with the resultant catch rates reflecting the abundance of rock lobsters in favoured habitat, rather

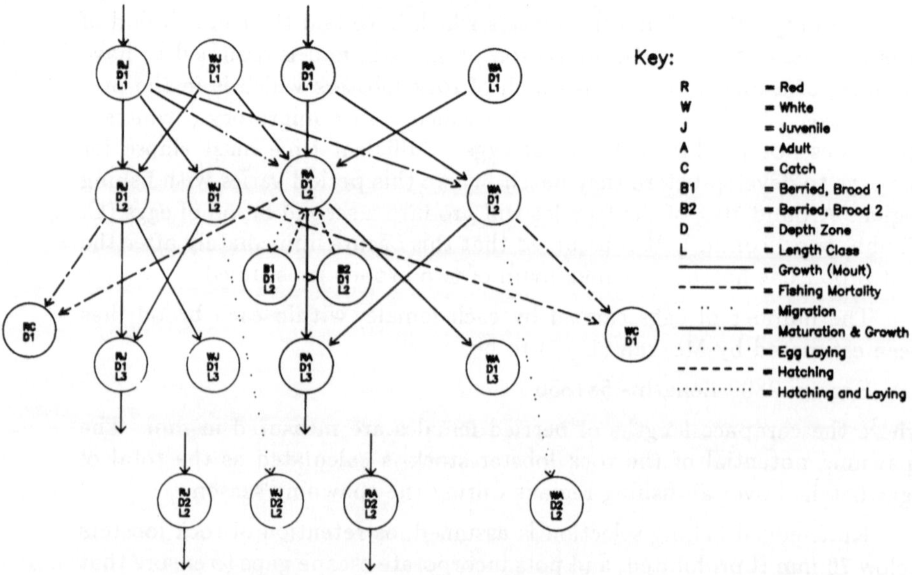

Figure 5. Flow for female rock lobsters of length-class, L2, within the shallowest depth zone, D1, where lobsters moult from length-class L1 to L2, and from L2 to L3. Natural mortality is not included in the diagram. Flows for other length-classes or depth zones are not completely represented.

than the average density over the entire depth zone. The assumption of constant catchability, although used in the model, will be the subject of future research.

The estimate of fishing mortality is multiplied by a factor reflecting the relative vulnerability of rock lobsters in a different stage of the moult cycle. Rock lobsters in the early intermoult stage reach a peak in relative vulnerability; this has been set to 1. During the remainder of the intermoult stage, the relative vulnerability is assumed to decrease to about 0.6, while in the pre-moult and post-moult stages, the relative vulnerability is assumed to be about 0.2. The estimate of fishing mortality must also be multiplied by a factor to adjust for the enhanced catchability of migratory rock lobsters. The effect of this factor is assumed to be greatest immediately after the moult at which the rock lobsters enter the migratory state, declining with elapsed time since the last moult, until the catchability is again equivalent to that of non-migratory rock lobsters.

The instantaneous coefficient of natural mortality, M, is usually assumed to be constant. However, the natural mortality of migratory rock lobsters is likely to be greater than that of the more sedentary rock lobsters, as a result of increased predation risk due to lack of shelter. The effect of predation is assumed to be greatest immediately after the moult at which time rock lobsters enter the migratory state. The effect is assumed to decline with elapsed time in a similar fashion to the enhanced catchability of migratory rock lobsters.

Only those rock lobsters in the "white" state are assumed to migrate to deeper water, and the instantaneous rate of migration from shallower water, X, is assumed to be greatest immediately after the moult at which they enter the migratory "white" state. The rate of migration declines with elapsed time after the moult, until it is again zero.

After adjusting F, M, and X for these factors for each batch of rock lobsters, the effects of mortality and migration at the time step may be approximated by:

$$\text{Proportion caught} = F[1 - \exp(-ZT)]/Z$$

$$\text{Proportion migrating} = X[1 - \exp(-ZT)]/Z$$

$$\text{Proportion at end of time step} = \exp(-ZT)$$

where T is the duration of the time step, and $Z = M + F + X$.

Live weight of the catch is calculated, and classified into a tail weight grade, using appropriate weight and length relationships.

The model is fitted to the observed data from the fishery using the simplex technique described by Nelder and Mead (1965), to minimize the sum of squared deviations of the natural logarithms of observed monthly catch in each depth zone from the logarithms of the estimates. In estimating parameters, some depth zones may be combined to reduce problem complexity, and each fishing region is processed independently. Parameters estimated include the catchability coefficient, the scaling factors for the newly introduced year-classes in the depth zones, the multipliers of catchability and mortality and the rate of movement of the migratory animals.

3. Results and Discussion

A more realistic approach to the modelling of crustacean populations has been taken by Caddy (1977, 1979, and 1987), Campbell (1985), and Ennis (1985, as reported by Botsford, 1987). They have incorporated into models of yield per recruit and eggs per recruit the processes of growth through moulting, and reproduction associated with the intermoult period, which characterise crustacean fisheries.

A similar model has been developed for the western rock lobster, which also reflects the effect of ecdysial growth on the recruitment process, catchability by pot fishing, and the duration of the paler colouration of the exoskeleton during the annual migration. These features of the system are of particular interest to the fishery managers and to the fishing industry, and must be incorporated in the model to allow exploration of alternative management strategies.

The level of complexity within our model reflects the demand for management advice on monthly catches resulting from alternate management strategies. Because market demand favours the smaller "red" rock lobsters over paler "whites" or larger lobsters, the composition of the catch resulting from the different strategies must also be determined. The need to communicate results to the fishing industry requires that our model reflects the mechanisms of growth and migration which are understood to operate within this fishery.

The structure of our model attempts to incorporate the current understanding of western rock lobster biology, however the model's ability to accurately predict future catches has yet to be fully determined. Initial estimates of the various fixed parameters and relationships used in the model are approximations, but are considered to be realistic; further analysis of available data will however be required to determine more precisely the form of these relationships, and the values of the associated parameters.

Data for the period from November, 1975, to December, 1985, for the north coastal region has been used in preliminary analyses. For these

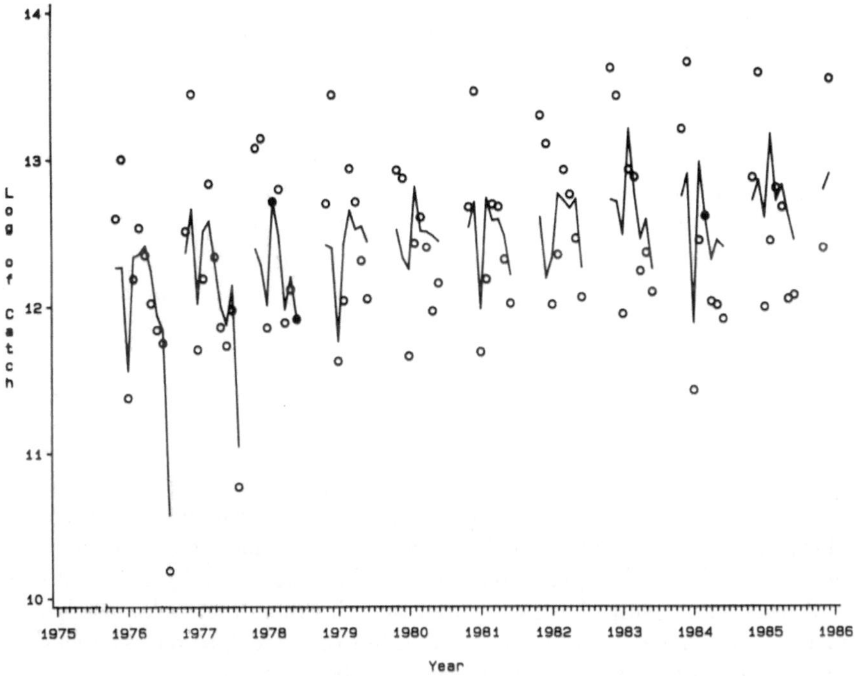

Figure 6. Monthly catches between November, 1975, and December, 1985, for the north coastal region, for depths less than 36 m. Circles represent observed catches, while the line joins catches estimated by the model.

data, a fit has resulted which explains approximately 93% of the sum of squares around the mean of the logarithms of monthly catches in depths from 0-36 m and over 36 m (Figures 6 and 7); lack of fit to the within season catch is being investigated as the study is continued. Estimations of unknown parameters are currently determined by producing estimates of monthly catches from the various depth zones which approximate the observed catches. Such an approach tends to ignore the other information which is available from the model regarding the grade (tail weight) composition of the catch. Details are available of the grade composition of the product passing through the processing establishments during each year. It is possible to utilize this information in parameter estimation by adding an auxiliary sum of squares term (Deriso *et al.*, 1985). The resulting estimates of the parameters would then reflect both the requirement that catches are estimated well, and also that the resulting grade composition of the catch agrees with the grade composition observed in processing establishments.

A stock-recruitment relationship is included in the model to relate total egg production to the subsequent recruitment index. However, parameters of this relationship reflect the fact that, at present, spawning stock is not considered to be limiting recruitment; management of the fishery has the objective of ensuring that the spawning stock is not reduced to such a level that the abundance of spawning females becomes the major determinant of subsequent recruitment. Factors presently limiting recruitment to the fishery are the survival of the larval and juvenile phases of the rock lobster. Where observed values of the recruitment index are available, or where predictions of the index are available from observed puerulus and juvenile indices, these are used in preference to the values estimated by the stock-recruitment relationship. It was required that this more precise knowledge of future recruitment strengths be used within the model, as management strategies must be based on these more accurate predictions.

The model allows a range of scenarios of effort redistribution to be examined, by specifying the monthly effort to be directed towards each depth zone, within each fishing region, over a number of fishing seasons. The model is also being extended to account for changes in fishermen's behaviour, by dynamically allocating effort to the various depth zones in accordance with the month and the abundance of legal-sized rock lobsters. Without considering effort redistribution, changes in management strategies such as a reduction in the level of fishing during the annual migration, an extension of the length of open fishing season, or a change in the minimum legal size for retention of rock lobsters, are unlikely to provide adequate predictions of the possible impact on the spawning potential and yield from the fishery.

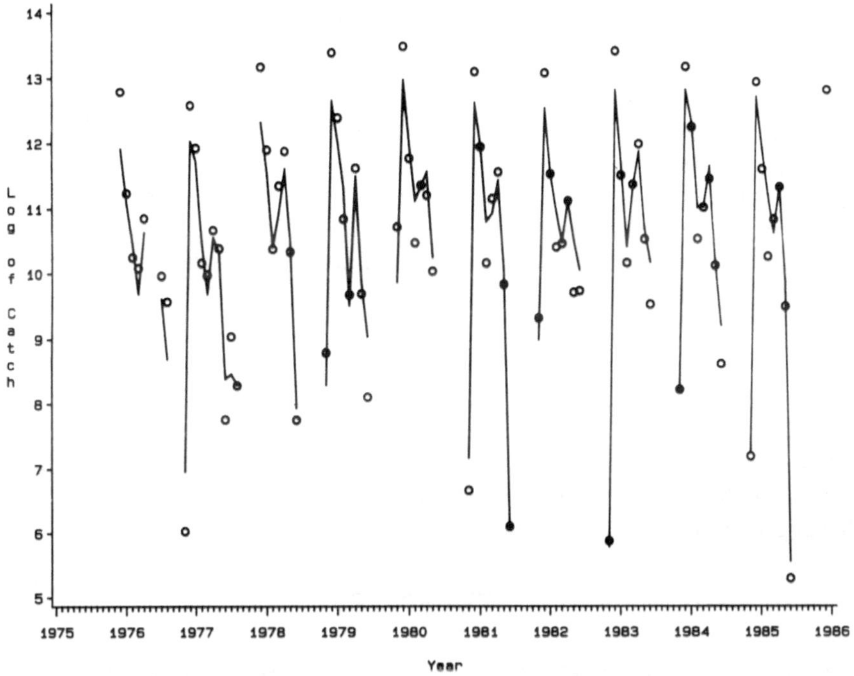

Figure 7. Monthly catches between November, 1975, and December, 1985, for the north coastal region, for depths greater than 36 m. Circles represent observed catches, while the line joins catches estimated by the model.

While the model is currently deterministic, the stochastic nature of the rock lobster fishery cannot be ignored. At a later stage, the model will need to be extended to reflect the variability inherent in the real world. This may be introduced by including random variation about the structural relationships. Variation in recruitment is likely to have the greatest impact on the results predicted by the model. This variation is likely to be related to environmental factors, and will not necessarily be completely random. As indicated earlier, use of indices of recruitment strength obtained from research observations of the stock ensures that this factor is considered within the model.

The model, in its current form, provides a good representation of the biological processes of the rock lobster and of the associated fishery. While further validation and analysis of data may result in minor changes to the form of the relationships, and to the values of fixed parameters, the basic structure of the model is unlikely to change. A significant benefit of the modelling process is that it focuses research attention on the critical areas; this benefit is already being realised.

Acknowledgements: The authors would like to thank M. Zingross for her help in preparing the figures, and C. Chubb and E. Barker for the help they gave in providing required information, and for their helpful suggestions. They wish also to express their appreciation to Dr J. Penn, and other Research Officers at the Western Australian Marine Research Laboratories for suggestions made.

REFERENCES

[1] Botsford, L. W. 1987. Crustacean egg production and fisheries management, *Crustacean Egg Production. A symposium at the Annual Meeting of the American Society of Zoologists*, New Orleans, December, 1987.

[2] Bowen, B. K. 1971. Management of the western rock lobster (*Panulirus longipes cygnus* George). *Proc. Indo-Pac. Fish. Counc.* **14(II)**, pp. 139-153.

[3] Bowen, B. K. 1980. Spiny lobster fishery management. *The biology and management of lobsters Vol. II*, pp. 243-264, J. S. Cobb and B. F. Phillips (editors), Acad. Press, N.Y..

[4] Bowen, B. K. and D. A. Hancock. 1989. Effort limitation in the Australian rock lobster fisheries. *Marine invertebrate fisheries: their*

assessment and management, pp. 375-393, Caddy, J. F. (editor), John Wiley and Sons, New York.

[5] Caddy, J. F. 1977. Approaches to a simplified yield-per-recruit model for Crustacea, with particular reference to the American lobster, *Homarus americanus*. *Fish. Mar. Serv. MS Rep.* 1445, 14p.

[6] Caddy, J. F. 1979. Notes on a more generalized yield per recruit analysis for crustaceans using size-specific inputs. *Fish. Mar. Serv. MS Rep.* 1525.

[7] Caddy, J. F. 1987. Size-frequency analysis for crustacea: moult increment and frequency models for stock assessment. *Kuwait Bull. Mar. Sci.*, **9**, pp. 43-61.

[8] Campbell, A. 1985. Application of a yield and egg-per-recruit model to the lobster fishery in the Bay of Fundy. *North American Journal of Fisheries Management*, **5**, pp. 91-104.

[9] Caputi, N. and R. S. Brown. 1986. Relationship between indices of juvenile abundance and recruitment in the western rock lobster (*Panulirus cygnus*) fishery. *Can. J. Fish. Aquat. Sci.* **43**, pp. 2131-2139.

[10] Caputi, N., R. S. Brown and B. F. Phillips. 1988. Forecasting rock lobster catches - check and double check. Fisheries Department, Western Australia. *FINS*, **21(2)**, pp. 18-22.

[11] Chittleborough, R. G. 1970. Studies on recruitment in the Western Australian rock lobster *Panulirus longipes cygnus* George: density and natural mortality of juveniles. *Aust. J. Mar. Freshwat. Res.*, **21**, pp. 131-148.

[12] Deriso, R. B., T. J. Quinn II, and P. R. Neal. 1985. Catch-age analysis with auxiliary information. *Can. J. Fish. Aquat. Sci.*, **42**, pp. 815-824.

[13] Ennis, G. P. 1985. An assessment of the impact of size limit and exploitation rate changes on egg production in a Newfoundland lobster population. *North Am. J. Fish. Management.* **5**, pp. 86-90.

[14] George, R. W. (1958). The biology of the Western Australian commercial crayfish *Panulirus longipes*. Ph.D. Thesis, Univ. Western Australia, Nedlands, 124p.

[15] Hancock, D. A. 1981. Research for management of the rock lobster fishery of Western Australia. *Proc. Gulf. Caribb. Fish. Inst.* **33**, pp. 207-229.

[16] Joll, L. M. and B. F. Phillips. 1984. Natural diet and growth of juvenile western rock lobsters *Panulirus cygnus* George. *J. Exp. Mar. Biol. Ecol.* **75**, pp. 145-169.

[17] Morgan, G. R. 1977. Aspects of the population dynamics of the western rock lobster and their role in management. Ph. D. Thesis, Univ. Western Australia.

[18] Morgan, G. R. 1980a. Population dynamics and management of the western rock lobster fishery. *Marine Policy*, January 1980, pp. 52-60.

[19] Morgan, G. R. 1980b. Increases in fishing effort in a limited entry fishery - the western rock lobster fishery 1963-1976. *J. Cons. int. Explor. Mer*, **39**, pp. 82-87.

[20] Morgan, G. R. and E. H. Barker. 1982. The western rock lobster fishery 1961-1971. *West. Aust. Dep. Fish. Wildl. Rep.*, **55**, pp. 41.

[21] Morgan, G. R., B. F. Phillips, and L.M.Joll. 1982. Stock and recruitment relationships in *Panulirus cygnus*, the commercial rock (spiny) lobster of Western Australia. *Fish. Bull.* 80(3), 475-486.

[22] Nelder, J. A. and R. Mead. 1965. A simplex method for function minimization. *Computer Journal*, **7**, pp. 308-313.

[23] Pearce, A. F. and B. F. Phillips. 1988. ENSO events, the Leeuwin Current, and larval recruitment of the western rock lobster. *J. Cons. int. Explor. Mer*, **45**, pp. 13-21.

[24] Phillips, B. F. 1981. The circulation of the southeastern Indian Ocean and the planktonic life of the Western Rock Lobster. *Oceanography of Marine Biology Annual Review*, **19**, pp. 11-39.

[25] Phillips, B. F. 1986. Prediction of commercial catches of the western rock lobster *Panulirus cygnus*. *Can. J. Fish. Aquat. Sci.*, **43**, pp. 2126-2130.

PARTICIPANT'S COMMENTS

Few of us at the conference live with the real-world consequences of applying our models to biological systems. This is not so for Hall 's *et al* model. Their model is used to make management decisions that directly influence the livelihood of the fisherman, and the abundance and persistence of rock lobster. In this way, their modelling objective is somewhat akin to Mike Fisher's which was to suggest better means of allocating insulin doses to diabetics. In contrast, other models presented at the conference were intended to: test ideas, illustrate phenomena (both biological and mathematical), provide heuristics, and develop mathematical tools. Hall *et al* are at once modellers and managers.

There are at least two demands placed upon Hall *et al* 's model. The first requires the model to accurately predict the future abundance of commercially valuable lobsters. Such predictions can assist fishermen in deciding whether future seasons will be good or bad. For instance, it would be of interest to know how well the model tracked the decline in lobster harvests from the 81/82 season through the 85/86 season (see Fig. 2).

The second demand requires that the model accurately incorporate the effects of fishermen as predators on the dynamics of their prey. Such accuracy assists in determining sustainable yields and in determining the risks to the fishery from over-harvesting. In asking how effective is the present model as a tool for steering the future course of the fishery, I am reminded of the childhood experience of "driving" cars at amusement parks. The cars, of course, were on tracks; and the cars, of course, for realism had steering wheels. I would faithfully steer my car along the "roadway" convinced that my driving prowess was averting accidents at every turn. In short, the true test of a management model's utility comes from its successful application.

Joel Brown

This is a fascinating account of the way in which a relatively simple mathematical model can be used to obtain reliable predictions, of paramount importance for management, of population dynamics in a major marine fishery (or crustaceary!). It illustrates how a predictive management model has to be tailored to match the kind and precision of field data that is available at reasonable cost. A particular feature of interest in this Australian rock lobster population model is that it is stage-structured, moving individual animals through classes that incorporate size, location, moult stage and reproductive state, instead of simply age.

Robert McKelvey

Legumes at Loggerheads: Modelling Competition Between Two Strains of Sub-Clover

ANTHONY G. PAKES

Abstract

In essence this paper is a précis of a substantial investigation of deterministic modelling of two strains of sub-clover which interact competitively at the time they set their seed.

1. Introduction

The annual legume subterranean clover (sub-clover) is of Mediterranean origin and is an important pasture crop in southern Australia and parts of the U.S.A. It propagates by means of 'soft' or germinable seeds, whilst retaining a store of 'hard' dormant seeds for germination in future years. Agronomists in Western Australia have conducted long term competition experiments (up to 7 years) in which binary mixtures of strains of sub-clover are grown together with the object of determining, *inter alia*, whether long term coexistence is possible; if not then which strain will win the competition and how long will it take to do so? Since data is accumulated so slowly, it is quite difficult to answer these questions, even if environmental conditions are assumed to be temporally unchanging.

An approach to these problems is to build analytically tractable mathematical models which capture important aspects of reality. By using field data to estimate model parameters it is possible to predict the behaviour of sub-clover mixtures, and this can aid agronomists in selecting strains best suited for pastoral use.

This paper summarises the main findings of work reported in Pakes and Maller (1989) which arose from requests to construct and analyse such models. Field aspects of sub-clover competition together with applications of the simplest model (called 1A below) have been published by Rossiter et al. (1985). Model formulation is discussed in Section 2, and in Section 3 we describe some of their properties. Their application to field data is covered in Section 4, and in Section 5 we outline directions for further work. Some mathematical issues were raised during the oral presentation and they are discussed in Section 6.

2. Outline of model formulation

The models are based on a straightforward life history analysis of sub-clover. Consider a new monoculture of sub-clover at the beginning of summer (during December in southwestern Australia) – see Figure 1. All plants have set their seed and died, and this seed is in a burr form which lies on, or just beneath, the soil during summer. A proportion P_S of it survives summer to remain as viable burr seed by the break of season, i.e. the end of summer, around April or May. Temperatures are lower by then, and there is occasional rain. [Summer is very dry].

Viable seed may be released from the burr to become 'free', with a probability $P(F|B)$, or it remains in burr form with probability $P(B|B) = 1 - P(F|B)$. Both burr and free seed are initially 'hard', i.e. impermeable to water and unable to germinate. Over summer this seed can soften, i.e. become permeable and hence germinable, and the probabilities of it doing so are $P(S|B)$ and $P(S|F)$ for burr and free seed, respectively. Residual hard burr and free seed is carried over to the next year, provided it survives the intervening summer.

A fraction P_E of the soft seed pool gives rise to established plants and a proportion P_A of these survive to adulthood; ungerminated soft seed is lost. By the start of the second summer adult plants have set their seed; M seeds per plant. At the end of this summer the seed bank contains surviving one year old burr seed (produced by the start of that summer) and two year old burr and free seed. The cycle continues in like manner with the seed bank accumulating hard burr and free seed of varying ages. The most general case allows transition probabilities from burr to free seed, burr to soft seed, and so on, to be age dependent. Where relevant we indicate this

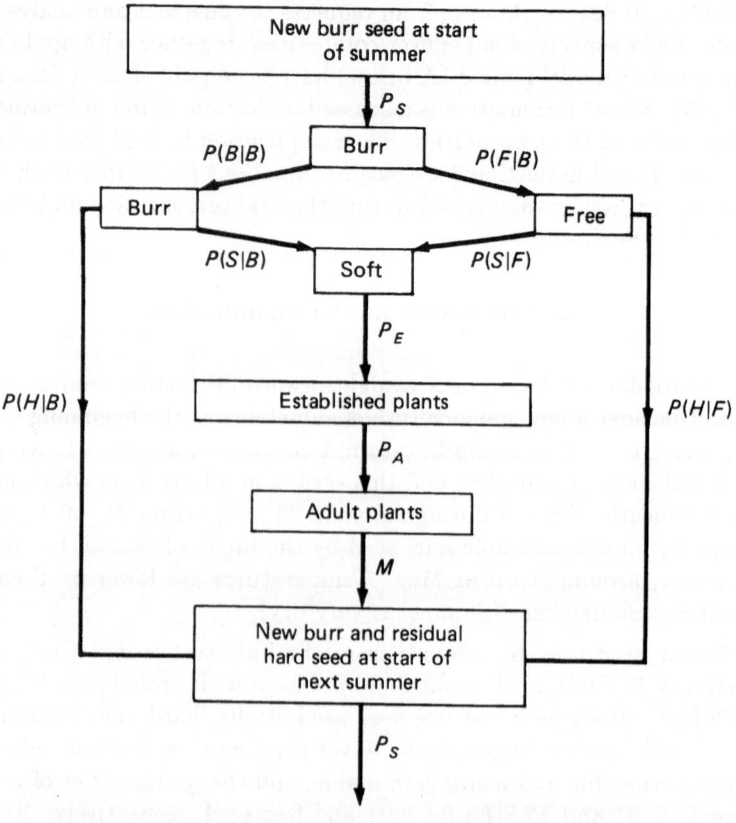

Figure 1. The life history over one year of a new monoculture.

with numerical subscripts; e.g., $P_i(S|B)$ is the softening rate of i-year old burr seed.

We now consider what happens when two strains are grown together. Experience has shown that their growth dynamics can be modelled as if they develop independently up to the time that mature plants set their seed. The numbers set depend on the densities of both strains at this time and, on advice from agronomists, we have chosen to handle this using the de Wit replacement model, see Braakhekke (1980), Chapter 4. While it receives good empirical support, until recently it had little theoretical support – but see Gates and Westcott (1988).

Let M_i be the monoculture seed yield of a typical strain-i plant ($i = 1, 2$) and let Y_i be the corresponding yield in the mixture. The de Wit model assumes that seeds of one strain replace those of the other in the sense that

$$\frac{Y_1}{M_1} + \frac{Y_2}{M_2} = 1.$$

It also assumes that the relative proportion of daughter seeds is proportional to the relative proportions of parent numbers. So if Z_i is the initial number of strain-i plants, adjusted for relative seed sizes, then

$$\frac{Y_1}{M_1} / \frac{Y_2}{M_2} = c\frac{Z_1}{Z_2}$$

where c is a 'de Wit' constant measuring relative crowding of Strain 1 by Strain 2 (and denoted by k_{12} in the botanical literature). These equations give the yield relationships

$$Y_1 = M_1 \frac{cz}{[(c-1)z+1]} \text{ and } Y_2 = M_2 \frac{d(1-z)}{[(d-1)(1-z)+1]}$$

where $z = Z_1/(Z_1 + Z_2)$ and d (or k_{21}) is determined by

$$cd = 1. \tag{1}$$

In using the de Wit model it is often convenient to assume that c and d do not satisfy (1). For example, we might set $cd > 1$ if the strains compete for different resources and mutually benefit each other, and then the total seed yield can exceed the larger of the monoculture yields. In such more general cases (1) is called the de Wit condition.

Two sub-clover strains grown together develop independently except for their interaction via the de Wit model when seed is set. The most general situation we consider is illustrated in Figure 2 which shows the flow of events for one year in the life of Strain 1, and it is completed by imagining a mirror image on the left hand side representing life history events for Strain 2. We denote quantities for Strain 2 by using the next letter of the alphabet from the corresponding Strain 1 quantity, e.g., $P_i(S|B)$ and $Q_i(S|B)$. In our experience age dependent softening rates are zero for seed

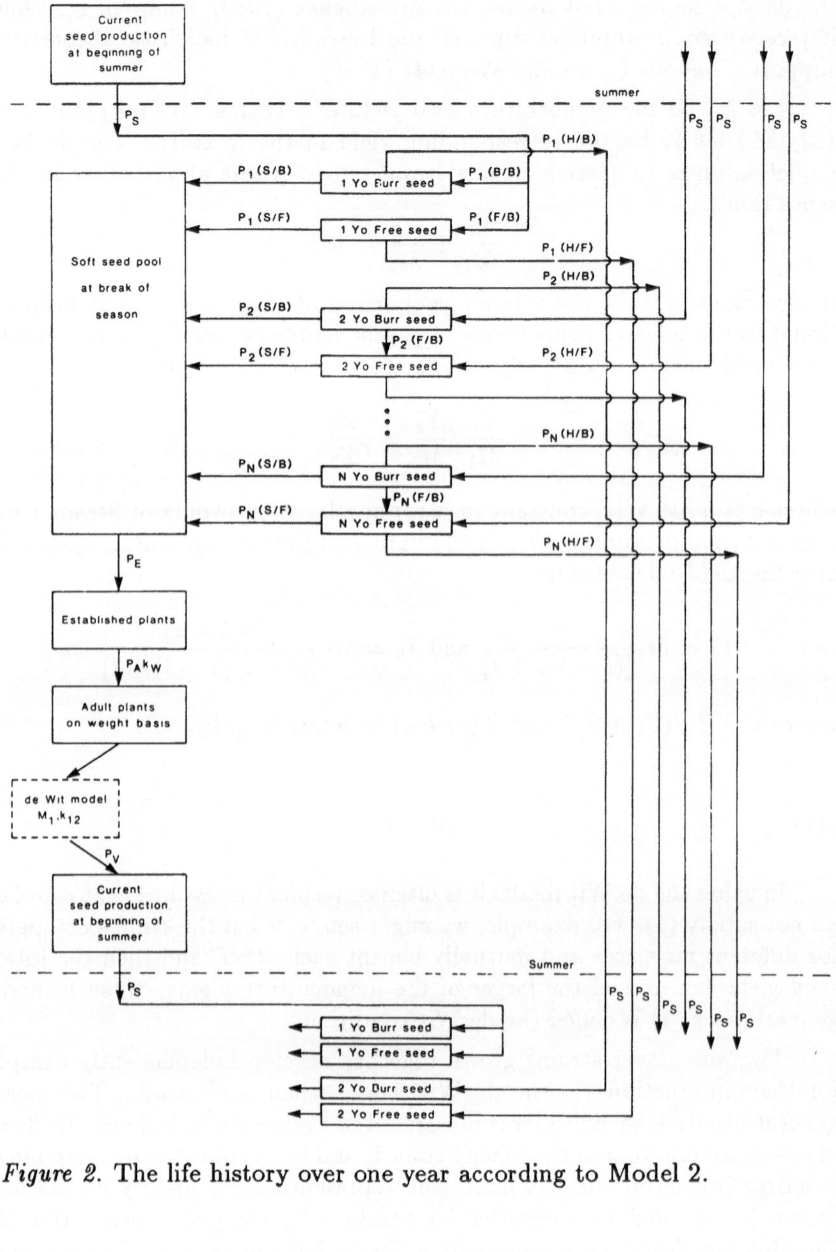

Figure 2. The life history over one year according to Model 2.

aged over seven years, and the resulting models are then said to have a finite memory.

The seed numbers of both strains are discounted by a factor P_V (typically time varying) to allow for possible intrusion by so called volunteer species whose presence reduces sub-clover densities. This factor decreases from unity over seven years and then remains constant. The finite memory and eventual constancy of P_V allows us to assess the asymptotic behaviour of our models.

We have considered two models which are differentiated by the finesse which can be achieved in censusing the seed bank. In Model 1 residual burr and free seed are differentiated only at the time of seed softening and hence transition rates from burr to free seed are independent of age. Model 1 has the property that the proportion of Strain 1 free seed is always exactly $P(F|B)$.

Model 2 applies where burr and free seed of all ages are separately counted. Figure 2 depicts this situation. The simplification offered by Model 1 is more apparent than real — the mathematical structure of both models is the same. Let x_n and y_n denote the numbers of new seeds produced by Strains 1 and 2, respectively, at the beginning of summer in year n. A careful accounting of the flow of seed from one year to the next (Pakes and Maller (1989), Chapter 2) yields the following pair of equations

$$(2a) \qquad x_n = \frac{Lc\sum_{1\leq i<n} a_i x_{n-i}}{Lc\sum_{1\leq i<n} a_i x_{n-i} + M\sum_{1\leq i<n} b_i y_{n-i}}$$

and

$$(2b) \qquad y_n = \frac{Md\sum_{1\leq i<n} b_i y_{n-i}}{L\sum_{1\leq i<n} a_i x_{n-i} + Md\sum_{1\leq i<n} b_i y_{n-i}}$$

Here $L = M_1 P_V$, $M = M_2 P_V$, $a_i = M_1 P_E P_A W S_i$ and $b_i = M_2 Q_E Q_A T_i$ and W is a factor accounting for the seed weights of each strain. The S_i and T_i are fairly complicated combinations of the survival and transition rates for each strain; they satisfy $S_i = 0$ implies $S_{i+1} = 0$, and similarly for the T_i. Define the memory length

$$N = sup\{i : S_i T_i > 0\}.$$

Other quantities can be derived from the the new seed numbers, e.g. the total hard seed of all ages in year n is

$$t_n = x_n + \sum_{1 \le i < n} P_S^i H_i x_{n-i}$$

where H_i is another parameter depending on the age dependent transition probabilities. A particular case of Model 1, called Model 1A, gives a useful mathematical simplification. For this we assume that all rates of intra-seed bank transitions are independent of seed age; see Figure 1. The S_i and T_i now form geometric sequences, whence the total hard seed numbers t_n and u_n are given by the first order difference equations

$$t_{n+1} = \alpha t_n + \frac{cUt_n}{(cXt_n + Yu_n)} \tag{3a}$$

and

$$u_{n+1} = \beta u_n + \frac{dVu_n}{(Xt_n + dYu_n)} \tag{3b}$$

Here U and V are positive constants depending on nearly all the model parameters, $X = U/L$, $Y = V/M$ and α and β depend on the transition rates. We always have $0 < \alpha, \beta < 1$. Within the rather broad limits of experimental variability of the field trials, it has been found that Model 1A gives an acceptable fit to the data available to us; see Rossiter, Maller and Pakes (1985) for details.

3. Basic properties of the models

First we consider Model 1A with initial conditions u_1, $t_1 > 0$. This is a dynamical system $\{(t_n, u_n) : n \in \mathbb{N}\}$ whose state space $\mathcal{S}^\circ$ is the interior of $\mathcal{S} = \mathbb{R}_+^2 \backslash \{0\}$. It has a pair of linear isoclines,

$$\mathcal{I}_t : cXt + Yu = \frac{cU}{(1-\alpha)} \quad for\ u,t \ge 0$$

and

$$\mathcal{I}_u : Xt + dYu = \frac{dV}{(1-\beta)} \quad for\ u,t \ge 0.$$

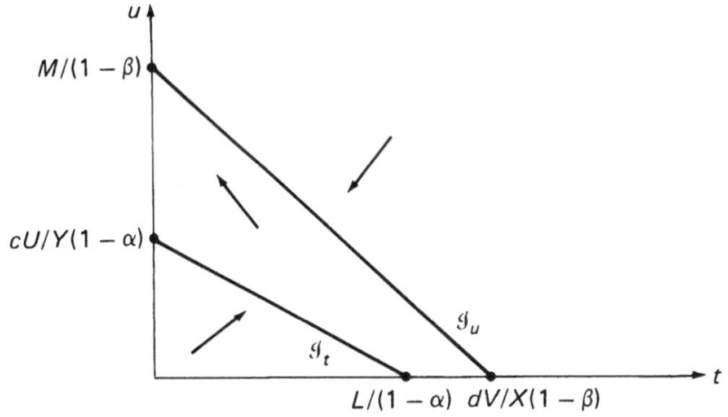

Figure 3. Typical configurations of the isoclines and directions of motion for Model 1A corresponding to Outcome B.

The subscripts are chosen so that if $(t_n, u_n) \in \mathcal{I}_t$ then $t_{n+1} - t_n = 0$ and if $(t_n, u_n) \in \mathcal{I}_u$ then $u_{n+1} - u_n = 0$. The system can be started from the positive parts of the co-ordinate axes and then the state space is $\mathcal{S}$. It is easily checked that the boundary points $(0, M/(1-\beta))$ and $(L/(1-\alpha), 0)$ are always equilibria of the system; see Figure 3 for a typical configuration. The isoclines can intersect in $\mathcal{S}^\circ$ only if the de Wit condition (1) is violated. More precisely let

$$\rho = \frac{V(1-\alpha)}{U(1-\beta)} \tag{4}$$

and when $cd \neq 1$ let,

$$\bar{p} = \frac{d(c-p)}{(cd-1)} \text{ and } \bar{q} = \frac{c(\rho d - 1)}{(cd-1)}$$

The isoclines intersect in $\mathcal{S}^\circ$ iff $\bar{p}, \bar{q} > 0$ and then this point is an equilibrium given by

$$(t_a, u_a) = \left(\frac{L\bar{p}}{(1-\alpha)}, \frac{M\bar{q}}{\rho(1-\beta)}\right).$$

Consider, e.g., Figure 3 in which $\mathcal{I}_u$ lies above $\mathcal{I}_t$ and the arrows indicate the general directions of flow of the system. The regions below $\mathcal{I}_u$, and above $\mathcal{I}_t$, are invariant, i.e. an orbit entering one of them cannot subsequently leave it, whence the trapezoidal region $\mathcal{T}$ between the isoclines is invariant. These properties ensure that the boundary equilibrium $(0, M/(1-\beta))$ is globally attracting and stable. Similarly, the possible long term outcomes predicted by Model 1A are as follows. We assume that at least one inequality in each of the following pairs is strict.

(A) $c/\rho \geq 1$ and $\rho d \leq 1$, and then Strain 1 wins; or

(B) $c/\rho \leq 1$ and $\rho d \geq 1$, and then Strain 2 wins; or

(C) $c/\rho \geq 1$ and $\rho d \geq 1$, and there is a globally attracting stable interior equilibrium (t_a, u_a); or

(D) $c/\rho \leq 1$ and $\rho d \leq 1$, and then (t_a, u_a) is an unstable interior equilibrium.

In Case (C) it is possible to exhibit a Liapunov function to prove the stability result. In Case (D) the boundary equilibria have basins of attraction which are open and simply connected and Liapunov functions can be used to obtain inner estimates of them. Many other qualitative properties of the orbits can be determined, see Pakes and Maller (1989), Chapters 4 and 5.

Only Case (B) is important for our field data, so we assume this case in the sequel. Let

$$\Delta = \alpha + \frac{Uc(1-\beta)}{V}.$$

The sign of $\Delta - \beta$ determines the rate at which orbits approach the stable equilibrium. Typically $\beta < \Delta$ and then $t_n \sim const.\Delta^n$ and $M/(1-\beta) - u_n \sim const.\Delta^n$ as $n \to \infty$; the unidentified constants are positive and depend on the entire history of the orbit. This result shows that the attracting boundary equilibrium is approached from *within* $\mathcal{T}$, i.e., all orbits enter $\mathcal{T}$ and approach $(0, M/(1-\beta))$ along a linear asymptote through this point and whose slope is $-X(1-\beta)/Yd(\Delta-\beta)$, between the slopes of the isoclines. Figure 4 shows typical orbit configurations for this case. The situation when $\beta > \Delta$ is briefly discussed near the end of Section 6.

Consider now the system defined by (2). Strictly speaking this does not define a two-dimensional dynamical system because (x_n, y_n) may depend on the entire past history $\{(x_j, y_j) : j < n\}$. In the practically relevant case of a finite memory N we can imbed our model in a dynamical system

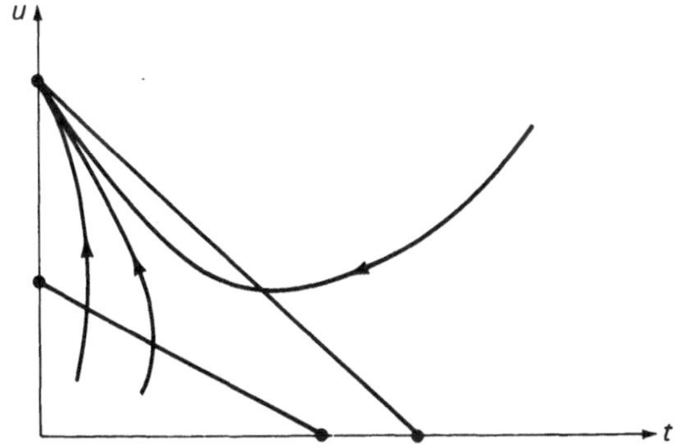

Figure 4. Typical orbital configurations corresponding to Figure 3 with $\beta < \Delta$.

whose state space is $\mathbb{R}_+^{2N} \setminus \{\underline{0}\}$. A very important invariance property of (2) is that for $n \geq 2$ the motion is confined to the one-dimensional manifold

$$y = \frac{M(1 - x/L)}{[1 - x(cd - 1)/Lcd]};$$

which joins the points $(L, 0)$ and $(0, M)$.

The boundary points $(0, M)$ and $(L, 0)$ are equilibria of (2), but there is never an interior equilibrium. Instead the following occurs. Let

$$A = \sum_{j \geq 1} a_j \text{ and } B = \sum_{j \geq 1} b_j,$$

suppose $cd \neq 1$ and let

$$x_a = \frac{d(ALc - BM)}{(cd - 1)A} \text{ and } y_a = \frac{c(BMd - AL)}{(cd - 1)B}.$$

If x_a and y_a are positive and if we have $x_1 = \ldots = x_N = x_a$ and $y_1 = \ldots y_N = y_a$ then $(x_n, y_n) = (x_a, y_a)$ for all n. Thus (x_a, y_a) is a kind of 'quasi-equilibrium' point of (2). The transformation which takes the system (2) into the above mentioned 2N-dimensional dynamical system takes this quasi-equilibrium into an actual interior equilibrium point, and it is the only such point.

When $N < \infty$ the long term behaviour of (2) can be resolved in all cases of practical interest by deriving bounds for x_{nN+j} $(j = 0, \ldots N-1)$ in terms of the n-fold functional iterate of a certain monotonic function. This approach also yields estimated convergence rates. If we redefine ρ by

$$\rho = \frac{B}{A}$$

then, in essence, Outcomes (A)-(C) above continue to hold for Model 2, but with (C) being weakened to: (x_a, y_a) is globally attracting for initial points in $\mathcal{S}^o$. Under the conditions for Outcome (D) it can be shown that both boundary equilibria are locally stable. Additional stringent conditions on the a_i and b_i allow the proofs to be pushed through when $N = \infty$. A local analysis giving convergence rates is possible without these conditions. However, a general global theory for the infinite memory case does not yet exist.

4. Application of the models.

Model behaviour has been examined using parameter values estimated from field data gathered over many years by Western Australian agronomists. We consider the so-called main mixtures:

Dwalganup and Daliak;
Yarloop and Seaton Park; and
Seaton Park and Midland B.

Each was sown as a 50-50 combination in 1970 and observed (and grazed) under field conditions until 1977, and the results of this gave estimates of summer survival, and establishment and softening rates. The de Wit constants and monoculture yields were derived from other experiments. Some of the transition rates, e.g. from burr to free seed, are little more than educated guesses by the agronomists.

It appears that $c = d = 1$ for each mixture, and Outcome B occurs in each case, i.e., all models predict that the second strain of each mixture eventually displaces the first. Table 1 below shows the most important derived parameters for Model 1A.

Table 1. Stability Analysis for Model 1A

	$\frac{c}{\rho}$	$d\rho$	Outcome	u_a	y_a	β	Δ
Dwal.-Daliak	.519	1.926	B	183.5	87.3	.524	.757
Yarloop-S. Park	.922	1.084	B	73.0	43.2	.408	.951
S.Park-Mid.B	.977	1.024	B	130.1	59.1	.546	.986
				(t_a, u_a)	(x_a, y_a)	R	
S.Park-Mid.B*	1.045	1.024	C	(47.9, 46.7)	(28.4, 21.7)	.992	

*$c = 1.07$, $d = 1$ and all other parameters values are standard data values for Seaton Park-Midland B.

Since $d = 1$ the third column actually lists the values of ρ. Note that $\beta < \Delta$ in each case, so the boundary equilibrium $(0, u_a)$ is approached from within $\mathcal{T}$; the isoclines are parallel by virtue of the de Wit condition. Both ρ and Δ are close to unity for the Yarloop-Seaton Park and Seaton Park-Midland B mixtures. The condition $\rho \cong 1$ means that the isoclines are very close and hence once an orbit enters $\mathcal{T}$ the relation between t_n and u_n is very nearly linear. The condition $\Delta \cong 1$ means that the orbit takes a very long time to make its excursion through $\mathcal{T}$, i.e., Strain 2 takes a long time to effectively displace Strain 1. Indeed a Seaton Park-Midland B mixture started with a small proportion of Midland B requires about 500 years to secure the effective elimination of Seaton Park.

Further insight can be gained by inspecting typical phase portraits for Model 1A. Phase portraits of scaled versions of $\{(t_n, u_n)\}$ (in fact of $\{(p_n, q_n)\}$ as defined in Section 6) were prepared with the package PHASER (Koçak (1986)), and they are shown in Figures 5 and 6. Starting positions for the orbits were taken around the edges of the window shown in the figures, except that initial positions near the axes had a value of .1 for at least one coordinate. Orbital configurations are as predicted, except for the time development of the Seaton Park-Midland B mixture. All orbits reach $\mathcal{T}$ in a few (≤ 8) years, and then abruptly change direction to move very slowly through $\mathcal{T}$ toward the attracting boundary point. We discuss this further in Section 6.

In Figure 7 we show the values of the maximum seed pools observed in the seven year mixture experiments together with values predicted from

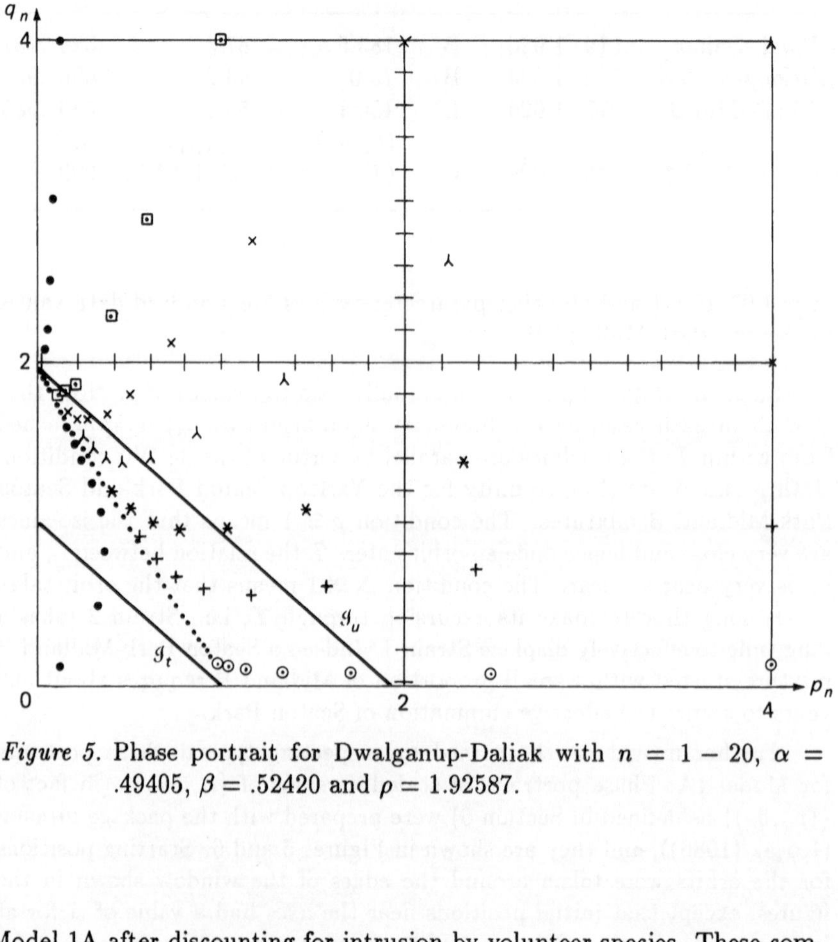

Figure 5. Phase portrait for Dwalganup-Daliak with $n = 1 - 20$, $\alpha = .49405$, $\beta = .52420$ and $\rho = 1.92587$.

Model 1A after discounting for intrusion by volunteer species. These comparisons are accepted as showing that Model 1A adequately describes the field situation-see Rossiter, Maller and Pakes (1985).

The Seaton Park-Midland B mixture is 'structurally unstable'; a small change of c from unity will cause the isoclines to intersect inside $\mathcal{S}^{o}$ at an internal equilibrium. In the last row of Table 1 we have set $c = 1.07$. The equilibrium (47.9,46.7) so created is globally attracting and stable, but

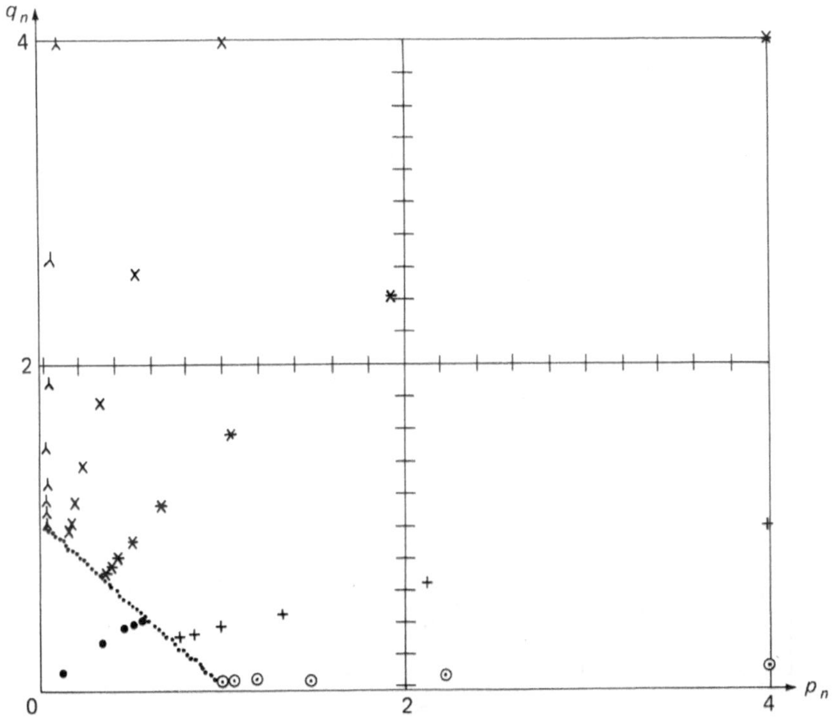

Figure 6. Phase portrait for Seaton Park-Midland B with $n = 1 - 300$, $\alpha = .40795$, $\beta = .54590$ and $\rho = 1.02388$.

again it is approached very slowly. This is indicated by the quantity R=.992 which has the property that $t_n - t_a \sim const.R^n$ and $u_n - u_a \sim const.R^n$. In Figure 8 we show some orbits which started in a neighbourhood of (t_a, u_a). Upon entering the wedges between the isoclines they move almost linearly, and very slowly, toward the equilibrium. For comparison, Figure 9 shows the phase portrait for Dwalganup-Daliak with $c = 3$.

Table 2 displays the important derived parameters for Models 1 and

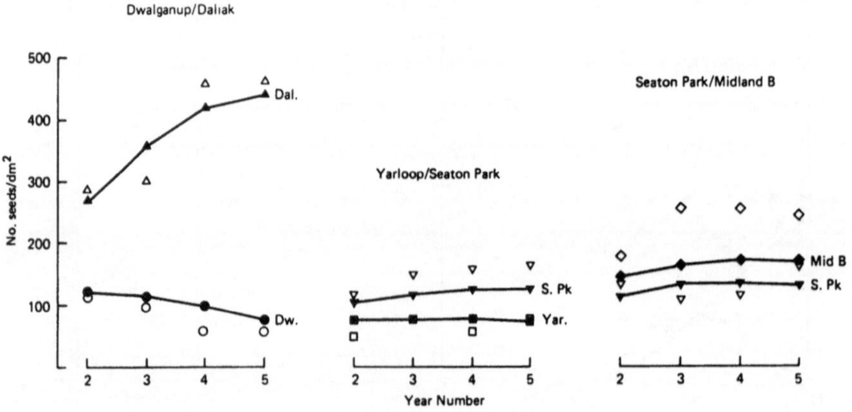

Figure 7. Maximum seed numbers at the end of years 2-5 after allowing for volunteer species intrusion. Lines and closed shapes indicate model predictions and open shapes represent observed values. (Adapted from Rossiter et. al. (1985)).

2. The figures in the first two panels show that these models predict that the second member of each mixture will eventually displace the first, but only after a very long time for the second and third mixtures. For a given mixture, all models predict the same limiting seed numbers, but Model 1A predicts a slightly slower convergence rate for Yarloop-Seaton Park

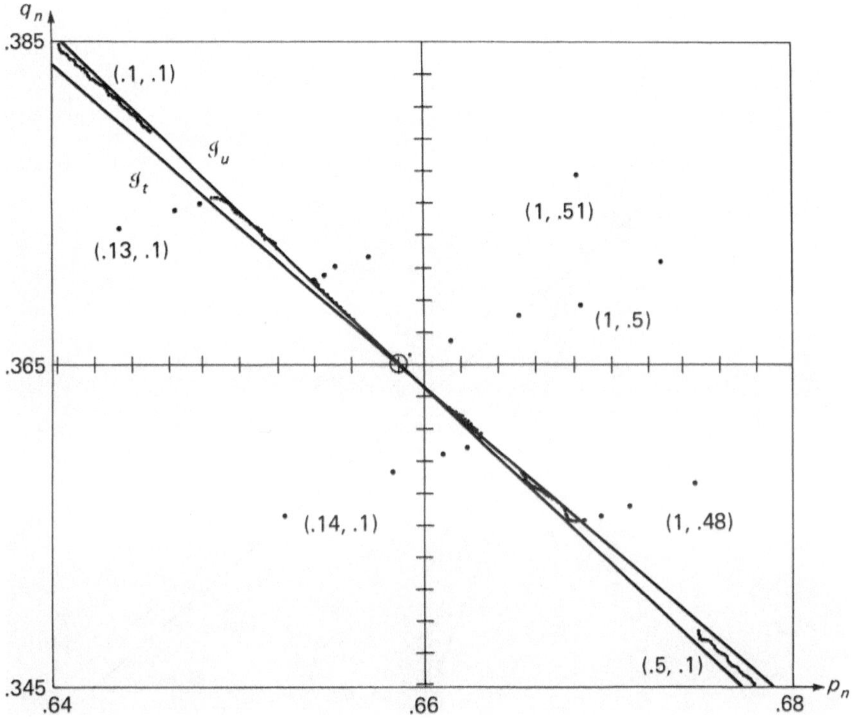

Figure 8. Phase portrait for Seaton Park-Midland B with $c = 1.07$ and all other parameters as for Figure 6. Orbits are labelled with their initial points and they are plotted after entering the portion of the phase space that is shown. The orbit starting at (.1,.1) was run for $n \leq 200$, $n \leq 400$ for (.5,.1), and $n \leq 60$ for all other orbits.

than do Models 1 and 2, but the reverse is true for the other mixtures. However, these differences are minor and we conclude that all models are in substantial agreement.

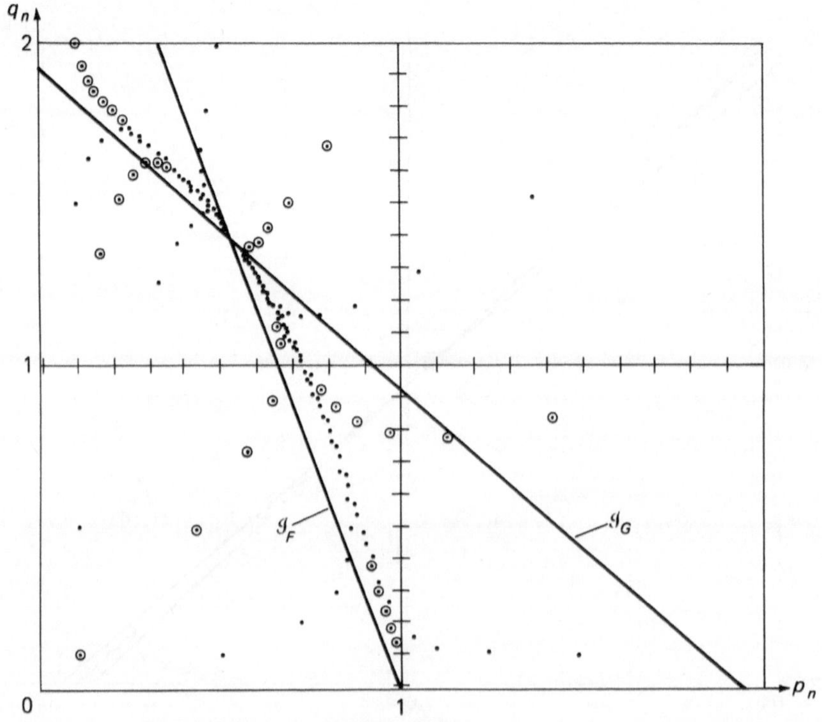

Figure 9. Phase portrait for Dwalganup-Daliak with the same parameters as for Figure 5, but with $c = 3$.

Table 2. Stability analysis for Models 1 and 2.

	$\frac{c}{\rho}$	$d\rho$	(x_a, y_a)	R
Model 1.				
Dwal.-Daliak	.5121	1.9527	(0,87.3)	.761
Yarloop-S. Park	.9145	1.0935	(0,43.2)	.949
S. Park-Mid. B	.9945	1.0055	(0,59.1)	.997
Model 2.				
Dwal.-Daliak	.5090	1.9645	(0,87.3)	.767
Yarloop-S. Park	.9115	1.0971	(0,43.2)	.949
S. Park-Mid. B	.9920	1.0081	(0,59.1)	.996
S. Park-Mid. B with c=1.07				
Model 1.	1.0641	1.0055	(39.8,4.9)	.998
Model 2.	1.0614	1.0081	(38.2,7.2)	.997

The last lines of Table 2 give the Model 1 and 2 predictions for Seaton Park-Midland B when c is increased to 1.07. As for Model 1A, Outcome B gives way to Outcome C. The change of .07 in c is comparatively large insofar as model behaviour is concerned, and hence the new seed productions are very different to those predicted when $c = 1$.

In a similar vein, comparison of the last lines of Tables 1 and 2 show a considerable difference in the limiting seed numbers predicted by Model 1A, and by Models 1 and 2. This arises because the estimated value of $\rho - 1$ for Model 1A is rather larger than for Models 1 and 2. The expressions for x_a and y_a show that when c and d are close to unity then these quantities are quite sensitive to small variations of ρ around unity.

This situation highlights the need for care in assessing model adequacy. When $c = 1$ all models predict Outcome B and hence variations in ρ which stay within the ambit of this outcome have no effect on limiting seed numbers. But this need not entail similarities of other aspects of the models. For example, the estimated values of S_i and T_i for Models 1 and 2 do not, for the most part, show a geometric variation with i. With respect to this characteristic then, Model 1A is not an acceptable surrogate of Models 1 and 2 and hence if estimated parameters indicate that Outcome C is occurring, then Model 1 or 2 should probably be used instead of Model 1A.

5. Some other problems

Model behaviour is discussed further by Pakes and Maller (1989), Chapter 7. For example, a sensitivity analysis identifies parameters whose variation most strongly affects predicted seed yield. This is potentially useful to agronomists concerned with breeding programs because it reveals characteristics whose enhancement will do most to improve yield.

Nonlinear effects arising from density dependence (self-thinning, inter-strain competition etc.) are regarded in the literature of plant population biology as significant practical concerns. Our advice has been that they are not significant aspects of the life cycle of sub-clover mixtures under grazed conditions for the environments of current interest. Instead, there are some other exigencies which could more usefully be included in mathematical model building. Examples are environmental and demographic stochasticity, and the behaviour of n-ary mixtures. Another aspect of a real clover's life is as follows.

In Western Australia, sub-clover is often grown in rotation with a crop pasture, such as wheat, thus improving soil quality. The pasture is grazed by sheep or cattle during sub-clover years, and then it is ploughed in preparation for cereal crops. Grazing and tillage redistribute clover seed within the soil and it is known that seed transition rates vary with depth. Pakes and Maller (1989), Chapter 8, have discussed some aspects of these regimens in the case of monocultures, but much more work remains before a general theory emerges.

6. An approximate analysis of Model 1A

Some auditors commented that Figure 6 illustrates well how the 'slow manifold' grabs the motion from the 'fast foliation'. These terms are explained by Mees (1981) in connection with the van der Pol oscillator, but they are not mentioned in most books on dynamical systems theory.

To understand the import of the comments we transform (3) into a more economically parametrised form. Recalling (4), let

$$p_n = \frac{(1-\alpha)Xt_n}{U} \text{ and } q_n = \frac{\rho(1-\beta)Yu_n}{V}.$$

Then (3) reduces to

$$p_{n+1} = \alpha p_n + \frac{(1-\alpha)cp_n}{(cp_n+q_n)} \tag{6a}$$

and

$$(6b) \qquad q_{n+1} = \beta q_n + \frac{(1-\beta)\rho d q_n}{(p_n + d q_n)}.$$

We suppose first that when $n = 1$ the first term on the right-hand side of (6a) and (6b) is much larger than the second — observe that $cp_n/(cp_n + q_n)$ and $dq_n/(p_n + dq_n) < 1$. Then $p_{n+1} \cong \alpha p_n$ and $q_{n+1} \cong \beta q_n$, whence

$$(7) \qquad p_n \cong p_1 \alpha^{n-1} \text{ and } q_n \cong q_1 \beta^{n-1}.$$

Thus p_n and q_n initially decrease geometrically fast. Elimination of n between the members of (7) shows that the orbits move down curves given approximately by $q = const.p^r$ where $r = \log\beta/\log\alpha$. By varying the constant we obtain a family of curves which intersect only at the origin and which together cover $\mathcal{S}^\circ$ as a layered structure, i.e. the family is a foliation. It is a 'fast' foliation if, e.g., α and $\beta < \Delta$ so that orbital points fall down the foliation more rapidly than their asymptotic approach to a boundary equilibrium.

Now consider the asymptotic regime by starting the system (6) within $\mathcal{T}$. To mimic the behaviour of the Seaton Park-Midland B mixture suppose that $c < \rho$ and $\rho d > 1$ — the second strain wins. Let the de Wit condition (1) hold (the isoclines are parallel), and let $\rho \cong c$ (the isoclines are close together). From (4) and (5) we obtain

$$\Delta = \alpha + \frac{(1-\alpha)c}{\rho}$$

whence $\Delta \cong 1$ no matter what the value of α or β.

The isoclines of the system (6) are

$$\mathcal{I}_p : cp + q = c \text{ and } \mathcal{I}_q : cp + q = \rho$$

(using (1)) and $\mathcal{I}_q$ lies above $\mathcal{I}_p$. The proximity of the isoclines entails $cp_n + q_n \cong \rho$ and hence (6a) yields $p_{n+1} \cong \Delta p_n$. Consequently $p_n \to 0$ and in addition $q_n \cong \rho - cp_1\Delta^{n-1}$. Thus orbits approach $(0, \rho)$, but very slowly. At this level of approximation we conclude that orbits attaining $\mathcal{T}$ move toward $(0, \rho)$ very slowly along a curve close to $\mathcal{I}_q$. Let us see if we can find a better estimate of this path.

The above approximate relation between p_n and q_n can be replaced by the exact connection

$$cp_n + q_n = \rho - \epsilon_n$$

where $\epsilon_n \to 0$ as $n \to \infty$, and $\epsilon_n > 0$ since the orbit lies within $\mathcal{T}$. Using (6) we have

$$\rho - \epsilon_{n+1} = cp_{n+1} + q_{n+1} = \alpha c p_n + \frac{(1-\alpha)c^2 p_n}{(\rho - \epsilon_n)} + \beta q_n + \frac{(1-\beta)\rho q_n}{(\rho - \epsilon_n)}$$

and linearization yields

$$\epsilon_{n+1} \cong \epsilon_n + c(1-\Delta)p_n - \epsilon_n\left[\frac{(\Delta - \alpha)cp_n}{\rho} + \frac{(1-\beta)q_n}{\rho}\right].$$

The dominant term in ϵ_n on the right is $\beta\epsilon_n$ and since $\beta < \Delta$ we might expect that the time variation of ϵ_n will be most influenced by p_n. To see that this is true write the linearised equation as

$$\epsilon_{n+1} = \beta\epsilon_n + c(1-\Delta)p_n + \epsilon_n R_n,$$

where $R_n \to 0$, whose solution is

$$\epsilon_n = \epsilon_1\beta^{n-1} + c(1-\Delta)\sum_{j=1}^{n-1}\beta^{n-j-1}p_j + \sum_{j=1}^{n-1}\beta^{n-j-1}\epsilon_j R_j.$$

The first sum is

$$\sum_{j=0}^{n-1}\beta^j p_{n-j-1} \sim p_n \sum_{j=0}^{\infty}\frac{\beta^j}{\Delta^{j+1}} = \frac{p_n}{(\Delta - \beta)}.$$

The remainder term is

$$R_n = -\left[\frac{(\Delta-\alpha)cp_n}{\alpha} - \frac{(1-\beta)(cp_n + \epsilon_n)}{\rho}\right].$$

which evidently is $O(\Delta^{2n})$. Consequently the second sum is $O(\Delta^{2n})$, whence

$$\epsilon_n \sim \left[\frac{c(1-\Delta)}{(\Delta-\beta)}\right]p_n.$$

We conclude from this that

$$\rho - q_n = cp_n + \epsilon_n \sim \frac{cp_n(1-\beta)}{(\Delta - \beta)}.$$

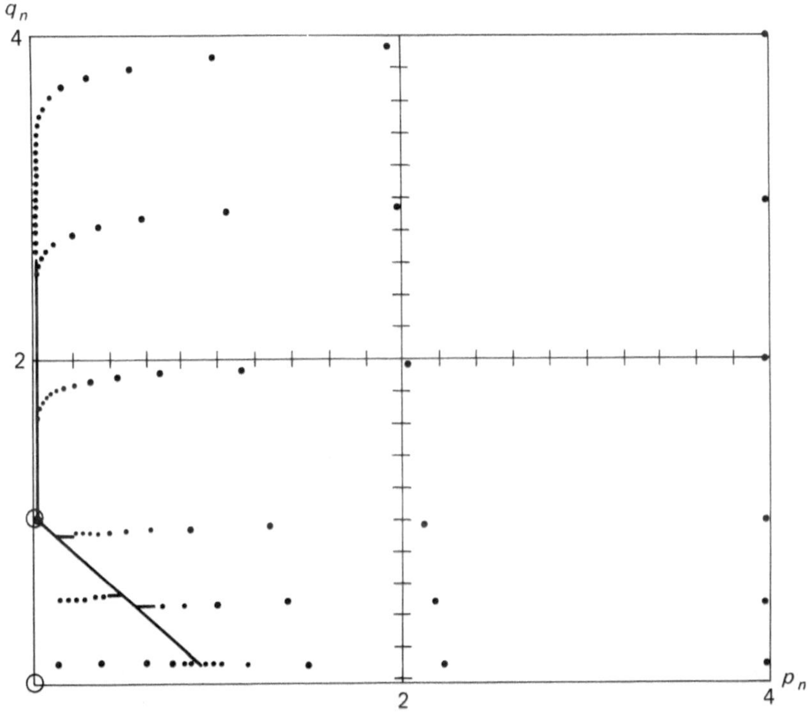

Figure 10. Phase portrait for Seaton Park-Midland B with the same parameters as in Figure 6, but with $\beta = .99$.

This finer analysis shows that orbits in T move slowly along a path, the slow manifold, that is approximately the straight line

$$q + c[(1-\beta)/(\Delta - \beta)]p = \rho. \tag{8}$$

This line ends at the attracting equilibrium point and it has a steeper slope than the isoclines. It cannot be visually distinguished from the isoclines in the Seaton Park-Midland B mixture.

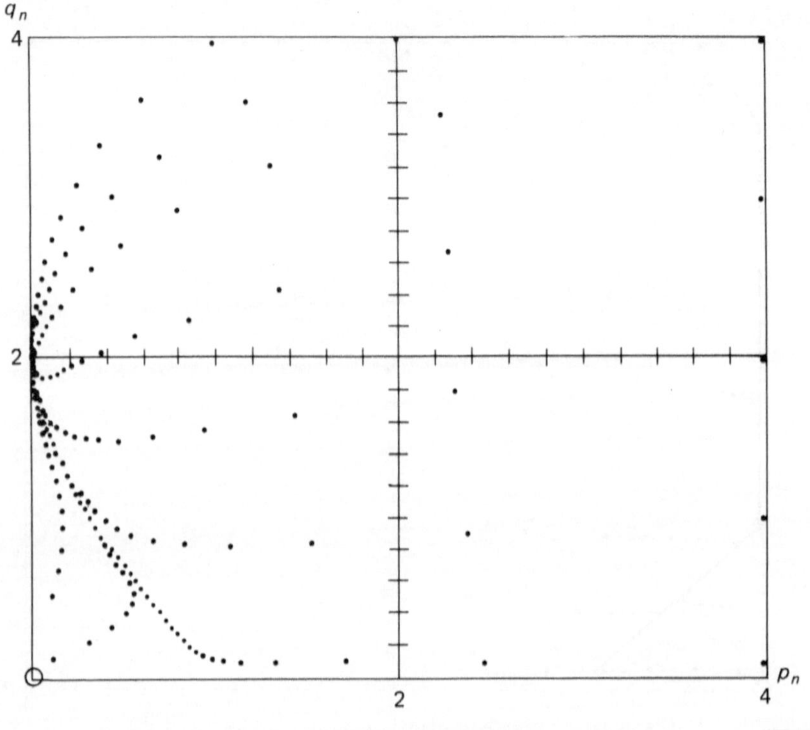

Figure 11. Phase portrait for Dwalgannup-Daliak with the same parameters as in Figure 5, but with $\beta = .85$.

These approximate results accord with those of the longer, exact and more comprehensive analysis of Pakes and Maller (1989), Chapter 4. The analysis here suggests that all orbits eventually enter $\mathcal{T}$ (so they can asymptote along the line (8)), but confirmation of this requires the exact analysis.

We have assumed $\beta < \Delta$ above, and in Section 3 we mentioned the possibility of a different outcome when $\beta > \Delta$. We always have $p_n \sim const.\Delta^{n-1}$, but when $\Delta < \beta < 1$, $\rho - q_n \sim const.\beta^{n-1}$. In this case the asymptotic approach to $(\rho, 0)$ involves two time scales and orbits always approach along curves having the vertical axis as an asymptote. This is illustrated in Figures 10 and 11. The parameter values used to obtain Figures 5 and 6 remain unchanged except that β is increased so that $\beta > \Delta$, leaving the isoclines unchanged. Indeed we set $\beta = .85$ for Dwalganup-Daliak and $\beta = .99$ for Seaton Park-Midland B. Figures 10 and 11 show that the boundary equilibrium is approached in directions depending on the initial conditions. Also, orbits not entering $\mathcal{T}$ reach $(0, \rho)$ much more quickly than those attaining $\mathcal{T}$. For example, the orbit starting at (4,4) in Figure 11 requires about 500 years to approach $(0, \rho)$, but any orbit starting near the horizontal axis requires about 5000 years.

REFERENCES

[1] Braakhekke, W.G. (1980). *On coexistence: a causal approach to diversity and stability in grassland vegetation.* Centre for Agricultural Publishing and Documentation, Wageningen.

[2] Gates, D.J. and Westcott, M. (1988). Stability of plant mixtures in the absence of infection or predation. *J. Theoret. Biology* **131**, pp. 15-31.

[3] Koçak, H. (1986). *Differential and Difference Equations through Computer Experiments.* Springer-Verlag, New York.

[4] Mees, A.T. (1981). *Dynamics of Feedback Systems.* Wiley, Chichester.

[5] Pakes, A.G. and Maller, R.A. (1989). *Mathematical Ecology of Plant Species Competition.* Cambridge University Press, New York. (To appear).

[6] Rossiter, R.C., Maller, R.A. and Pakes, A.G. A model of changes in the composition of binary mixtures of subterranean clover strains. *Aust. J. Agric. Res.* **36**, pp. 119-143.

PARTICIPANT'S COMMENTS

This paper illustrates one of the important uses of modelling, that of predicting long term trends. In this case, the trends being assessed are in the field of agronomy, and the problem is to determine the likely outcome of competition between two strains of sub–clover. While some "long term" information (7 years) has been collected in the field, this is still not of a sufficient time frame to answer some of the questions of the agronomists. The author has clearly made good use of the feedback he obtained at the

Workshop in Albany by devoting a section of his paper (section 6) to re-examining an aspect of his model using a different mathematical approach.

Nick Caputi

Quite complicated biological situations can be modelled by relatively simple mathematical formulations and this paper is an excellent example of such an approach. The subtle competitive interaction between plant strains, involving the added complication of time lags in seed survival and persistence, has been modelled by Pakes as a system of difference equations. This model is easy to understand and describes the competition in terms of seed production and survival in a seed population which persists in the soil from year to year.

The model succeeds in fitting field data, from different locations, quite well and can also be used to make predictions. This is obviously valuable for future development of clover strains in any pasture breeding programme. Relatively simple descriptions give a strong insight into the way probable biological mechanisms interact and provide a comprehensible theoretical framework to test hypotheses and the sensitivity and importance of various measurable field parameters.

Unfortunately, it takes a fair number of years to evaluate the predictive capabilities of such models because they describe a microcosm of Evolution and are thus often, of their very nature, long term descriptors and predictors. Nevertheless, the class of models analysed by Pakes is pretty convincing as to the probable competitive mechanisms at work. The de Wit replacement principle is neatly linked into the dynamics and the fit with extant data and intuition persuades one that the models capture important aspects of the real world situation.

Phil Diamond

Two Dimensional Pattern Formation In a Chemotactic System

M.R. MYERSCOUGH, P.K. MAINI, J.D. MURRAY,

K.H. WINTERS

Abstract

Chemotaxis is known to be important in cell aggregation in a variety of contexts. We propose a simple partial differential equation model for a chemotactic system of two species, a population of cells and a chemoattractant to which cells respond. Linear analysis shows that there exists the possibility of spatially inhomogeneous solutions to the model equations for suitable choices of parameters.

We solve the full nonlinear steady state equations numerically on a two dimensional rectangular domain. By using mode selection from the linear analysis we produce simple pattern elements such as stripes and regular spots. More complex patterns evolve from these simple solutions as parameter values or domain shape change continuously. An example bifurcation diagram is calculated using the chemotactic response of the cells as the bifurcation parameter. These numerical solutions suggest that a chemotactic mechanism can produce a rich variety of complex patterns.

1. Introduction

There are numerous biological phenomena which involve organisation or pattern formation in one, two or three spatial dimensions. Examples of these include spatial distribution of species in ecology, the spatial spread of an epidemic and pattern formation (morphogenesis) during development.

Mathematical models describing these phenomena will necessarily have a spatial component.

Spatial models for developing systems should produce one of two types of behaviour. The model may either be robust or produce a variety of spatial patterns for comparatively small changes in initial conditions or model parameters (sensitive). Robust models are required to describe the morphogenesis of systems such as the skeletal system where essentially the same pattern is always produced in every individual. Models which produce a variety of pattern are suitable, for example, for describing pigmentation pattern. There can be a wide variation of pigment markings between individuals in the same species or closely related species but the patterns are all generated by the same mechanism. We investigate in this paper a simple mechanism which can produce a variety of complex patterns in a population of motile cells.

Migration, localisation and aggregation of cells leading to spatial pattern play an important role in many developing systems. Examples of such systems include skeletal structures in the vertebrate limb (Hinchliffe and Johnson 1980), slime mould aggregration (Loomis 1975) and the neural system and pigmentation cells (Le Douarin 1982). We consider here a simple model for motile cells whose aggregation is driven by chemotaxis and investigate what type of patterns such a system can produce in two dimensions.

Chemotaxis is the process whereby motile cells migrate in response to a gradient of some chemical substance. The cells may either migrate towards high concentrations of this particular substance (chemoattraction) or away from high concentrations (chemorepulsion). Chemotaxis is known to operate in the aggregation of slime mould amoebae and in the localisation of leukocytes in tissue where bacterial inflammation is occurring (Alt and Lauffenburger 1987). We consider here a population of motile cells responding to a chemoattractant. This chemoattractant is produced by the cells themselves and so promotes aggregation and localisation of cells in clusters of high cell density.

2. Mathematical Model

We propose a simple model based on the models for slime mould aggregation which were first proposed by Keller and Segel (1970). The model comprises a population of motile cells of density $\bar{n}$ and a chemoattractant of concentration $\bar{c}$. The cells undergo both random and chemotactic motion and are able to divide and, where cell density is high, to differentiate, die or be removed from the population in some other way. The chemoattractant is produced by the cells themselves, diffuses through the cells' environment and decays linearly. These factors may be captured in a mathematical formulation as follows:

Equation for cell density

$$\frac{\partial \bar{n}}{\partial \bar{t}} = \underbrace{D_n \bar{\nabla}^2 \bar{n}}_{\substack{random \\ motion}} - \underbrace{\bar{\alpha} \bar{\nabla} \cdot (\bar{n} \bar{\nabla} \bar{c})}_{\substack{chemotactic \\ motion}} + \underbrace{\bar{r} \bar{n} (\bar{N} - \bar{n})}_{\substack{replication \\ and\ removal}} \tag{1a}$$

Equation for chemoattractant concentration

$$\frac{\partial \bar{c}}{\partial \bar{t}} = \underbrace{D_c \bar{\nabla}^2 \bar{c}}_{diffusion} + \underbrace{\frac{\bar{S} \bar{n}}{\beta + \bar{n}}}_{\substack{production \\ by\ cells}} - \underbrace{\gamma \bar{c}}_{\substack{linear \\ decay}} \tag{1b}$$

where the operator $\bar{\nabla}^2$ represents $\frac{\partial^2}{\partial \bar{x}^2} + \frac{\partial^2}{\partial \bar{y}^2}$, D_n and D_c are diffusion coefficients for the cells and the chemoattractant respectively, $\bar{\alpha}$ is the chemotactic coefficient, $\bar{r}$ is the mitotic rate of the cells, $\bar{N}$ the carrying capacity of the cells' environment, $\bar{S}$ and β determine the rate of synthesis of the chemoattractant and γ its rate of decay. We can write these equations in non-dimensional form by setting

$$\begin{array}{llll} n = \frac{\bar{n}}{\beta}; & c = \frac{\gamma \bar{c}}{\bar{S}}; & t = \gamma \bar{t}; & \mathbf{x} = \bar{\mathbf{x}} \sqrt{\frac{\gamma}{D_c}}; \\ D = \frac{D_n}{D_c}; & \alpha = \frac{\bar{S} \bar{\alpha}}{D_c \gamma}; & r = \frac{\bar{r} \beta}{\gamma}; & N = \frac{\bar{N}}{\beta} \end{array} \tag{2}$$

which gives

$$\begin{aligned} \frac{\partial n}{\partial t} &= D \nabla^2 n - \alpha \nabla \cdot (n \nabla c) + r n (N - n) \\ \frac{\partial c}{\partial t} &= \nabla^2 c + \left\{ \frac{n}{1+n} - c \right\}. \end{aligned} \tag{3}$$

We consider a finite domain where neither cells nor chemoattractant can cross the boundary of the domain. Hence Neumann boundary conditions apply, namely

$$\mathbf{s}(\mathbf{x}) \cdot \nabla c(\mathbf{x}, t) = \mathbf{s}(\mathbf{x}) \cdot \nabla n(\mathbf{x}, t) = 0, \quad \mathbf{x} \in \partial \mathcal{D} \tag{4}$$

where $\mathbf{s}(\mathbf{x})$ is the outward unit normal to the boundary $\partial \mathcal{D}$.

Equations (3) have one homogeneous steady state,

$$n = n_0 = N \text{ and } c = c_0 = N/(1+N). \tag{5}$$

Setting $n = n_0 + u$, $c = c_0 + v$ with $|u|$, $|v| << 1$ in equations (3) we get the linearised system,

$$\begin{aligned} \frac{\partial u}{\partial t} &= D\nabla^2 u - \alpha N \nabla^2 v - rNu \\ \frac{\partial v}{\partial t} &= \nabla^2 u + \left\{ \frac{u}{(1+N)^2} - v \right\}. \end{aligned} \tag{6}$$

We set $(u, v) \propto e^{i\mathbf{k}\cdot\mathbf{x}+\sigma t}$ in (6) and obtain the dispersion relation as the appropriate solution of

$$\sigma^2 + ((D+1)k^2 + rN + 1)\sigma + Dk^4 + \left\{ rN + D - \frac{N\alpha}{(1+N)^2} \right\} k^2 + rN = 0. \tag{7}$$

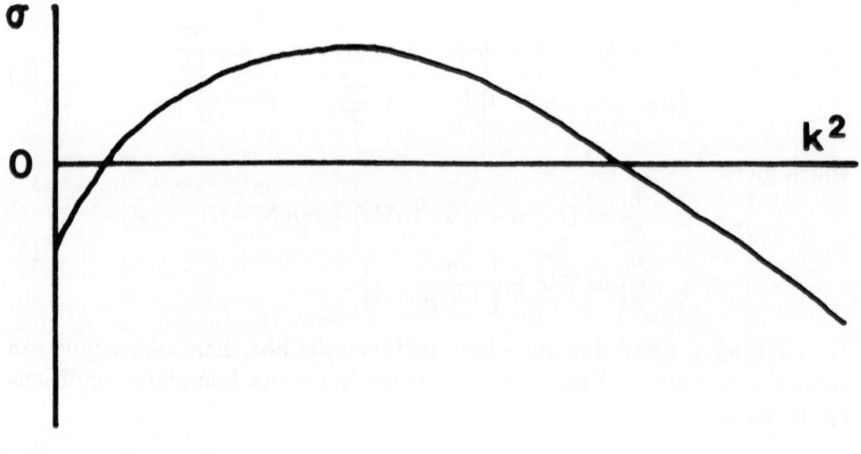

Figure 1. Dispersion curve $\sigma(k^2)$.

The dispersion curve is illustrated in Figure 1. When $\sigma(k^2) > 0$ the homogeneous solution is unstable to small perturbations provided the solution domain allows disturbances of wave number $k = |\mathbf{k}|$. The homogeneous solution is stable for k^2 very small or very large. For k^2 small the perturbation from the homogeneous steady state has a long wavelength. The gradient of chemoattractant is so low in this case that cell death and reproduction smooths out any deviation from the homogeneous steady state before chemotactic cell migration can take effect. For very large k^2 the perturbations have a small wavelength and on such a small length scale random cell motion quickly disrupts cell pattern and hence eliminates any chemotactic effect. When $\sigma = 0$, k^2 satisfies

$$Dk^4 + \left\{ rN + D - \frac{N\alpha}{(1+N)^2} \right\} k^2 + rN = 0. \tag{8}$$

For unstable modes to exist at least one root of (8) must be real and non-negative. This implies that

$$\begin{aligned} rN + D &< \frac{N\alpha}{(1+N)^2}, \qquad \text{and} \\ \left\{ rN + D - \frac{N\alpha}{(1+N)^2} \right\}^2 &> 4rND. \end{aligned} \tag{9}$$

On a finite domain a number of discrete modes given by k^2 will be unstable. By a suitable parameter choice we can cause only one mode to be linearly unstable (see Figure 2). Patterns of this wave number will then dominate the subsequent evolution of the system. To isolate the mode k_i^2 we choose our parameters so that the maximum value of $\sigma(k^2)$ occurs at k_i^2 and is zero, i.e equation (8) holds and has equal roots in k^2. Therefore we require

$$rN + D - \frac{N\alpha}{(1+N)^2} = -2\sqrt{(rND)}. \tag{10}$$

We take the negative square root in agreement with (9) so that k_i^2 is non-negative. Substituting (10) into (8) we have

$$k_i^2 = \sqrt{(rN/D)}. \tag{11}$$

On a rectangular domain of dimensions $L_x \times L_y$ we can use (10) and (11) to select a pattern with ℓ half wavelengths in the x direction and m half wavelengths in the y direction by choosing the parameters such that (10) is satisfied and

$$\pi^2 \left\{ \frac{\ell^2}{L_x^2} + \frac{m^2}{L_y^2} \right\} = \sqrt{(rN/D)}. \tag{12}$$

We refer to this pattern as the (ℓ, m) mode.

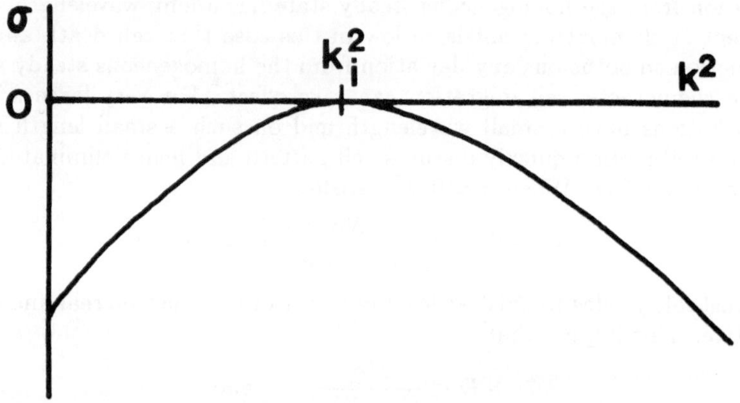

Figure 2. Mode isolation. Dispersion curve showing isolated mode k_i^2.

3. Steady State Pattern

We solved the steady state equations

$$\begin{aligned} D\nabla^2 n - \alpha\nabla\cdot(n\nabla c) + rn(N-n) &= 0 \\ \nabla^2 c + \left\{\frac{n}{1+n} - c\right\} &= 0 \end{aligned} \tag{13}$$

on a 1×4 rectangular domain $\mathcal{D}$. A norm $||n||$ was defined on the solutions to (13) by

$$||n|| = \frac{\int_{\mathcal{D}} |N-n|\, dx\, dy}{\int_{\mathcal{D}} dx\, dy}. \tag{14}$$

Using this norm the trivial solution is the homogeneous steady state solution $n = N$, $c = N/(1+N)$. We solved equations (13) numerically using the finite element approximation with nine- noded rectangular elements in a standard Galerkin formulation. To investigate bifurcation behaviour and follow solutions we evaluated the Jacobian of (13) on the trivial branch using α as the bifurcation parameter. At points on the trivial axis where bifurcation from the uniform steady state occurs the Jacobian is singular. Thus we could identify bifurcation points from the uniform steady state. We then computed the eigenvector corresponding to the zero eigenvalue of the Jacobian matrix at these bifurcation points and used it as a first guess for the solution on the non-trivial branch. Continuation proceeds along the branch. We are also able to step off a branch of solutions in the $\alpha - ||n||$ plane and by continuation in another parameter such as S, r, N or domain size or shape to investigate how the solutions change with respect to this new bifurcation parameter. This is illustrated in Figure 3. Computations were performed using the ENTWIFE finite element package on the CRAY-1S at Harwell. Further details of the numerical methods are given in Riley and Winters (1987).

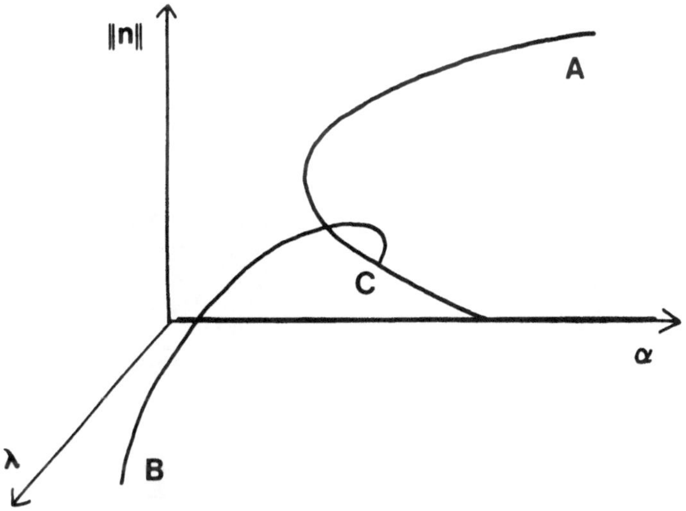

Figure 3. Bifurcation branches in three dimensions: A, bifurcation branch in the $\alpha - ||n||$ plane; B, bifurcation branch in the $\lambda - ||n||$ plane obtained by stepping off A at point C. Here λ represents any one of the parameter set.

To produce various patterns we selected each mode using (10) and (12) to give a good estimate of the appropriate parameter values. By allowing

α to vary we found numerically the exact point where bifurcation from the homogeneous steady state into each mode occurred. Using numerical continuation we stepped along the non-trivial branch until the solution was sufficiently far from the uniform steady state for the pattern to be well developed. Figures 4, 5 and 6 show some of the patterns we obtained in this way. The modes where $\ell = 0$, e.g. $(0,2)$ and $(0,3)$ all give lateral stripes (Figure 4). Longitudinal stripes are given by the modes with $m = 0$, e.g. $(2,0)$ (Figure 5).The modes $(0,4)$ and $(1,0)$ are degenerate as they have the same value of k^2 on a 1×4 domain. In order to select these modes we had to alter the domain size slightly so that the modes lost their degeneracy and could be separated. A variety of regular spots and blotches were produced in the modes where ℓ and m are both non-zero (Figure 6).

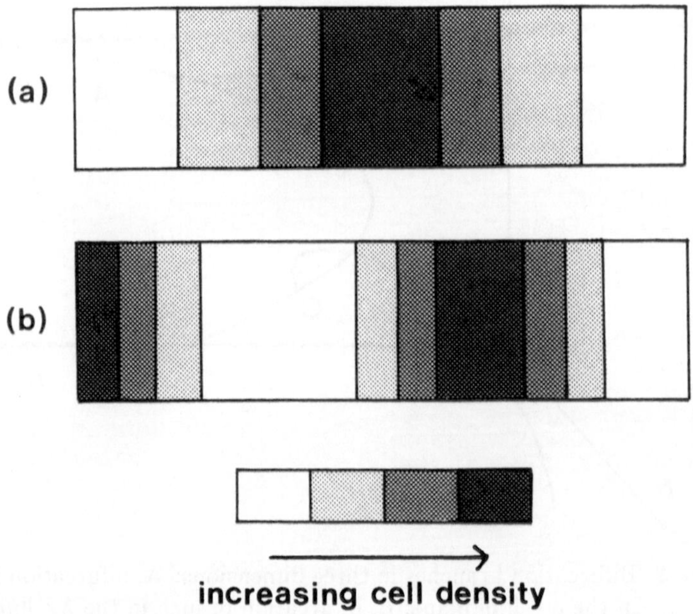

Figure 4. Lateral stripe patterns: (a) $(0,2)$ mode; (b) $(0,3)$ mode.

Because we are calculating the steady state solution we can say nothing about the stability of the patterns in Figures 4, 5,and 6. However the

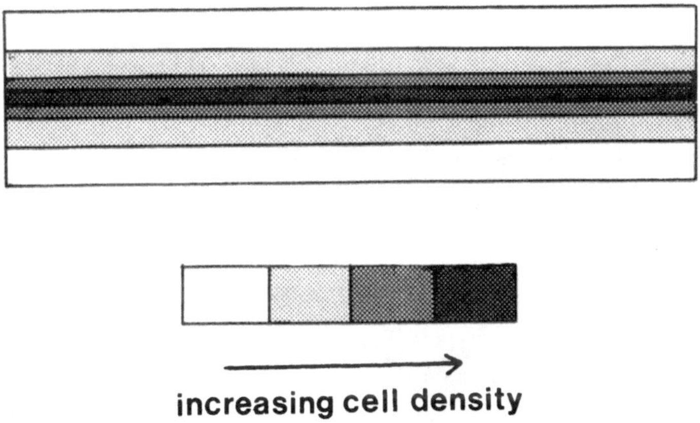

Figure 5. Longitudinal stripe patterns: (2, 0) mode.

results do suggest that a chemotactic model can produce basic pattern elements. These basic elements correspond to the patterns produced by the eigenfunctions of the linearised system. The model is also able to produce more complex patterns. For example if we start with the (1, 2) mode and continuously increase α, areas of high cell density form into spots and each spot divides to give a pair of spots. Changing the domain size and shape also produces interesting patterns. We elongated the domain in the y direction by stepping off the (0, 4) branch and continuing with L_y as the bifurcation parameter. Secondary peaks appeared in the interstices of the original peaks (Figure 7). These interstitial peaks grew until, when the domain length had doubled, they were the same height as the original peaks. Growth of the domain can also change the symmetry of the pattern and transform asymmetric spots into symmetric spots or bands. If we start with a pattern on the (1, 2) solution branch and step off that branch in the $\alpha - ||n||$ plane using L_x as the continuation parameter the asymmetric spots move to the centre of the domain and the pattern becomes symmetric (Figure 8). These examples of changing domain shape illustrate the important effect of growth on pattern and suggest that if growth takes place while the pattern formation mechanism is still operative but on a timescale

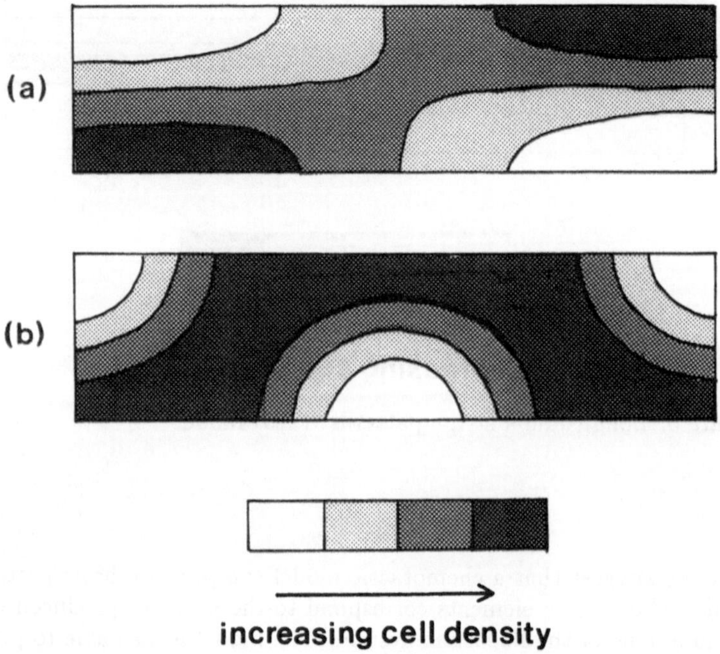

Figure 6. Spot patterns: (a) $(1,1)$ mode; (b) $(1,2)$ mode.

slower than cellular motile behaviour, such as chemotaxis, it will have an important bearing on final pattern.

In this section we have described some of the steady state patterns produced by the model. In the next section we investigate systematically the richness of possible pattern as just one parameter α is varied.

4. An Example of a Bifurcation Diagram

We found the first few primary bifurcation points as α increased for the parameter set given in Table 1. We continued along each of these branches.

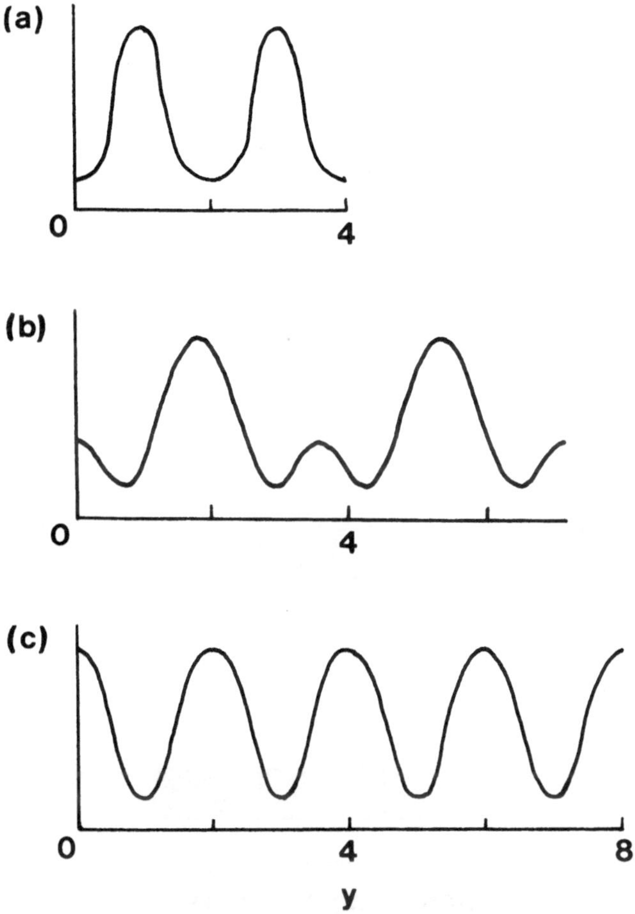

Figure 7. Schematic diagram of the cross-section in the y direction of cell density patterns as L_y increases: (a) $L_y = 4$; (b) $L_y = 7.6$; (c) $L_y = 8$.

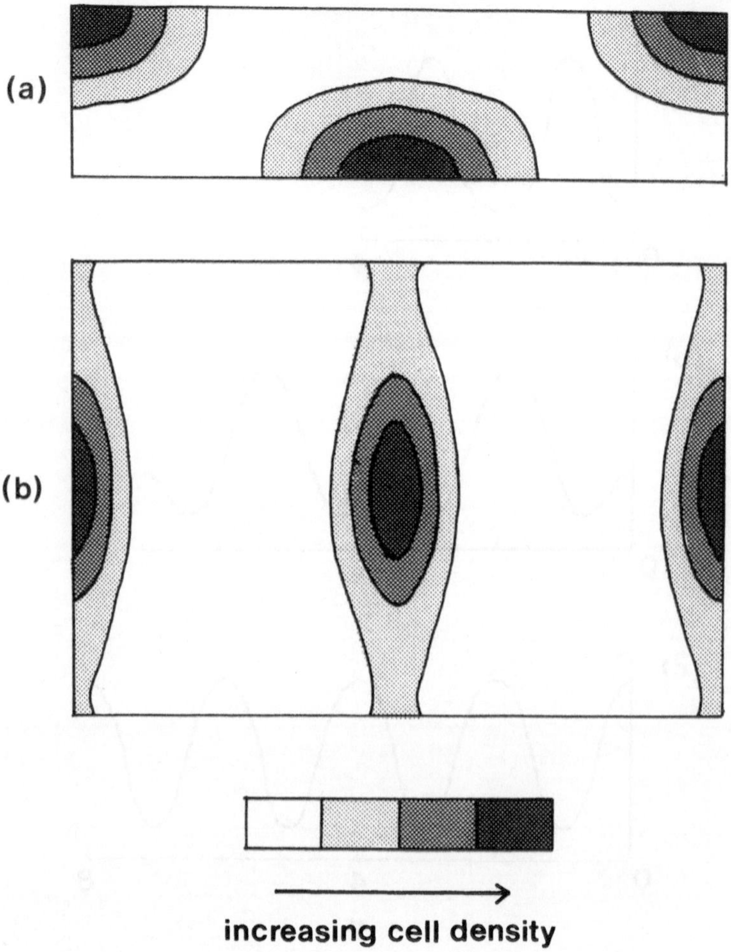

Figure 8. Effect of changing domain width L_x on $(1, 2)$ mode: (a) $L_x = 1$; (b) $L_x = 2.7$.

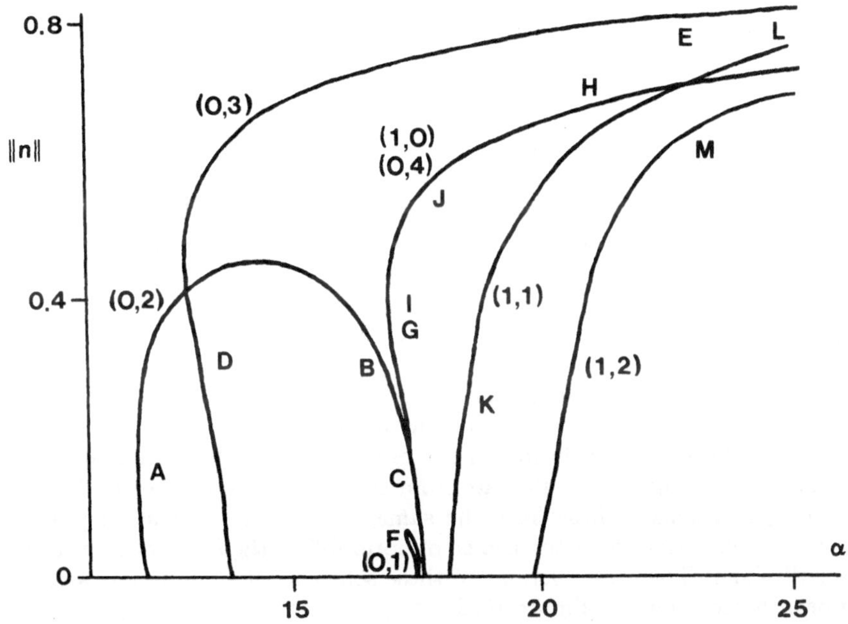

Figure 9. Bifurcation diagram. Ordered pairs refer to the mode on that branch. Letters A-M refer to Figure 10 which shows solutions on the branches.

The results are illustrated in Figure 9. The first branch to bifurcate from the trivial solution was the $(0,2)$ mode solution. The cross-sections of this solution parallel to the y direction are shown in Figure 10(a). As the solution reaches the region marked B interstitial peaks start to form. This solution comes closer and closer to the $(0,4)$ branch and finally joins onto it via a secondary bifurcation.

Table 1. Parameter values for the bifurcation diagram.

Cell carrying capacity N	1
Diffusion coefficient D	0.25
Linear growth rate r	1.52

As α increases the next branch to bifurcate from the trivial solution after the $(0,2)$ branch is the $(0,3)$ branch. As α continues to increase the solutions on this branch form sharper and sharper peaks (Figure 10(b)). This is what we might anticipate as α controls the aggregative force on the cells.

The next primary bifurcation gives the $(0,1)$ branch. This branch loops round and joins back to the trivial branch. A different choice of norm makes it clear that this is in fact a pitchfork bifurcation where the two branches of the pitchfork join on to each other. It is not a double bifurcation as it appears to be using the norm defined by equation (14). The general form of the solution on the branch is shown in Figure 10(c).

The next branches to bifurcate from the trivial branch are the $(1,0)$ and $(0,4)$ solutions. These have the same bifurcation point as they are degenerate and because of the choice of norm follow the same path in the $\alpha - \|n\|$ plane. The cross-sections of these solutions at some points along the branch are shown in Figure 10(d).

The last two primary bifurcation branches which we investigated were the $(1,1)$ branch and the $(1,2)$ branch. Like the $(0,3)$ branch these just give steeper and sharper peaks in cell density as α increased (Figure 10(e) and (f)).

We have given no indication of which branches on the bifurcation diagram are unstable and which are stable. At least some of the branches will be unstable. We shall investigate this question in a future publication. In other parameter regimes they may produce stable pattern. This diagram is also incomplete in that secondary bifurcations may occur which we have not investigated. Nevertheless this bifurcation diagram highlights and gives a useful insight into the complexities of pattern produced by this model and the way the different modes may interact with themselves and each other.

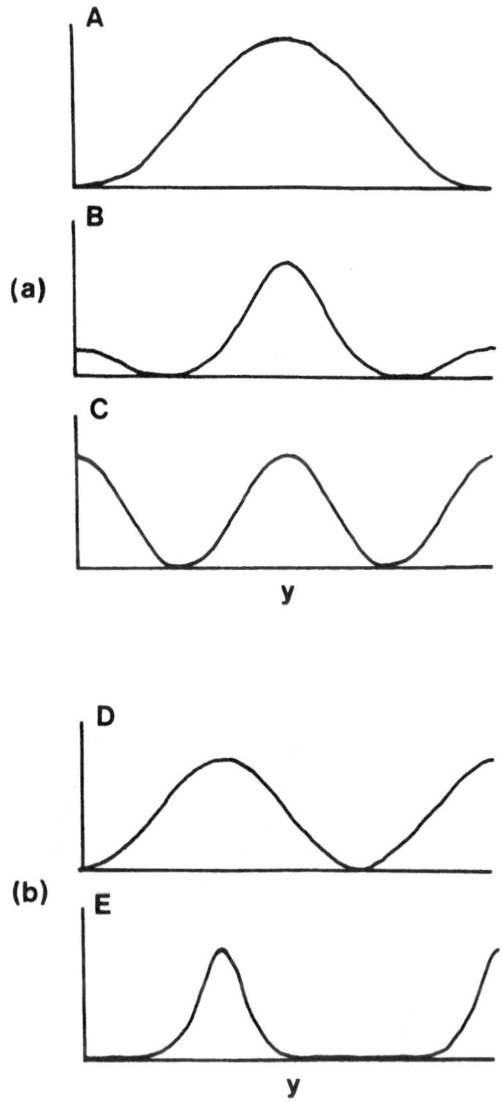

Figure 10. Cross-sections and shaded diagrams of solutions on the non-trivial branches in Figure 9: (a) (0, 2) branch; (b) (0, 3) branch; (c) (1, 0) branch; (d) (0, 4) branch; (e) (1, 0) branch; (f) (1, 1) branch; (g) (1, 2) branch.

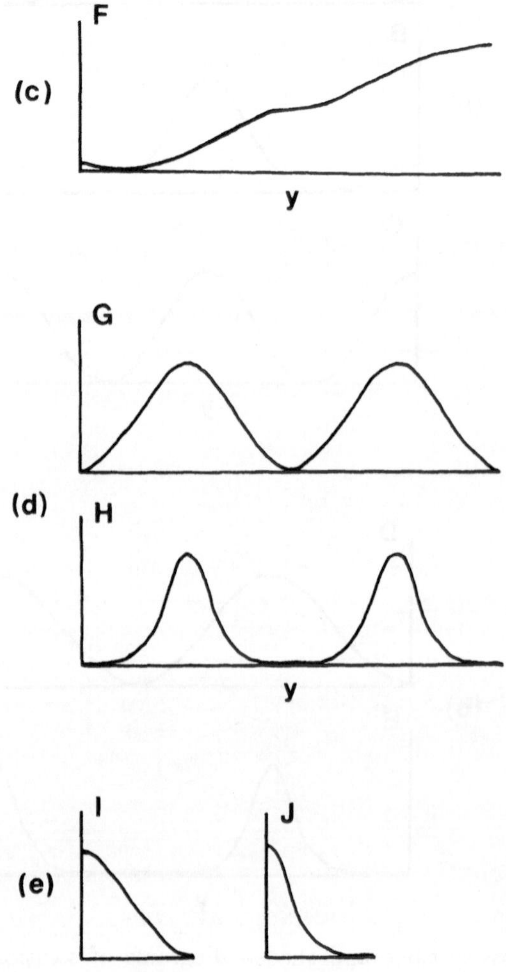

Figure 10. (contd).

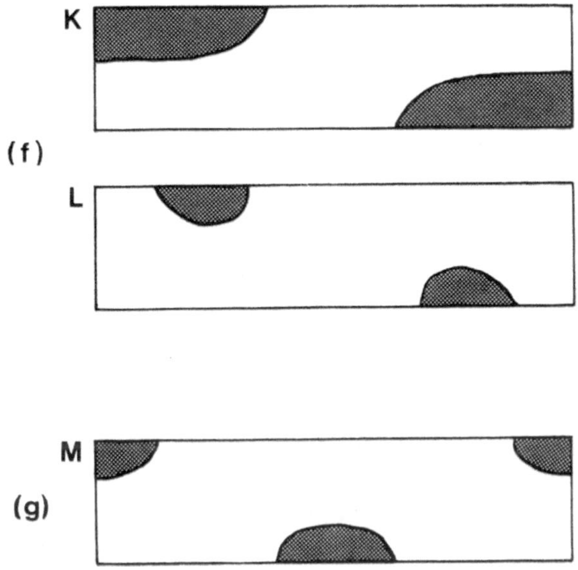

Figure 10. (contd).

5. Conclusion

The model for cell aggregation by chemotaxis which we have used here is a very simple one. We have shown that it produces a number of patterns in cell density. Some of these patterns, for example those in Figure 8 we cannot predict using linear analysis. It seems likely that this system of equations can generate other complex patterns in other parameter regimes and that even this very simple model can produce a rich variety of two dimensional pattern.

Acknowledgements: Part of this work was funded by the Underlying Programme of the U.K. Atomic Energy Authority. MRM acknowledges an Overseas Research Studentship (ORS) Award.

REFERENCES

[1] Alt, W. and Lauffenburger, D.A. 1987 Transient behaviour of a chemotaxis system modelling certain types of tissue inflammation. *J. Math. Biol.* **24**, pp. 691-722.

[2] Hinchliffe, J.R. and Johnson, D.R. 1980 *The Development of the Vertebrate Limb.* Clarendon Press, Oxford.

[3] Keller, E.F. and Segel, L.A. 1970 Initiation of slime mould aggregation viewed as an instability. *J. theor. Biol.* **26**, pp. 399-415.

[4] Le Douarin, N.M. 1982. *The Neural Crest.* Cambridge University Press.

[5] Loomis, W.F. 1975 *Dictyostelium discoideum: A Developmental System.* Academic Press, New York.

[6] Riley, D.S. and Winters, K.H. 1987 The onset of convection in a porous medium: a preliminary study. Harwell Report AERE R 12586.

PARTICIPANT'S COMMENTS

I liked this paper very much, and have practically nothing to contribute to it. I should only point out that the approach opens some interesting possiblities for ecologists: the patchy distribution of flora and fauna has intrigued ecologists ever since. Here is a method which may, in the least, describe, and reproduce known distributions of animals and plants. Wouldn't it be wonderful to recreate a coral reef with such a method?

Y. Cohen

This paper continues the development of mathematical models of pattern formation started by Turing (1952) with his seminal paper on chemical morphogenesis. That work provided the opening for one of the most exciting areas of mathematical biology today and Harrison (1987) provides an excellent review of the field. Turing's initial work has been developed by a number of researchers into a more complete mathematical theory and such topics as animal coat markings and the formation of cartilage condensations during early limb morphogenesis are currently areas to which considerable research effort is being directed.

This paper predicts that biological systems in which chemotaxis plays a major rôle can, under certain very reasonable assumptions, display regular patterns of cell aggregation. The fact that the equations are so relatively simple in character has meant that the authors have been able to carry out a detailed analysis of the bifurcation diagram which enables a much fuller understanding of the way in which one pattern changes into another when the dimension of the region of interest change.

One of the most exciting aspects of mathematical biology is the the interaction between the mathematical modeller and the experimental biologist. Most modelling is carried out after a biologist has completed an experiment and is looking for a mechanism to describe the observations. This paper seems to point the way to a set of experiments on certain pattern formation in chemotactic systems and as such is in the tradition of good Applied Mathematics where, the mathematics enriches the experimental discipline, in this case, biology.

Harrison, L.G. (1987). What is the Status of Reaction-Diffusion Theory Thirty-four years after Turing? Mini Review. *Journal of Theoretical Biology* **125**, pp. 369-384.

Turing, A. (1952). The Chemical Basis for Morphogenesis, *Philosophical Transactions of the Royal Society, London* **B237**, pp. 37-72.

Sean McElwain

Mathematical Modelling of the Control of Blood Glucose Levels in Diabetics by Insulin Infusion

MICHAEL E. FISHER

Abstract

Mathematical optimization techniques are applied to two simple mathematical models of blood glucose dynamics to derive insulin infusion programs for the control of blood glucose levels in diabetic individuals. Based on the results of the mathematical optimization a semiclosed loop algorithm is proposed for continuous insulin delivery to diabetic patients. The algorithm is based on three hourly plasma glucose samples. A theoretical evaluation of the effectiveness of this algorithm shows that it is superior to two existing algorithms in controlling hyperglycemia.

A glucose infusion term representing the effect of glucose intake resulting from a meal is then introduced into both model equations. A theoretical analysis is then undertaken to determine the most effective insulin infusion programs for the control of plasma glucose levels following a meal.

Key Words: *Optimal insulin infusion, mathematical modelling, blood glucose levels, diabetes*

1. Introduction

In this paper, we undertake a theoretical analysis of the control of plasma glucose levels in diabetic individuals using two well known simple mathematical models of the dynamics of glucose and insulin interaction in the blood system. The first model is the linearised model of Ackerman *et al* (1965) and the second is the nonlinear, although still very simple, model of Bergman *et al* (1981, 1985).

Mathematical optimization techniques are used to calculate optimal insulin infusion programs for the correction of hyperglycemia based on the two theoretical models. The relative merits of various insulin infusion programs such as single injections, continuous infusion and closed loop infusion for the control of blood glucose levels in diabetic individuals are then assessed. Based on the results of the mathematical optimization for the Bergman model a semiclosed loop algorithm is proposed for continuous insulin delivery to diabetic patients. The algorithm is based on three hourly plasma glucose samples. Computer simulation is then used to theoretically evaluate the effectiveness of this algorithm and compare it with two existing algorithms proposed by Chisolm *et al* (1984) and Furler *et al* (1985).

A glucose infusion term representing the effect of glucose intake resulting from a meal is then introduced into both model equations. The previously mentioned analysis is then repeated to determine effective insulin infusion programs for the control of plasma glucose levels in diabetics following a meal.

2. Two Mathematical Models

The first deterministic model of the dynamics of blood glucose and insulin interaction we shall study is the linearized model of Ackerman *et al* (1965). This is one of the earliest and simplest such models but one which still retains some compatibility with known physiological facts. In this model, $G(t)$ and $I(t)$ represent the differences of blood glucose level and net blood-glycemic hormone level, respectively, from their basal values. I is a composite hormone representing the circulating levels of all the hormones regulating glucose. The model equations, hereafter referred to as Model A, are:

$$\begin{aligned} \dot{G} &= -m_1 G - m_2 I + P(t) \\ \dot{I} &= -m_3 I + m_4 G + u(t) \end{aligned} \tag{2.1}$$

with initial conditions $G(0) = G_0$ and $I(0) = I_0$. Various studies (for example, Gatewood *et al*, 1970) have shown that changes in I can be associated almost completely with changes in plasma insulin. We will therefore refer to I, from now on, as the plasma insulin concentration. In (2.1), m_1, m_2, m_3 and m_4 are assumed to be constants and $P(t)$ and $u(t)$ are the rates of infusion of exogenous glucose and insulin per unit volume, respectively.

The units of measurement will be taken as mmol per litre of blood, for G, and net immunoreactive insulin mU (10^{-3} units) per litre of plasma, for I.

There have been various attempts to estimate the values of m_1, m_2, m_3 and m_4 from experimental data for both diabetic and non-diabetic subjects. Experimental evidence (see Gatewood *et al* , 1970, and Campbell and Abbrecht, 1968) suggests that it is not unreasonable to assume that the m_i's are indeed constant, provided that neither G nor I changes too rapidly. For normal individuals, all four parameters tend to have values around 0.05 (Yipintsoi *et al* , 1973). For diabetics, these parameter values can vary quite widely from subject to subject. Yipintsoi *et al* (1973) have shown that in a study of both diabetic and non-diabetic subjects, the values of m_1, m_2 and m_3 are shown to be always positive while for subjects with relatively severe diabetes, m_4 is effectively zero. From now on we will assume that $m_4 = 0$ in (2.1). As an illustration we will use the set of parameter values

$$m_1 = 0.0009, \qquad m_2 = 0.0031, \qquad m_3 = 0.0415, \tag{2.2}$$

which is given in Yipintsoi *et al* for a diabetic subject with relatively severe diabetes. These same parameters have also been used in Kikuchi *et al* (1980) and Swan (1983).

The second model we shall study is the so-called "minimal model" of Bergman *et al* (1981, 1985). The model consists of a single glucose compartment, whereas plasma insulin is assumed to act through a "remote" compartment to influence net glucose uptake. Bergman *et al* claim that amongst the models studied which satisfied certain validation criteria this model has the minimum number of parameters . While still classified as a very simple model of the dynamics of interaction of blood glucose and insulin in diabetics, this model is considered by many to be more realistic than Model A. The model equations, hereafter referred to as Model B, are:

$$\begin{aligned} \dot{G} &= -p_1 G - X(G + G_B) + P(t) \\ \dot{X} &= -p_2 X + p_3 I \\ \dot{I} &= -n(I + I_B) + u(t)/V_I \end{aligned} \tag{2.3}$$

with initial conditions $G(0) = G_0$, $X(0) = X_0$ and $I(0) = I_0$. G, I and P are as before and X is proportional to the insulin in the remote compartment. Here, u represents the rate of infusion of exogenous insulin measured in mU per minute as distinct from Model A where it represents the rate per litre of plasma. The constants G_B and I_B are basal values of plasma glucose and plasma insulin, respectively. V_I is the insulin distribution volume and n is the fractional disappearance rate of insulin. It should be noted that Model B is meant to represent the diabetic state as the production of insulin in the model equations is independent of the plasma glucose concentration.

Furler *et al* (1985) discuss appropriate values for the parameters together with values for p_1, p_2 and p_3 based on data from Bergman *et al* (1981). Values which they use for a person of average weight are

$$V_I = 12\text{L}, \quad n = (5/54)/\text{min}, \quad G_B = 4.5\,\text{mmol/L} \quad I_B = 15\,\text{mU/L}. \tag{2.4}$$

The somewhat unusual value of n is chosen so that a steady state insulin infusion rate of 1 U/hour is required to maintain the plasma insulin at its basal level, that is,

$$n = \frac{1000}{60 I_B V_I}.$$

Furler *et al* give three alternative sets of values for the remaining parameters p_1, p_2 and p_3 and values which they use for diabetic subjects are

$$p_1 = 0, \quad p_2 = 0.025, \quad p_3 = 0.000013. \tag{2.5}$$

There are fundamental differences between the two models which should be noted. Obviously Model A is linear while Model B is nonlinear. In Model A the steady state in which the plasma glucose and insulin levels are at their basal levels corresponds to zero external infusion of insulin. The same steady state in Model B corresponds to a constant insulin infusion rate of $u = nV_I I_B$ mU/min which, for the parameter values (2.4), corresponds to 1 U/hour. This is more consistent with observations of the infusion rates that are required to maintain steady state plasma glucose levels of severe diabetics at the basal values for normal subjects.

3. Optimal Control in the Absence of Glucose Infusion

In this section we examine various insulin infusion programs, based on models A and B, for correcting an initial state of hyperglycemia. As a means of comparing the effectiveness of the various programs we will use the performance criterion

$$J(u) = \int_0^T G^2(t)\,dt \tag{3.1}$$

where $[0, T]$ is the time interval under consideration. The problem is then to minimize $J(u)$ subject to the system equations (2.1) or (2.3) and their corresponding initial conditions and the constraints

$$0 \leq u(t) \leq u_{max}, \qquad \text{for all} \quad t \in [0, T], \tag{3.2}$$

where u_{max} is a constant.

Closed loop solutions for Model A have been obtained, in the absence of (3.2), (Kikuchi *et al* , 1980, Swan, 1983) using the performance criterion

$$\int_0^\infty \left[G^2(t) + \rho u^2(t)\right] dt, \tag{3.3}$$

where $\rho > 0$ is a scalar weighting factor. The introduction of the second term in the objective function appears to be purely for mathematical convenience. These solutions, however suffer from several drawbacks. For example, they do not necessarily satisfy the nonnegativity constraint (3.2). A possible remedy to this might be to set $u(t) = 0$ when the closed loop solution gives a negative value for $u(t)$. We will refer to this as the modified closed loop control. This is discussed in Fisher and Teo (1989) and is shown to be inferior to the optimal solution based on the objective (3.1). Other drawbacks, which are a feature of all closed loop infusion programs, are the requirement to continuously monitor plasma glucose levels and the implementation of the feedback control. These are complex and highly expensive procedures which have met with only limited success.

The problem of minimizing (3.1) subject to (3.2) for models A and B cannot, in general, be solved analytically. However, open-loop solutions can be obtained using numerical techniques. One numerical algorithm which can be used to solve these problems is the general optimal control program MISER developed by Goh and Teo (1987). The solution procedure used by this program partitions the interval $[0, T]$ into equal subintervals and approximates the control $u(t)$ by a constant function on each subinterval of the partition.

We have used the MISER program to solve the problem for Model A for the parameter set (2.2) together with a variety of initial conditions. The numerical results show that, as the partition of $[0, T]$ becomes finer and finer, if we remove the upper bound u_{max} on $u(t)$, the optimal control for this problem approaches an impulse control at $t = 0$. This suggests that the impulse control is in fact the true optimal control for the optimal relaxed control problem in which the control is not bounded above. For a discussion of optimal relaxed control problems the reader is referred to Gamkrelidze

(1975). The impulse control is superior to both the modified closed loop control based on (3.3) and controls which are held constant for periods of time. In Table 1 we compare the values of the performance criterion J over a 12 hour period resulting from three insulin infusion programs with initial plasma glucose levels of 10 mmol/L ($G_0 = 5.5$) and 15 mmol/L ($G_0 = 10.5$). These three programs are (i) a single injection at $t = 0$, chosen optimally, (ii) optimal infusion, constant over hourly periods and (iii) the modified closed loop control corresponding to the performance criterion (3.3) with $\rho = 1$. The insulin entries in Table 1 correspond to the total insulin infused over the 12 hour period for each of the three programs. In the modified closed loop program, the insulin infusion rate becomes zero after 42 minutes. Figure 1 shows the plasma glucose responses for these three infusion programs for an initial glucose level of 15 mmol/L together with the plasma glucose response when no insulin is administered.

	$G_0 = 5.5$ mmol/L		$G_0 = 10.5$ mmol/L	
	insulin (U)	$J(u)$	insulin(U)	$J(u)$
(i) Single injection	1.34	353	2.55	1288
(ii) Constant infusion (hrly)	1.32	1098	2.51	4002
(iii) Modified closed loop	1.44	729	2.74	2658

Table 1. Comparison of three insulin infusion programs for Model A

The optimal control problem for model B was also solved numerically for a variety of initial conditions. Once again, the numerical results show that the optimal solution requires a large infusion of insulin at $t = 0$. Unlike Model A, however, this initial impulse is then followed later by a period of continuous infusion. In Table 2 we compare the values of the performance criterion J over a 12 hour period resulting from two insulin infusion programs for the initial plasma glucose levels of 10 mmol/L and 15 mmol/L. These two programs are (i) a single injection at $t = 0$, chosen optimally, followed by optimal hourly infusion of insulin and (ii) optimal infusion, constant over hourly periods. The insulin entries in Table 2 correspond to the initial injection plus the total amount infused following the injection over the 12 hour period for program (i) and the total amount infused over the 12 hour period for program (ii).

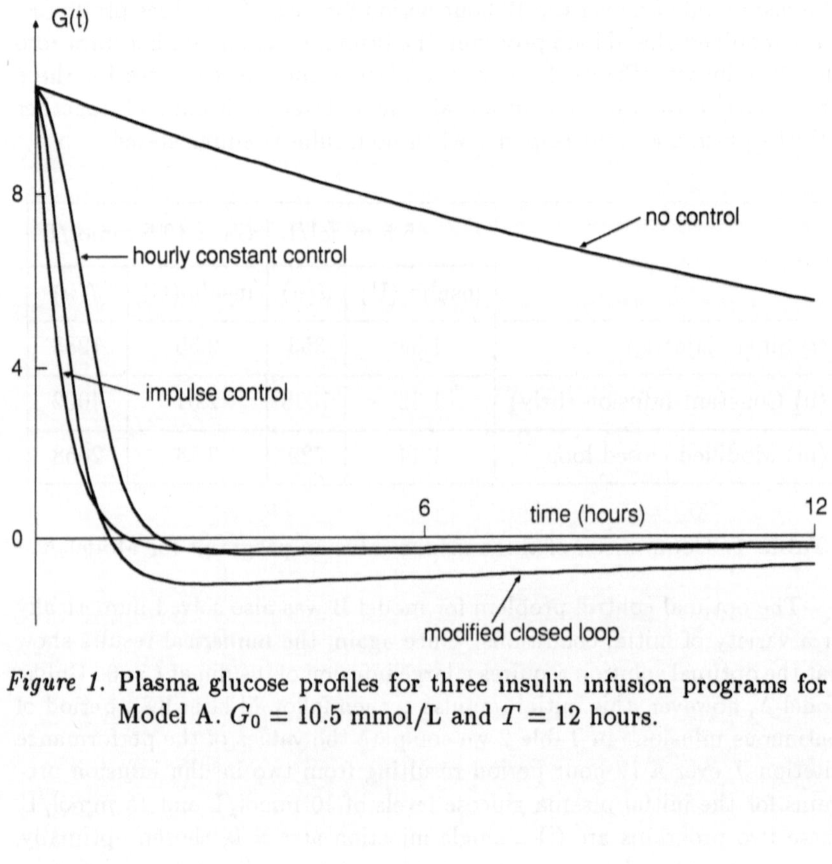

Figure 1. Plasma glucose profiles for three insulin infusion programs for Model A. $G_0 = 10.5$ mmol/L and $T = 12$ hours.

Figure 2(a) shows the plasma glucose responses for these two infusion programs for an initial glucose level of 15 mmol/L together with the plasma glucose response when only the basal insulin level (1 U/hour) is administered. Figure 2(b) shows the insulin infusion rate profile for program (i) where the initial glucose level is 15 mmol/L.

	$G_0 = 5.5$ mmol/L		$G_0 = 10.5$ mmol/L	
	insulin (U)	$J(u)$	insulin(U)	$J(u)$
(i) Injection/hrly infusion	$3.29 + 10.41$	455	$4.56 + 10.01$	1599
(ii) Constant infusion (hrly)	13.71	1168	14.57	4065

Table 2. Comparison of two insulin infusion programs for Model B

4. Semiclosed loop control based on Model B

Because of the complexity and expense of fully closed loop insulin infusion devices, the development of a simple semiclosed loop system based on, say, three hourly plasma glucose readings is especially appealing. Such systems have been developed by Chisolm *et al* (1984) and Furler *et al* (1985) with the insulin delivered by a computer-assisted insulin infusion system. Ollerton (1988) has also used theoretical techniques to derive closed loop insulin infusion algorithms.

The systems of Chisolm *et al* and Furler *et al* have been used with some success for short term therapy of diabetic inpatients. The algorithms on which these systems are based supply a constant insulin infusion rate over a three hour period calculated from a simple piecewise linear graph on the basis of a three hourly plasma glucose reading. The first algorithm (Chisolm *et al*), which we shall refer to as Algorithm 1, delivers insulin at the rate of 0.5 U/hour for plasma glucose levels below 4 mmol/L and 2.5 U/hour for levels above 8 mmol/L, with a linear transition between these rates for plasma glucose levels from 4 to 8 mmol/L. The second algorithm (Furler *et al*), which we shall refer to as Algorithm 2, delivers insulin at

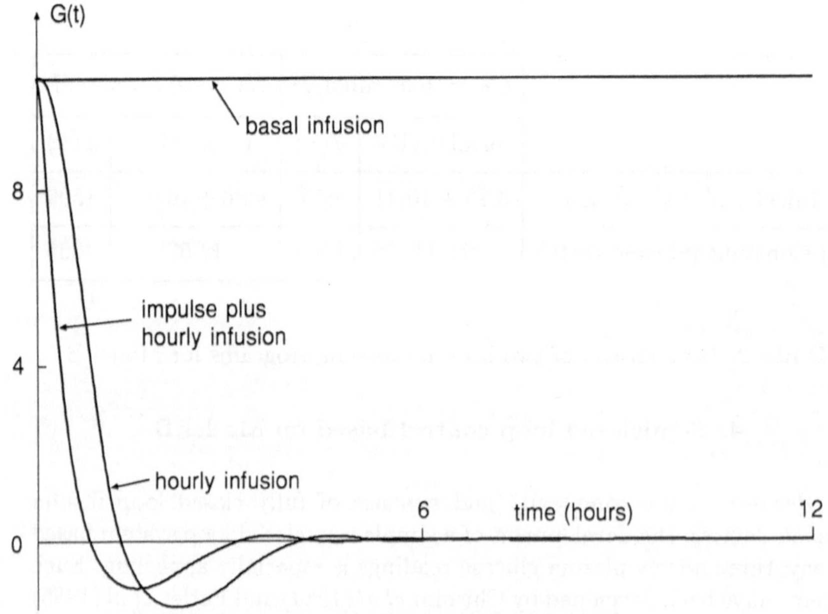

Figure 2. (a) Plasma glucose profiles for two insulin infusion programs for Model B. (b) Insulin infusion rate profile for program (i) for Model B. In both cases, $G_0 = 10.5$ mmol/L and $T = 12$ hours.

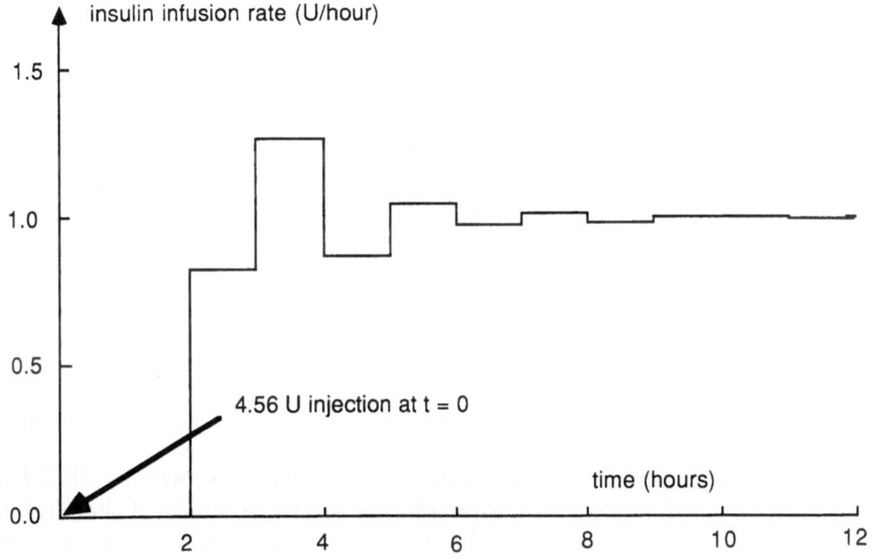
insulin infusion rate (U/hour)
1.5
1.0
0.5
0.0
4.56 U injection at t = 0
time (hours)
2
4
6
8
10
12

Figure 2(b).

the rate of 0.5 U/hour for plasma glucose levels below 2 mmol/L and 2.5 U/hour for levels above 12 mmol/L, with a linear transition between these rates for plasma glucose levels from 2 to 12 mmol/L. These two algorithms have been evaluated theoretically using Model B and Algorithm 2 was found to be superior in controlling hyperglycemia.

The results of solving the optimal control problem formulated in the previous section clearly show that an impulse control or a combination of an impulse with infusion held constant over fixed time periods results in a far superior correction of a hyperglycemic state than control based purely on constant infusion. Based on the results from solving this optimal control problem for various initial values of plasma glucose levels we have constructed a simple semiclosed loop algorithm for plasma glucose control. This algorithm, which we shall call Algorithm 3, depends on the glucose reading at the beginning of each three hour period and consists of two parts. Firstly, if the plasma glucose reading is greater than or equal to 6 mmol/L (a somewhat arbitrary amount) then an injection is given followed by constant infusion for the three hour period at the rate required to maintain plasma insulin at its basal value in the steady state, that is, $u = nV_I I_B$ mU/min. The size of the injection is given by the formula

$$u = G(0.41 - 0.0094G) \text{ U}. \tag{4.1}$$

This formula was obtained by a least squares fit of the expression

$$u = G(a - bG) \text{ U} \tag{4.2}$$

to the data obtained from solving the optimal control problem for Model B for various initial values of plasma glucose levels. Secondly, if the plasma glucose reading is below 6 mmol/L then insulin is delivered at a constant rate over the next three hours which is

$$u(t) = nV_I I_B(1 + \frac{G}{G_B}) \text{ mU/min}. \tag{4.3}$$

This formula is very close to that obtained from linear regression based on solving the optimal control problem for Model B for various initial values of plasma glucose levels.

The effectiveness of Algorithm 3 was then assessed from a theoretical viewpoint using Model B and the parameter values (2.4) and (2.5) which correspond to a patient with severe diabetes. The performance criterion (3.1) was used as a means of comparing the performances of this algorithm with those of Chisolm *et al* and Furler *et al* . The time period considered was 24 hours. In all cases, Algorithm 3 was found to be vastly superior to the other two, not only in the values of J obtained, which are shown in Table 3, but also in the increased stability (speed with which any oscillations are damped out) of plasma glucose levels. Figures 3a and 3b show the theoretical plasma glucose profiles resulting from the three algorithms for initial glucose values of 10 mmol/L and 15 mmol/L, respectively.

	Initial Plasma Glucose Level (mmol/L)				
	8	10	12	15	20
Algorithm 1	5284	7600	6563	8347	17092
Algorithm 2	1220	2859	5041	8134	16790
Algorithm 3	294	672	1197	2213	4375

Table 3. Values of the performance criterion $J(u)$ for Model B for Algorithms 1, 2 and 3

5. Control of Plasma Glucose Levels Following a Meal

Here we examine means of controlling the blood glucose level following an infusion of exogenous glucose. In the models we shall assume that oral glucose infusion commences at $t = 0$ prior to which plasma glucose and insulin are at their fasting levels.

The term $P(t)$ in (2.1) and (2.3) will be taken to represent the rate at which glucose enters the blood from intestinal absorption following a meal. There have been various studies which have attempted to find an appropriate functional form for $P(t)$, particularly in relation to Model A. In oral glucose tests with normal subjects, the aim is for the model to produce the desired effect of the plasma glucose level rising quite rapidly (from the rest level) to a maximum in less than 30 minutes and then falling to the base level after about two to three hours. There is some evidence (Gatewood et al., 1968, 1970) to suggest that the exact form of $P(t)$ for non-diabetics is not important provided the previously stated aim is met. A function which produces the desired behaviour is

$$P(t) = B\exp(-kt), \qquad t \geq 0, \tag{5.1}$$

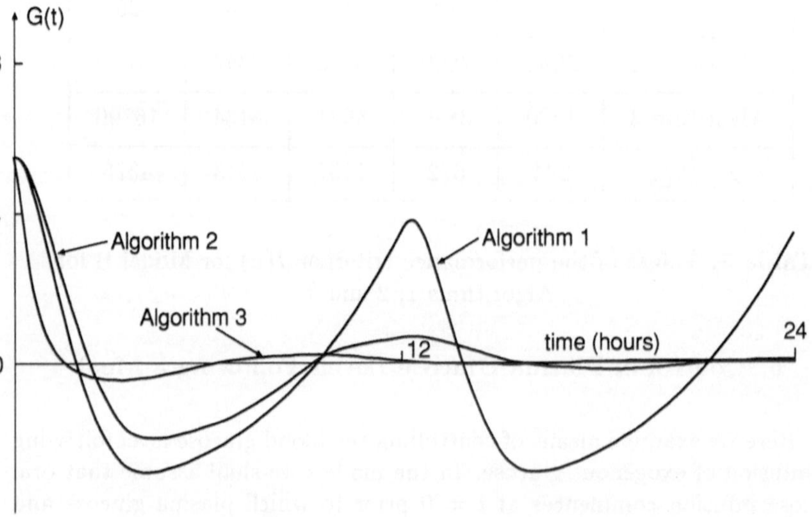

Figure 3. Plasma glucose profiles over a 24 hour period for Model B for Algorithms 1, 2 and 3. (a) G_0 = 5.5 mmol/L. (b) G_0 = 10.5 mmol/L.

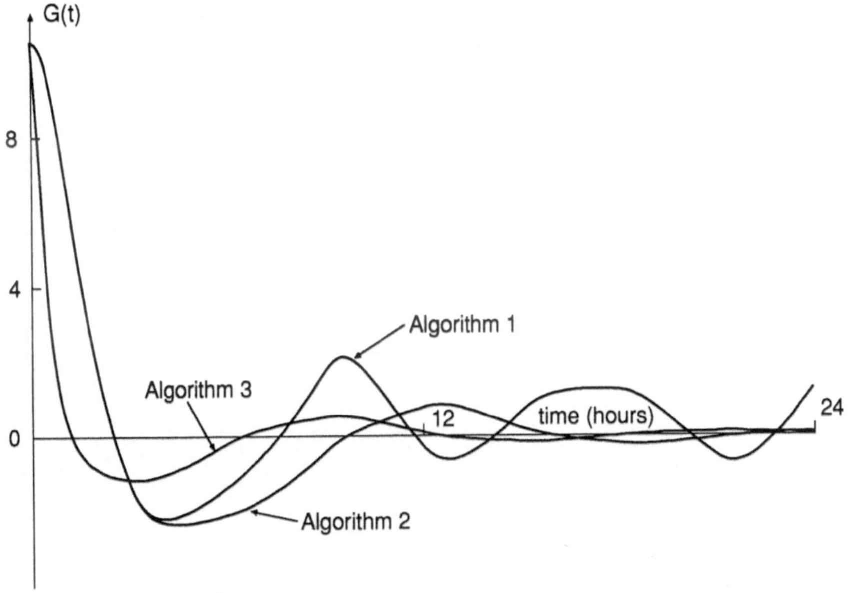

Figure 3b.

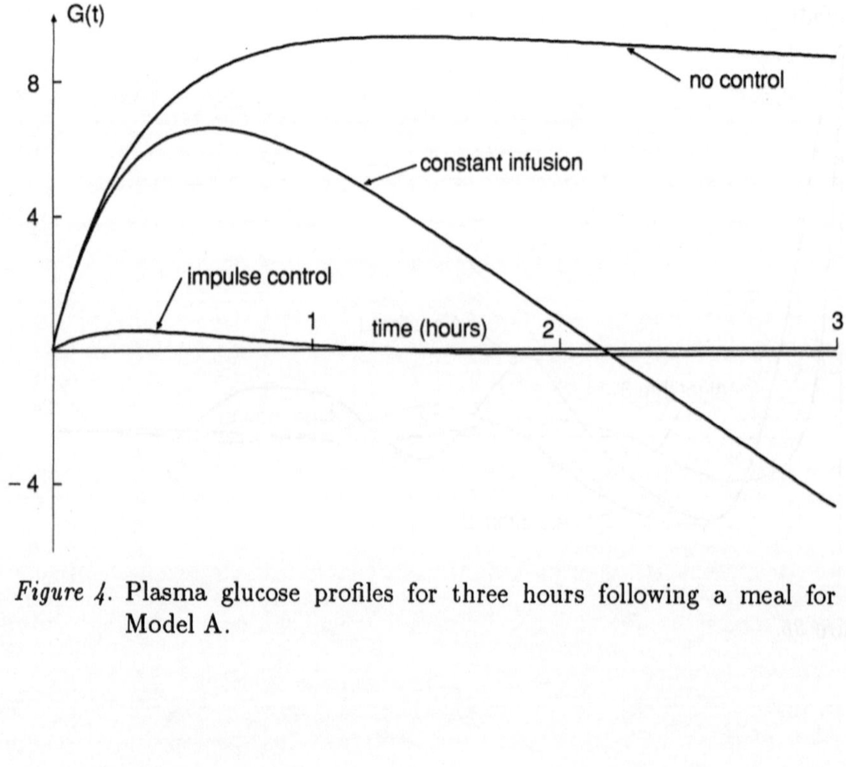

Figure 4. Plasma glucose profiles for three hours following a meal for Model A.

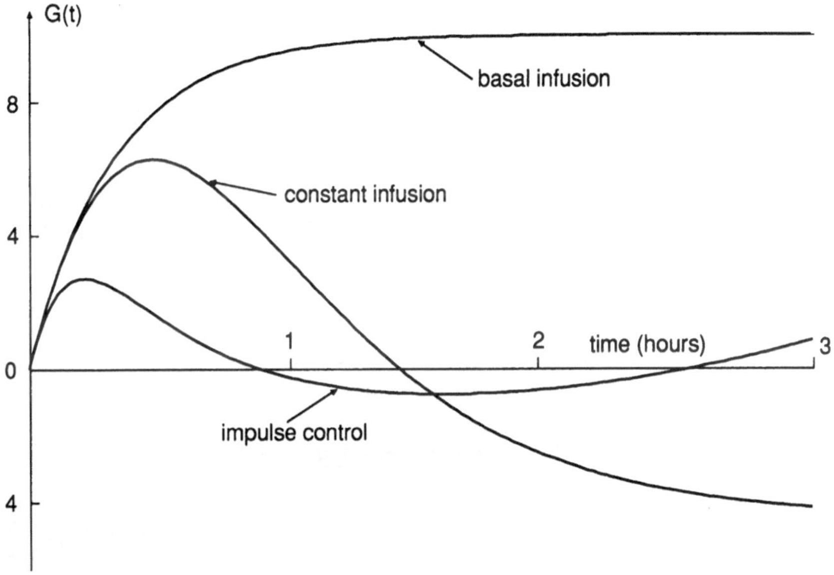

Figure 5. Plasma glucose profiles for three hours following a meal for Model B.

provided appropriate values are chosen for the constants B and k. If we use the values

$$B = 0.5 \qquad k = 0.05 \tag{5.2}$$

in models A and B, then, with parameter values chosen to represent normal subjects, we obtain the desired effect of the plasma glucose rising to a maximum of about twice the basal level in 20 to 30 minutes and falling to near the base level in two to three hours.

The optimal control problem to minimize (3.1), subject to (3.2) was now solved numerically using the basal values of plasma glucose and insulin as the initial conditions. The parameter values were again those corresponding to a subject with severe diabetes.

Figure 4 illustrates the plasma glucose profile for three hours following a meal for Model A with the parameter values (2.2). The three curves correspond to no insulin infusion, constant infusion over the three hour period and a coincident injection with no other infusion. Figure 5 illustrates the plasma glucose profile for three hours following a meal for Model B with the parameter values (2.4) and (2.5). The three curves correspond to insulin infusion at the basal rate (1 U/hour), constant infusion over the three hour period and a coincident injection with no other infusion. The results for both models show that the most effective short term control (one to three hours) is achieved by an insulin infusion program which includes an injection at the same time as the meal.

REFERENCES

[1] E. Ackerman, L.C. Gatewood, J.W. Rosevear, and G.D. Molnar, (1965). Model studies of blood glucose regulation. *Bull. Math. Biophys.* **27**, pp. 21-37.

[2] R. N. Bergman, L. S. Phillips, and C. Cobelli, (1981). Measurement of insulin sensitivity and β-cell glucose sensitivity from the response to intraveous glucose. *J. Clin. Invest.* **68**, pp. 1456-1467.

[3] R. N. Bergman, D. T. Finegood, and M. Ader, (1985). Assessment of insulin sensitivity in vivo. *Endocrine Reviews* **6**, pp. 45-86.

[4] J.A. Campbell and P.H. Abbrecht, (1968). An analysis of the control of insulin secretion. *Proc. 21st Annual Conference Eng. Med. Biol.* **10**, pp. 29.

[5] D. J. Chisolm, E. W. Kraegen, D. J. Bell, and D. R. Chipps, (1984). A semi-closed loop computer-assisted insulin infusion system. *Med. J. Aust.* **141**, pp. 784-789.

[6] M. E. Fisher and K. L. Teo, (1989). pp. Optimal insulin infusion resulting from a mathematical model of blood glucose dynamics. *IEEE Trans. Biomed. Eng.* **36**, pp. 479-486.

[7] S. M. Furler, E. W. Kraegen, R. H. Smallwood, and D. J. Chisolm, (1985). Blood glucose control by intermittent loop closure in the basal mode: Computer simulation with a diabetic model. *Diabetes Care* **8**, pp. 553-561.

[8] R.V. Gamkrelidze, (1975). *Principles of Optimal Control Theory*, Plenum Press, New York.

[9] L.C. Gatewood, E. Ackerman, J.W. Rosevear, and G.D. Molnar, (1968). Simulation studies of blood-glucose regulation: Effect of intestinal glucose absorption. *Comp. Biomed. Res.* **2**, pp. 15-27.

[10] L.C. Gatewood, E. Ackerman, J.W. Rosevear, and G.D. Molnar, (1970). Modeling blood glucose dynamics. *Behav. Science* **15**, pp. 72-87.

[11] C.J. Goh and K.L. Teo, (1987). *MISER, An Optimal Control Software*, Applied Research Corporation, National University of Singapore.

[12] M. Kikuchi, E. Machiyama, N. Kabei, A. Yamada, Y. Sakurai, Y. Hara, and A. Sano, (1980). Adaptive control system of blood glucose regulation. In *MEDINFO* **80**, pp. 96-100. ed. A.B. Lindberg and S. Kaihara.

[13] R.L. Ollerton, (1988). *An application of optimal control theory to diabetes mellitus*, preprint.

[14] G.W. Swan, (1983). *Applications of Optimal Control Theory in Biomedicine*, Marcel Dekker, New York .

[15] T. Yipintsoi, L.C. Gatewood, E. Ackerman, P.L. Spivak, G.D. Molnar, G.W. Rosevear, and F.C. Service, (1973). Mathematical analysis of blood glucose and plasma insulin responses to insulin infusion in healthy and diabetic subjects. *Comput. Biol. Med.* **3**, pp. 71-78.

PARTICIPANT'S COMMENTS

Michael Fisher's paper examines a problem of obvious interest to diabetics and to the medical community. His study demonstrates that the method of analysis used produces more efffective programs for the control of the blood glucose level of diabetics than other algorithms or infusion programs considered. The conclusions from this theoretical assessment are well suited to subsequent empirical validation.

For the two simple models considered, similar results were obtained to the optimal control problems which were examined. The models differ slightly in their representation of the blood glucose system for diabetic patients; for one model, no external infusion of insulin is required to keep the subject at the basal level of blood glucose, while for the other a constant infusion rate is required. Uncertainty in model structure or in parameter values is an element of most models of biological systems. Use of a range of models, each providing a slightly different representation of the system under consideration, is therefore highly desirable when evaluating any control problem.

N.G. Hall

This paper is an excellent example of the way in which mathematical modelling can be useful and illuminating in medical and physiological applications. The models for insulin and glucose interactions and the evaluation, using optimization techniques, of various insulin infusion strategies to control hyperglycemia give some indication of which strategy may be best *in vivo*. The results in this paper, for example, show that one of the best strategies for insulin infusion, both in terms of keeping glucose near to normal and using as little insulin as possible, is a large single injection rather than a constant infusion rate. This may seem counter–intuitive if it were not for the mathematical analysis to back it up. Moreover the results of the mathematical modelling and optimization make it clear that the choice of overall strategy as well as dosage is very important in obtaining the best possible results.

Two models are investigated. Model A which is linear with no infusion of insulin required to maintain basic glucose levels and Model B which is nonlinear, has an implicit delay in the form of the remote insulin compartment and requires a constant infusion of insulin to keep glucose levels at steady state. Model A is mathematically simple and has clearly been used by many workers in the past. Linear models in biology should, however be viewed with some caution as most biological processes are intrinsically nonlinear. For this and other reasons, such as the requirement of constant insulin infusion Model B appears to be the better model of the two. It is still a relatively simple model and easy to understand despite the extra equations and the nonlinearities.

This work has direct and obvious application to clinical work in insulin infusion programs for diabetic individuals. In particular the results in section 4 using a semi-closed loop control based on Model B and relying on only 3 hourly testing provide an uncomplicated and easily implemented control strategy.

M.R. Myerscough

Part II

Tools

Modelling Complex Systems

ALISTAIR I. MEES

Abstract

Traditionally, mathematical models of biological systems have mostly been phenomenological because of the difficulty of deriving a set of equations from first principles in the way that can typically be done in the physical sciences. Sometimes, attempts are made to fit a parametrised model to data, but the fit is usually very bad and all that is hoped for is a qualitative understanding. Recent work in dynamical systems theory has made it possible to construct non-parametric models directly from data. This paper describes a particular approach – embedding followed by tesselation – which appears able to capture the main features of some biological systems, and to provide a degree of quantitative predictive power.

1. Modelling biological systems

This paper is about modelling dynamical systems directly from data. Although the methods apply to all kinds of systems for which a deterministic model is appropriate, I shall concentrate on modelling chaotic systems because they are particularly interesting and because they have led to the important realisation that complex behaviour need not imply that the underlying system is complex. I shall assume that data comes from a deterministic dynamical system, and try to model it. Any modelling errors will be ascribed to noise. How to distinguish between the case where a system's behaviour is random and that where it is chaotic, possibly with some level of noise, is a question that has no entirely satisfactory answer at present. Workers in dynamical systems would probably say that the question is misguided, in that (except in quantum mechanics) randomness is another name for complexity which we cannot describe concisely. In dynamical systems terms, asking for a concise description is close to asking whether the dimension of the system is low. Statisticians might have a completely different viewpoint. The question of whether a mathematical model is legitimate is traditionally answered in science by testing whether it has any predictive power, and this is the attitude I shall adopt.

A dynamical system consists, as far as this paper is concerned, of a state space which is a finite-dimensional manifold M, and a (deterministic) map f taking any point z in M to another point $f(z)$ in M. Often z cannot be observed directly; instead, there is another map g which is the *readout map* or *observer*, and at any instant we can actually observe $x = g(z)$ where x is a single real number. This represents the case where all that can be measured is a single variable such as the population of a particular species. If more than one variable can be tracked the problem does not change in any important way, so for simplicity I shall only discuss the single variable case. Values x_t are given for x at different values of t, and t is assumed to take on equally spaced discrete values which we may as well take to be integers. To keep the description reasonably concise, I shall deal only with this model, which is essentially a difference equation. If a differential equation model is more appropriate than a difference equation, the fact remains that measurements will only be available at discrete time intervals, and the difference equation setup will still work as described. Summing up, we have

Definition. A dynamical system is a triple $\{M^p, f, g\}$ where M^p is a p dimensional manifold and $f : M^p \to M^p$ and $g : M^p \to \mathbb{R}$ are C^1 maps.

Even under the assumption that a time series of measured data is from a deterministic system and is sufficiently free of noise to be useful, there are many problems in using it. For example, the true system $\{M, f, g\}$ is usually unknown, and although all we are given is values of x we must somehow deduce values of z, or of something equivalent to z, if we are to be able to make any use of dynamical systems theory. This is the problem that embedding addresses. It reconstructs a time series which turns out to be as good as z for modelling purposes.

Once z has been reconstructed, then what do we do? Most work to date has concentrated on examining indicators such as fractal dimension which say something about the geometry or even the dynamics, without having to find M or f. The novel contribution of this paper is a method for reconstructing both the manifold M (or some submanifold which contains the attractor for the dynamics) and the map f. They are built from the embedded data, or from the true statespace data if we are lucky enough to have it. This gives a good test for chaos, in that it may discover a low-dimensional chaotic dynamical system which has good predictive power.

It may also be useful for interpolation when additional experiments are difficult or expensive, but interpolated data is needed for, say, control purposes. It is important to realise, though, that numerical predictive power will be inherently limited by one of the defining features of chaos, namely sensitivity to initial conditions. We can only expect to predict well in the short term even if our models are very good.

2. Takens embedding

A brief description of embedding is helpful to set the scene for what follows. It is often called Takens embedding since Takens (1981) seems to have been the first to put it on a firm mathematical footing, but it was invented independently by others (Packard et al, 1980).

As we have seen, in producing a dynamical model it is necessary somehow to go from the data x to the system state z; that is, from measurement space to statespace. We are given a time series of data points

$$x_0, x_1, \ldots, x_N$$

together with a lag τ and an embedding dimension k which are chosen according to methods to be discussed later. The embedding process constructs the k dimensional vectors

$$\begin{aligned} v_0 &= (x_0, x_\tau, \ldots, x_{(k-1)\tau}) \\ v_1 &= (x_1, x_{1+\tau}, \ldots, x_{k\tau}) \\ &\vdots \\ v_t &= (x_t, x_{t+\tau}, \ldots, x_{t+(k-1)\tau}) \\ &\vdots \end{aligned}$$

The claim is that v_t can be used in place of the system state z_t, in the sense that any calculations that can be done with z can also be done with v. Of course there are some conditions: we need k to be large enough, we require any noise not to be too great, we require the system to be "well-behaved" in certain ways, and we require there to be no unfortunate symmetries which mask modes of z behaviour from x. The issues are discussed widely in the literature. They are essentially the same as in the well-known control theoretic concept of observability (which is about provision of adequate instrumentation) except that control theory's checkable observability condition (Mees, 1981) is replaced by an essentially uncheckable genericity condition. There are great compensations, though: the embedding theorem, unlike the standard linear observability theorem, allows nonlinearity and does not require the system map and state space to be known.

Formally, the theorem statement is as follows.

Takens' Theorem *Given a p dimensional manifold M^p and C^1 maps $f : M^p \to M^p$ and $g : M^p \to \mathbb{R}$, then whenever $k \geq 2p+1$, the map $V : M^p \to \mathbb{R}^k$ is generically an embedding, where*

$$V(z) = (g(z), g(f(z)), g(f^2(z)), \ldots, g(f^{k-1}(z))).$$

An embedding is a proper injective immersion (Guillemin and Pollack, 1974). Intuitively, it is a satisfactory representation of the manifold in the same way that the surface of a ball is a satisfactory representation in 3 dimensional space of the two dimensional manifold S^2, called the two-sphere. As can be seen from this example, it is sometimes possible to relax the requirement that the embedding space dimension k be at least $2p+1$, where p is the manifold dimension. Indeed, $k = p+1$ is very often adequate. Nevertheless, the theorem gives no guarantees in such cases and care must be taken.

The lag τ introduced in the definition of v_t involves a generalization: if $\tau = 1$ then

$$v_t = V(z_t)$$

whereas $\tau \neq 1$ requires the theorem to be modified slightly. The lag is introduced because it proves to have practical advantages. For discussion in this paper, the possibility that $\tau \neq 1$ will be ignored, though some of the examples used nonzero lags in their embeddings. Finding a good lag is discussed by Rapp *et al* (1988) and by Fraser (1988).

The value of p is usually unknown. It may even be infinite, though most definitions of deterministic chaos would insist on a finite dimensional attractor which would therefore be containable in a manifold with some finite value of p. If p is not known, how may we find k? The usual way is to try several values of k and perform a variety of tests: projections of the data should "look right", there should be agreement over several values of k in fractal dimension estimates, and so on. A new and very interesting approach to this is calculation of mutual information (Fraser, 1988). We shall see later that finding tesselations is a good way to determine good values of k.

From now on, I assume that either true state space data or embedded data is available. Except during discussions such as choice of dimension, I shall not distinguish between these two types of data, though one may well feel more confident about calculations performed with the first.

3. Tesselation and the geometry of embedded phase portraits

The data is now in the form of an ordered set of vectors in k-dimensional space. Its representation in the space is a cloud of points, and if we can obtain reasonable projections into the plane, we may be able to gain some feel for what the attractor looks like. This uses the excellent pattern recognition powers of the human brain, and is most useful when it is possible to discern folds or branches, which are both good chaos indicators (Judd, this volume). If k is large it is unlikely that this will work, but even when k is small, looking at projections of the cloud is often uninformative, either because it is difficult to find a good projection, or because it is difficult to see the local relationships between points in the cloud. There are methods for trying to find a good projection, such as projection pursuit (Silverman, 1986), so let us concentrate on how to reveal the geometric structure inherent in the data. In other words, can we help in showing what the attractor really looks like?

There are at least two features missing from any picture of a projection of the point cloud. It does not show what points are near to one another, and it does not show the ordering in the time series. The remainder of this section deals with the first problem, and the next section deals with the second.

The neighbourhood structure can be shown by drawing lines from nearby points to one another and displaying the projections of these lines. Still better, if the points all lie on a manifold M we can try to use them to triangulate the manifold: that is, to cover it with simplices (which are triangles if it is two dimensional) with disjoint interiors, and whose union covers a connected component of M.

Rather than start by considering how to triangulate a manifold from points on it, look at the easier problem of constructing a triangulation of $\mathbb{R}^k$ from the points, without regard to the manifold. Even in this case, there is no unique way to triangulate, but there is a way which has certain optimality properties and which proves to have other properties which are useful later. This is the Delaunay triangulation (Green and Sibson, 1978) in which the triangles are as nearly equilateral as possible (subject to a proper definition of this implied optimality). Figure 1 shows how the Delaunay triangles have the data points as their vertices, and are dual to a tiling of $\mathbb{R}^k$ called the Dirichlet tesselation. Duality means here that data points are defined to be neighbours if they are linked by edges in the triangulation, or if they are on opposite sides of a polytope's face in the tesselation.

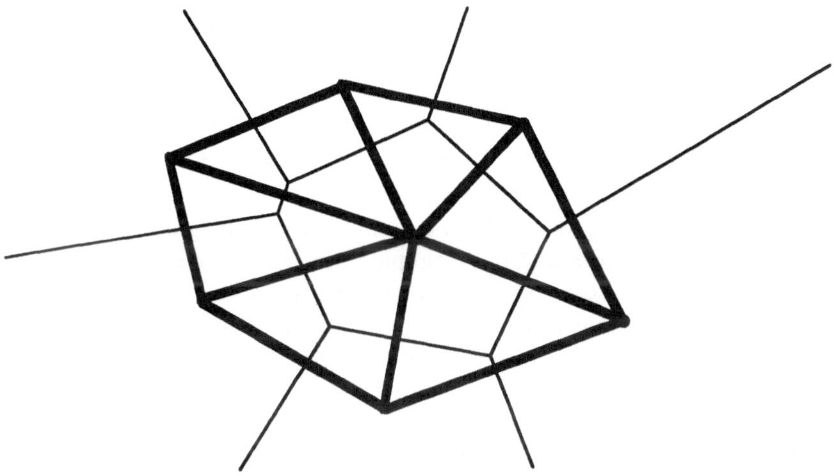

Figure 1. Data points v_i are vertices of the Delaunay triangulation (solid lines). Connected points are neighbours; the perpendicular bisectors of the line segments joining neighbours define a tiling of the plane (dashed lines) called the Dirichlet tesselation.

The Dirichlet tesselation of $\mathbb{R}^k$ by a set of distinct points v_i, $i = 1, \ldots, N$, is the set of polytopes (or *tiles*)

$$T_i = \{u \in \mathbb{R}^k : |u - v_i| < |u - v_j| \quad \text{for all} \quad j \neq i\}.$$

Thus the closure of $\cup_i T_i$ is $\mathbb{R}^k$, and two points v_i and v_j are neighbours if the intersection of the closures of T_i and T_j is nonempty.

The tesselation is sometimes called the Voronoi tesselation (Preparata and Shamos, 1985) and the polytopes are sometimes called Thiessen polygons. There are many other names associated with both the triangulation and the tesselation, presumably because they has been rediscovered a large number of times in one context or another.

Constructing either one of the triangulation or the tesselation allows the other to be constructed easily. Constructing either efficiently is, however, non-trivial. Green and Sibson (1978) and Preparata and Shamos (1985) both have good discussions of suitable algorithms for the $\mathbb{R}^2$ case, but there are rather few methods available for $\mathbb{R}^k$ when $k > 2$. The best I have discovered is by Bowyer (1981) and his algorithm was used as a starting point for much of what follows. The problem is that the number of neighbours of a point grows at least exponentially with k (Preparata and Shamos, 1985). With Bowyer's algorithm and 1988 workstations, a reasonable limit on the dimension of problem that can be tackled is $k \leq 5$, with $k \leq 3$ required for good response times with data sets of several thousand points.

As stated above, an important problem is to triangulate M, not $\mathbb{R}^k$. Sometimes M is known and has a simple form, such as a 2-sphere, in which case it is not hard to find a special purpose method for triangulating it. For other cases we have to find some general method for triangulating M rather than $\mathbb{R}^k$, even when the dimension p of M is not known *a priori*. Provided the embedding is good and there are enough data points, points which are nearby in $\mathbb{R}^k$ will usually be nearby on M. (In the limit of infinitely many points, this is always true by the definition of embedding.) We can therefore identify p and triangulate M simultaneously, by first triangulating the points as a subset of $\mathbb{R}^k$, then examining the simplices that result. They are likely each to have at least one point that comes from a different part of M, and it can be removed. This point can be identified by the fact that it is further from the centroid than the others. By removing one point from every simplex, then repeating the whole process as long as most simplices are strongly asymmetrical, we reduce the dimension of the triangulation in steps. If the process is taken too far, the tendency will be to disconnect the simplices from one another, but when all goes well the end result will be a cover of (part of) M with p dimensional simplices.

For example, triangulating a set of points in $\mathbb{R}^3$ which lie on a torus T^2 results in tetrahedra each of which has one face F that is an approximation to the local tangent space of T^2. The other 3 faces result from joining the points defining F to a point Q on some other part of T^2. Typically, Q will be much further from the points of F than are the points of F from one another. By examining all of the tetrahedra, and discarding from each the point furthest from the centroid, we are left with triangles the union of which should be an approximation to T^2. Figure 2 shows this in operation; the points used were from a dense trajectory on the torus, and the trajectory can even be seen in the edges of the triangles. There were too few points to give a perfect triangulation: a close examination will reveal that occasionally a triangle is produced which does not well-approximate the tangent space.

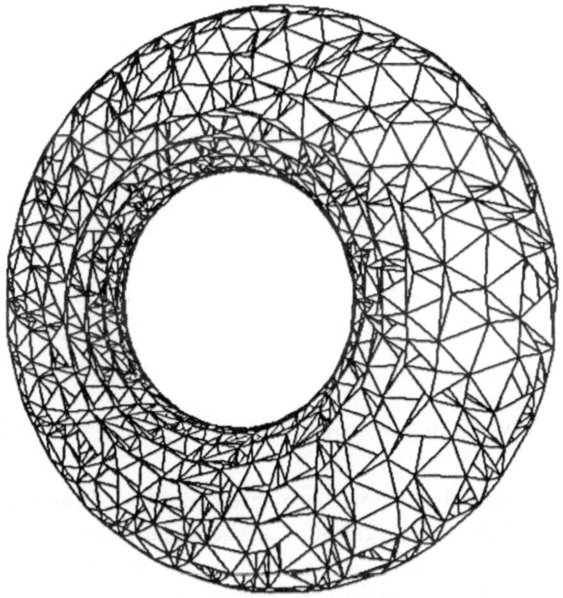

Figure 2. Points on a torus with their triangles formed as described in the text. A poor man's hidden line removal has been used: triangles were drawn, with black edges and white interior, in decreasing order of the distances of their centroids from the observer. The torus is nearly covered by the triangles but a few triangles do not lie on the surface.

Figure 2 shows an additional useful feature of this approach. Since the triangles can be drawn with hidden surface removal, the torus is much clearer to the eye in this representation than it would be in the point-cloud representation.

If the process is automated in the suggested way, removing points iteratively from simplices until simplices are on average approximately equilateral, we have a method of identifying p. How well this works will depend on how many points there are, as well as on how nearly the attractor (which typically has fractional dimension) fills a submanifold of the embedding space. When it does, we also have a way of finding k, since a successful identification of a topological dimension p for the manifold also identifies the current embedding dimension k as satisfactory. In the torus example, we would be confident both that $p = 2$ is the manifold dimension, and that this particular two dimensional manifold can be embedded in 3 dimensions.

Figure 3 shows the process applied to points from a trajectory of the Lorenz system (Sparrow, 1982). The attractor has dimension about 2.07 but the tesselation process has identified it as two dimensional, with some obvious difficulties near the join of the two lobes of the attractor. This is either good or bad, depending on one's point of view. It is bad in that the smallest manifold dimension that can hold the attractor is 3, not 2 as the tesselation implies. It is good in that it suggests that the data can fit onto a branched two dimensional manifold, which is a well-known model of the Lorenz system, providing a good approximation to the dynamics (Sparrow, 1982).

4. The dynamics of embedded phase portraits

One feature missing from a projected picture of the point cloud is a clue to which points are neighbours in the original space. Another is the ordering of the original time series: that is, the dynamics rather than the geometry. With almost no effort we can get some dynamical information, and with rather more effort we can get a great deal.

We make the following assumptions, which are largely a repeat of the assumptions of the previous section. The system has state $z_t \in \mathbb{R}^k$;

$$z_{t+1} = f(z_t)$$

with $f \in C^1$. We measure

$$x_t = g(z_t)$$

where $g \in C^1$. Embedding "works" and gives a map V which implies that the sequence $\{v_t\}$ is essentially diffeomorphic to $\{z_t\}$. (The map involved is not really a diffeomorphism but an immersion.)

Figure 3. Points on a Lorenz trajectory, triangulated as in Figure 2. The attractor is slightly more than 2 dimensional but this is not evident to the algorithm, except insofar as it experiences difficulty near the join of what it sees as two leaves of the attractor.

The last assumption, that V is an embedding, implies there is a C^1 map ϕ for which

$$v_{t+1} = \phi(v_t).$$

If we can estimate ϕ from the data we have rather a complete representation of the system. One method of estimating ϕ has been described by Farmer and Sidorowich (1987) and another by Casdagli (1989) but the method I shall describe is different and was developed independently.

Suppose $\mathbb{R}^k$ has been tesselated using the embedded data. In this section we do not need to worry about identifying p or M since we can think of ϕ as a map on $\mathbb{R}^k$, which just happens to hold invariant a certain submanifold of $\mathbb{R}^k$.

As a first application of this, let us use the fact that continuity of ϕ ensures that points which are close together will tend to remain so after being mapped by ϕ. Thus points which were neighbours in the tesselation will be close after one step of the mapping. Moreover, the edges of the small triangles (or the facets of the simplices) will, to a first approximation, be mapped into straight lines (or portions of affine submanifolds). If we take the original neighbours, find where they are mapped to, then draw the line segments as if they were mapped in the same way, we will see an approximation to how the part of the manifold covered by the data is affected by the map. Figure 4 shows this for a few points from a Poincaré section of a piecewise linear approximation to the forced Duffing equation. Notice how the original tesselation in Figure 4a appears in Figure 4b to be forced into a double-folded shape, highly suggestive of what the map actually does to the plane. It should be emphasised here that the tesselation program knows nothing whatever of the dynamics, nor anything about manifolds except that they are locally Euclidean.

We can do still better, though. Because the time series tells us the mappings of the data points we can approximate ϕ separately on each simplex, or on each tile. If the simplices or tiles are small we should be able to use relatively unsophisticated approximations – even linear ones. The main problem is to make sure that the overall approximate map is reasonably well-behaved.

Taking the simplex idea first, suppose $x \in S$ where S is a simplex with vertices $v_i, i \in I(S)$ where $I(S)$ is the index set for the vertices. A little work with Taylor series suffices to prove that since

$$\phi(v_i) = v_{i+1}$$

we can approximate ϕ in S by

$$\hat{\phi}(x) = \sum_{i \in I(S)} \lambda_i(x) v_{i+1}$$

where $\lambda_i(x)$ are the barycentric coordinates of x, i.e.

$$\sum_{i \in I(S)} \lambda_i(x) = 1$$

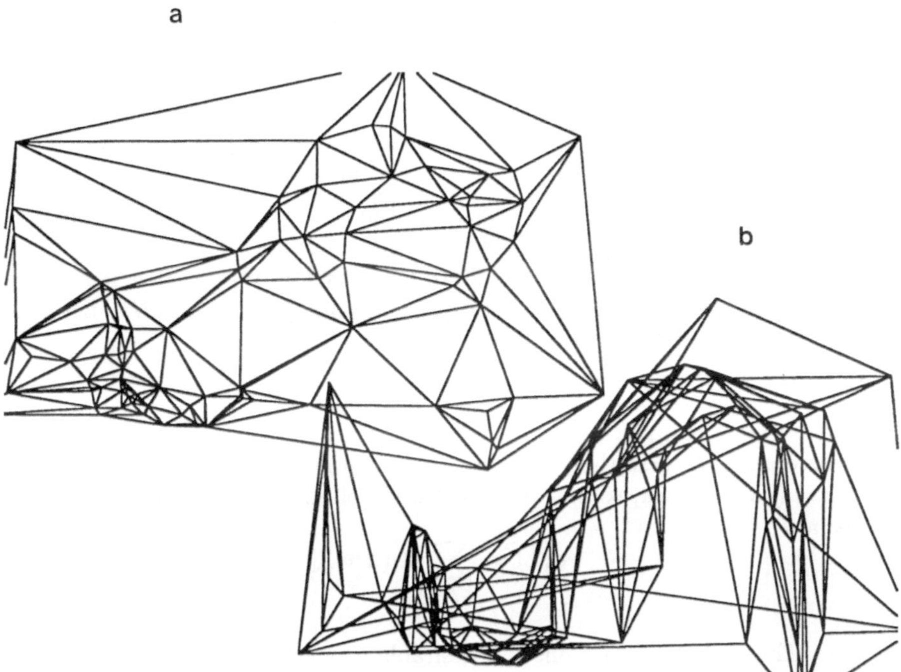

Figure 4. (a) A triangulation of some points in the plane which have come from a known map. (b) The approximate images of the triangles after one time step. Note the appearance of a double fold, which does correspond to the action of the real map.

and

$$\sum_{i \in I(S)} \lambda_i(x) v_i = x.$$

These are $k+1$ linear equations in $k+1$ unknowns, and are readily solvable. The reason for using barycentric coordinates is that (as the reader can easily check) they ensure that $\hat{\phi}$ is exactly correct at the given data points, and that affine functions of x are reproduced exactly. It is easy to show also that $\hat{\phi}$ is continuous as a function of x.

This method is straightforward but uses few neighbours, so it may give a poor result in regions where there are no data points v_i nearby. A better way is to work from the tesselation; the methodology is fully described elsewhere (Mees, 1989) but I shall outline the results here.

The approximation is now based on the tile $T(x)$ obtained by inserting x into the tesselation. $T(x)$ will gain territory from other tiles T_i (with zero territory gained from most, and finite territory always gained as long as x lies inside the convex hull of the original points $v_i, i = 1, \ldots, N$). Let the territory gained from T_i be $T_i(x)$. Define

$$\lambda_i(x) = \rho\, \mu(T_i(x))$$

where ρ is chosen so that $\sum \lambda_i(x) = 1$, and μ is the measure of the subtile $T_i(x)$ (i.e. the area, volume or generalised volume). The approximate map is now

$$\hat{\phi}(x) = \sum_{i \in I(T)} \lambda_i(x) v_{i+1}.$$

This calculation retains the desirable property that $\hat{\phi}$ is exact at the data points and is exact for affine ϕ. The new feature is that $\lambda_i(x)$ may be shown (Sibson, 1980) to be C^1 functions of x everywhere within the convex hull of the data, except possibly at the data points themselves, where only C^0 is guaranteed.

The main difficulty, both in programming the tesselation approximation and in calculations using it, is finding the measures of the subtiles. This is at least as computationally complex as tesselating in the first place, and forms the limiting factor in the calculation. Inventing and implementing a good algorithm was the most time consuming part of this research.

Both of the approximations to ϕ are better than might be expected; they are not merely piecewise linear as functions of x. For example, the tesselation version of $\hat{\phi}$ has components which are rational functions of the components of x and which join smoothly at the edges of tiles. It would

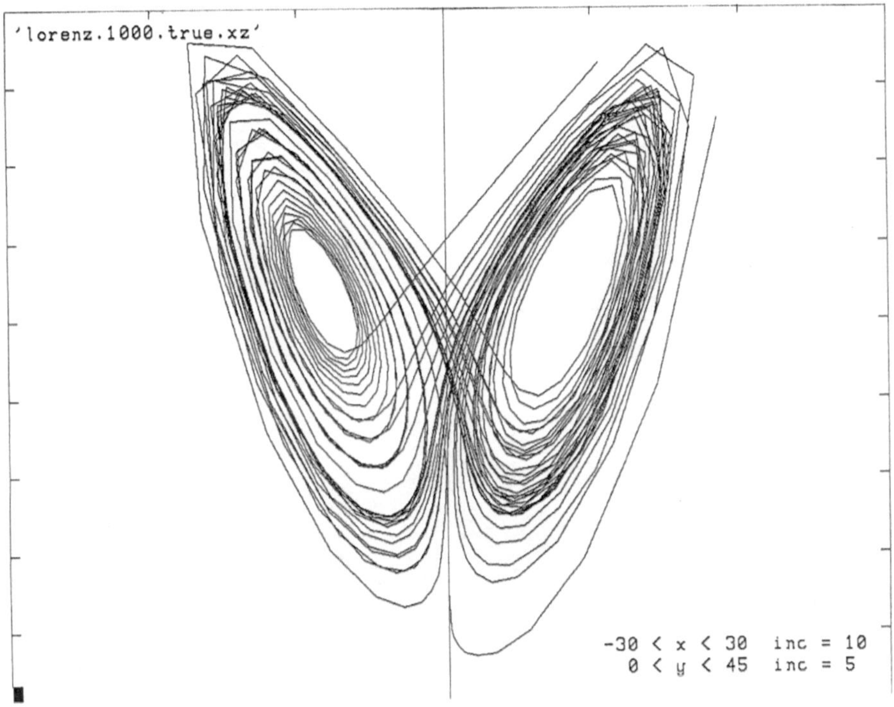

Figure 5. (a) A Lorenz trajectory from a numerical integration. (b) A trajectory from the modelling algorithm, deduced from the first half of the original trajectory.

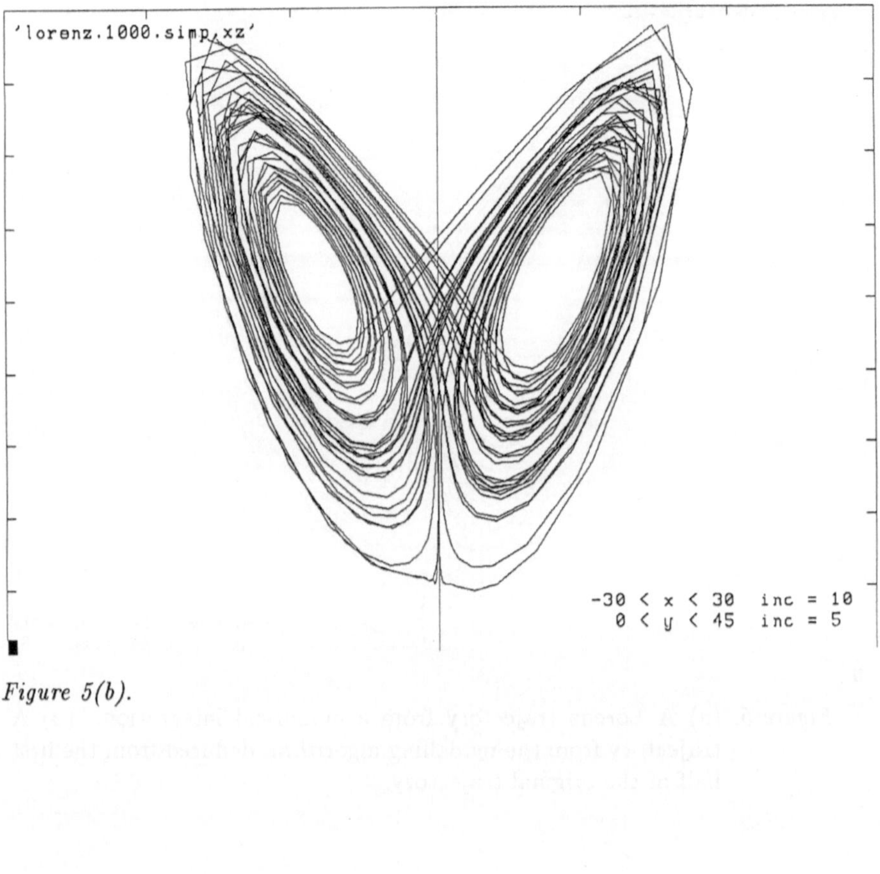
'lorenz.1000.simp.xz'
-30 < x < 30 inc = 10
0 < y < 45 inc = 5

Figure 5(b).

be possible to improve approximation of ϕ in terms of the known $\phi(v_i)$, with respect to which the present version is indeed only piecewise linear, but this becomes less important as the number of data increases since the tiles become smaller and smaller. Preliminary experiments with piecewise spherical quadratic fitting bear this out: in most cases the improvement is not significant.

Figure 5a shows a Lorenz trajectory of about 2000 points. The first 1000 points were used to construct a tesselation and the tesselation version of $\hat{\phi}$ was used to predict the next 1000. In Figure 5b it can be seen that the Lorenz attractor is very well represented: $\hat{\phi}$ seems to have similar properties to the true Lorenz map. (Here, the Lorenz map means the map that takes a sample point one more time step along the trajectory.) It is also clear that the trajectory is not the same as the original one, but given that the Lorenz attractor has a positive Lyapunov exponent, it is impossible in any finite accuracy calculation to follow another trajectory for an indefinite length of time. For this algorithm, qualititive agreement is excellent but quantitative agreement is only good for a modest length of time. Figures 6 and 7 show original and modelled system trajectories for lynx and for blowfly population data. The tesselation method has easily picked out the approximate periodicity, but since the data is certainly noisy and there are few data points, it is unsurprising that quantitative agreement is less good. The main errors in Figure 6 are in peak amplitude, while in Figure 7 peak amplitude is reasonably well predicted but the phase slips after the first cycle. Predictions for systems such as these, which are probably subject to external disturbances, can be improved greatly if instead of trying to predict into the far future we use a Kalman-filter-like predictor which only predicts one timestep, or a few timesteps, ahead, and updates the tesselation model as new data come in. This "recursive" predictor is described elsewhere (Mees, 1989).

5. Future work

The work described in this paper is preliminary. I have shown how it is possible to get information about both the geometry and the dynamics of attractors and manifolds for experimental data, with explicit reference to detection of folds and branches, and to prediction and interpolation. There are many more possible applications, including a quite remarkable degree of claimed noise reduction (Farmer and Sidoriwich, 1987) but a good deal of development is needed. The limitation of using tesselations for modelling is the exponential growth of the time required as the embedding dimension k increases. The best alternative seems to me to be generalised smoothing splines (Friedman, 1988) and this is now under investigation.

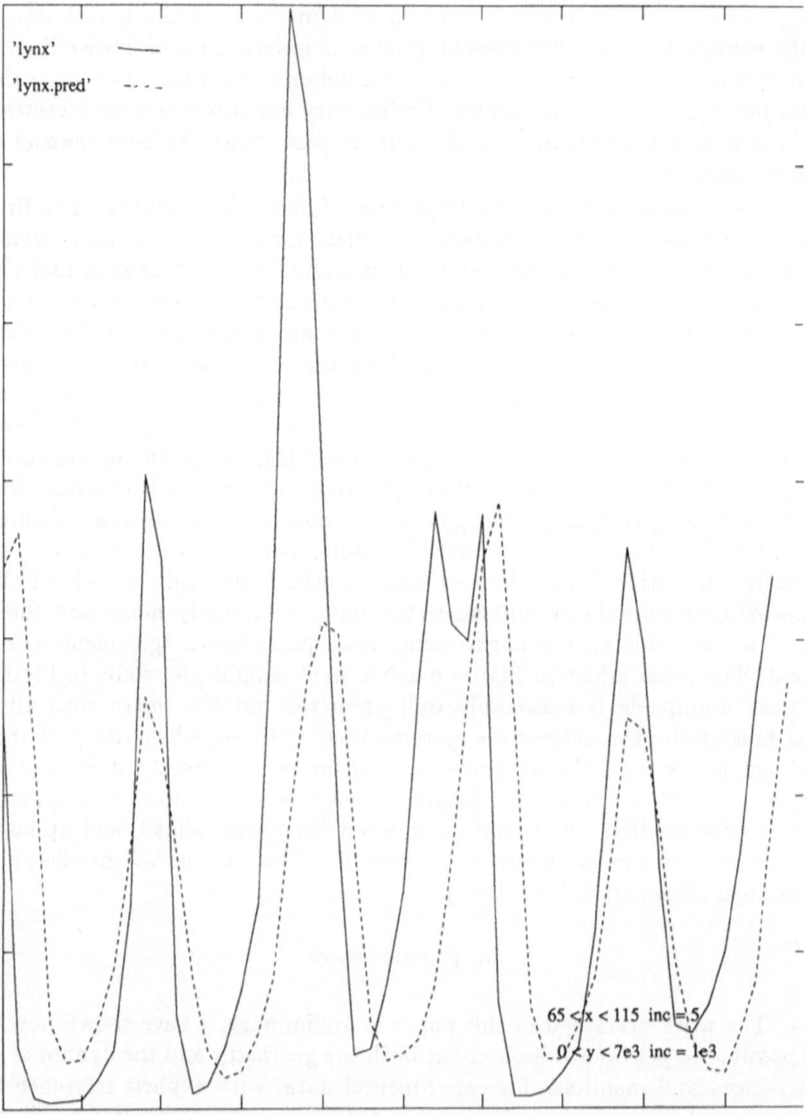

Figure 6. Solid lines: time series for lynx population: times 65 to 113 of the original data are shown. Dashed lines: time series generated by the modelling algorithm; the first 64 points of the original time series were read, then the tesselation model was built and used to predict the next 60 steps.

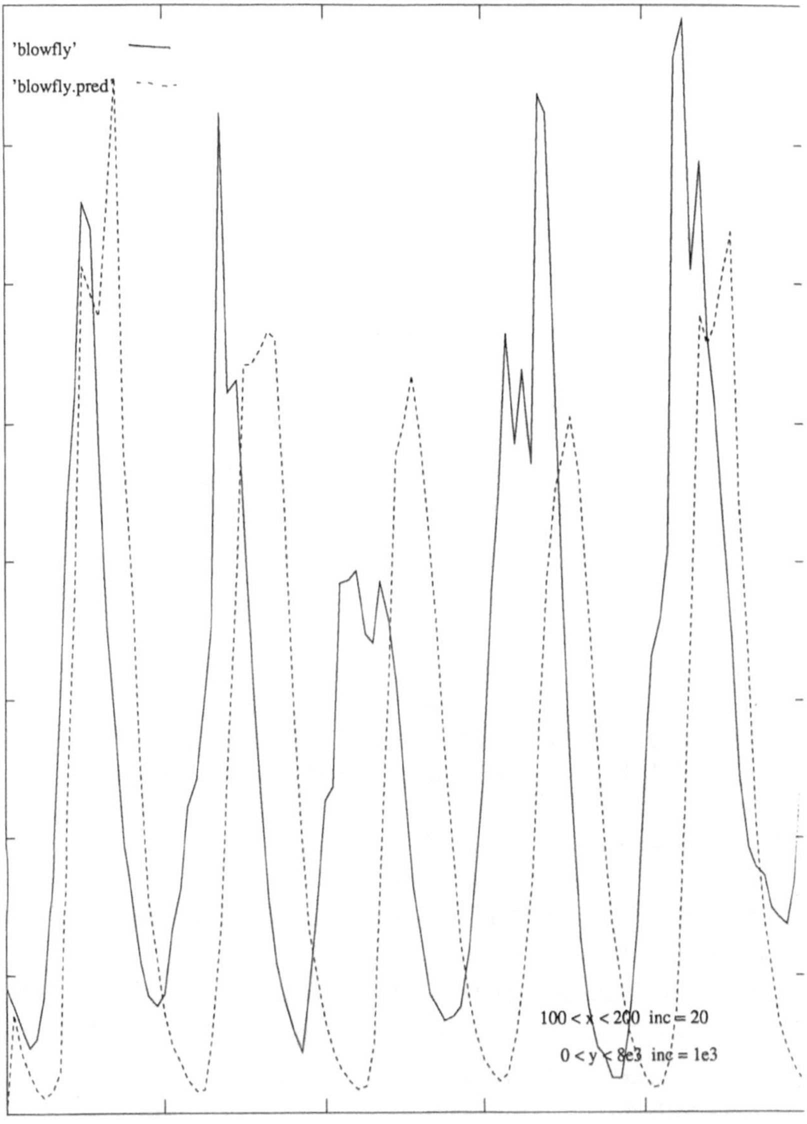

Figure 7. Solid lines: time series for blowfly population: times 100 to 200 are shown. Dashed lines: time series generated by the modelling algorithm; the first 100 points of the original time series were used to build tesselation, then the next 100 were predicted.

Even within the limitations of the approach as it has been implemented so far, there are very encouraging indications, including an intriguing success with stock market index data as well as with the biological data referred to here.

Acknowledgements: I thank Adrian Bowyer for providing a copy of a program implementing his algorithm for k dimensional tesselation. I have benefited from conversations with Lyle Noakes and Colin Sparrow. This work would not have been possible without the support and hospitality of the Dynamical Systems Research Centre at King's College, Cambridge, and that of the Departments of Mathematics and of Aerospace and Mechanical Engineering at the University of Arizona.

REFERENCES

[1] A. Bowyer, Computing Dirichlet tesselations, *The Computer Journal,* **24(2)**, pp. 162-166 (1981).

[2] M. Casdagli, *Phys. Rev. Letters*, to appear. (1989).

[3] J.D. Farmer and J.J. Sidorowich, Predicting chaotic time series, *Phys. Rev. Letters* **59(8)**, pp. 845-848 (1987).

[4] A. Fraser (1988), Information and entropy in strange attractors, *IEEE Trans. Information Theory*, in press.

[5] J.H. Friedman, Multivariate adaptive regression splines, *Technical Report* 102, Laboratory for Computational Statistics, Stanford University (1988).

[6] P.J. Green and R. Sibson, Computing Dirichlet tesselations in the plane, *The Computer Journal* **21**, pp. 168 (1978).

[7] V. Guillemin and A. Pollack, *Differential Topology*, Prentice-Hall New Jersey (1974).

[8] A.I. Mees, *Dynamics of Feedback Systems*, Wiley, New York, (1981).

[9] A.I. Mees, *Tesselations and Dynamical Systems*, in preparation (1989).

[10] N.H. Packard, J.P. Crutchfield, J.D. Farmer and R.S. Shaw, *Phys. Rev. Lett.* **45**, pp. 712 (1980).

[11] F.P. Preparata and M.I. Shamos, *Computational Geometry: An Introduction*, Springer, New York, (1985).

[12] P.E. Rapp, A.M. Albano and A.I. Mees, Calculation of correlation dimensions from experimental data: progress and problems. In *Dynamic*

patterns in complex systems, eds. J.A.S. Kelso, A.J. Mandell and M.F. Schlesinger, pp. 191-205. World Scientific, Singapore (1988).

[13] R. Sibson, A vector identity for the Dirichlet tesselation, *Math. Proc. Camb. Phil. Soc.,* **87**, pp. 151-155 (1980).

[14] B.W. Silverman, *Density Estimation for Statistics and Data Analysis*, Chapman and Hall, London (1988).

[15] C.T. Sparrow, The Lorenz Equations: Bifurcations, Chaos and Strange Attractors, *Applied Mathematical Sciences* 41, Springer, New York (1982).

[16] F. Takens, in Dynamical Systems and Turbulence, Vol 898 of *Lecture Notes in Mathematics*, ed. D.A. Rand and L.-S. Young, page 366, Springer, Berlin (1981).

PARTICIPANT'S COMMENTS

This paper forwards some intriguing possibilities and promise. Although laconically written, there is sufficient information in it for the diligent reader to pursue. I certainly hope that the "non-parametric" identification of strange attractors (as statisticians would most likely term this approach) would get off the ground soon. Perhaps the time when we could predict stock market crashes is nearing — with all of the ensuing lucrative results for dynamical systems experts.

Yosef Cohen

Mees's paper offers a re-interpretation of, and new computational methods for analysing time series data. His embedding technique provides a way of drawing out possible underlying higher dimensional nonlinear dynamics, while at the same time providing a way of visualising models of invariant sets and strange attractors which determine complex dynamical behaviour.

The numerical work captures the "essence of chaos". Sensitivity to initial conditions and positive Liapunov exponents make it impossible, in any practical sense, to predict quantitative behaviour very far into the future. However, a qualitative description of a model of the underlying system behaviour of complicated biological and physical systems can be formulated by the methods Mees describes.

This is very important for practical decision making at both the workface and political levels. Such a qualitative description lets a decision maker know when the system is in a fragile state and is likely to (apparently) jump sharply in the time series data as it moves along the embedded manifold. Moreover, acceptance of the chaotic paradigm and the intuitive visual models of an attractor provides a framework in which commonsense allowances for, and reservations about future behaviour, can be taken into account.

However, it must be remembered that these methods still only give a fit for plausible embeddings and attractors and should be seen as providing *models*. The very dimension of the underlying system model is itself only estimated. Consequently there may be several models which fit the *past* time series data in Mees's sense, yet whose *future* behaviour may be qualitatively different at some point of their time evolution. These techniques are plausible as well as being powerful and managers of biosystems should resist the temptation to use them without fairly long time series runs and very thorough examination of the qualitative picture which emerges.

Despite this slight reservation, I personally find this whole approach very exciting. Especially where insight and intuition are overwhelmed by inherent complexity, and where the data seems worse than noisy, this paper offers tools to reinterpret and heuristically analyse the situation.

Phil Diamond

Detecting Folds in Chaotic Processes By Mapping the Convex Hull

J.N. GLOVER

Abstract

An approximate tesselation map is used to detect global foldings in reconstructed time series by mapping the boundary of the convex hull of the reconstruction. This technique is applied to the Hénon map and the Lynx population data and in the latter case shows that a model of the system requires more than two degrees of freedom.

Introduction

An experiment in which a single output from a closed system is measured is often, mistakingly, considered to be inadequate for determining global phase space properties of the original system. This is equivalent to saying that all the possible systems which could give rise to the same output signal are unlikely to be related in a simple or useful way. In 1980 Takens[1] proved a theorem that suggests that the opposite is true and that a single output signal is sufficient to reconstruct an embedding of the phase space trajectory from which the output was measured. This reconstructed trajectory contains all the metric invariant and topological information of the original and can be used to infer properties of the system such as dimension, entropy, Lyapunov exponents etc. For further details see Schuster [2] and Mees[3].

This paper addresses a problem raised in [3] namely how to represent the data in order to determine qualitative properties of the map which generated it. In particular we wish to detect the folding and stretching of phase space which typically gives rise to chaotic dynamics and strange attractors as discussed by Judd [4]. Judd (private communication) suggests a procedure by which a triangulation of the data is mapped in the following manner for a two dimensional reconstruction; if $l_{i,j}$ denotes a line segment connecting the ith and jth points and is in the triangulation then its image is taken to be $l_{i+1,j+1}$. This procedure has been implimented by Mees [3] and for processes in which the folding is simple it provides a quick method for detecting folds that does not require much computing power. Its principle drawback is the crudeness of the approximating procedure which is equivalent to a piecewise linear approximation to the map and this can introduce spurious details that are not present in the system. An improvement to this procedure would be to use a better approximating map such as the the tesselation map discussed by Mees [3]. The number of points needed to adequately represent the image of the triangulation is quite high and hence it is desirable to only consider a subset of the triangulation, namely the set of edges which form the boundary of the convex hull of the data. This is the same as the set of edges which connect adjacent extreme points of the data set and can, once the triangulation is constructed, be obtained cheaply as a byproduct of the tesselation procedure. In practice this is done by constructing a triangulation for the set consisting of the data and a simplex containing the data. A point is taken to lie on the convex hull of the data if it is joined to one of the vertices of the simplex in the triangulation. Provided the simplex is sufficiently large this procedure seems to work well.

If the underlying map that generates the data is a diffeomorphism (i.e. a smooth invertible map) then it can be shown that the boundary of the image of the convex hull is the same as the image of the boundary. It should then be sufficient to map only the boundary of the convex hull in order to detect global folds. This reduces the number of points where the map needs to be approximated at since this is an $n-1$ dimensional set rather than an n dimensional set i.e. the boundary of the convex hull and the convex hull respectively. A discrete set of data will satisfy these conditions if it is, for example, generated by sampling a continuous trajectory of a differential equation. It does not automatically follow however that an approximation to a diffeomorphism will itself be a diffeomorphism even if there is a large number of uniformally spaced data points. In some sense though our approximation should still reveal the presence or absence of a global folding that is close to the fold in the underlying map. The substantial saving in time should offset any worries that might arise as to the exactness of the procedure in some cases. An additional 'problem' that arises in estimating probabilty distributions of high dimensional data turns out in this case to be an advantage. Data points in high dimensions tend to lie close to their convex hull , a property which is sometimes stated as 'the curse of dimensionality'. This will be true in the case of reconstructed phase portraits as well so mapping the boundary of the convex hull would be the natural approach to take in high dimensional reconstructions irrespective of the considerations above.

In the following we use the tesselation map [2] to approximate the underlying map but a property of the procedure we describe is common to all weighted average maps. These can be written as

$$\tilde{f}(x) \equiv \sum_{i=1}^{n-1} \lambda(x, y_i) f(y_i) \tag{1a}$$

$$1 = \sum_{i=1}^{n-1} \lambda(x, y_i) \tag{1b}$$

where $\tilde{f}$ is the approximate map to f and only the data points $\{y_i\}_{i=1}^{n}$ are known and $\tilde{f}(y_i) = f(y_i) \equiv y_{i+1}$. If we denote the data set by U and the convex hull by $Co(U)$then

$$\tilde{f}(U) \subseteq U \tag{2a}$$

$$Co(\tilde{f}(Co(U))) \supseteq Co(U) \tag{2b}$$

This means that the global length scales of U and $\tilde{f}(U)$ are the same and that they occupy the same region of phase space. If the last data point lies on the convex hull then this result is modified slightly but for a sufficiently large number of data points the conclusion is the same, the convex hull of the data is the largest set that the approximation can be considered to be meaningful.

The Hénon Attractor

To illustrate the above procedure we use the well known Hénon map [5] given by

$$\begin{aligned} x_{k+1} &= 0.3y_k - 1.4x_k^2 + 1 \\ y_{k+1} &= x_k. \end{aligned} \tag{3}$$

Our observation is the sequence $\{x_i\}_{i=1}^{n}$ for $n = 200$. The reconstructed sequence is $\{(u_k, v_k)\}_{i=1}^{n-1}$ where $u_k = x_k = y_{k+1}$ and $v_k = x_{k+1}$. A map which generates this sequence can be obtained by substituting the change of coordinates $u = y, v = x$ into (1) which gives

$$\begin{aligned} u_{k+1} &= v_k \\ v_{k+1} &= 0.3u_k - 1.4v_k^2 + 1. \end{aligned} \tag{4}$$

In general neither the original map nor the change of coordinates relating it to the reconstruction is known. In this case it is and the accuracy of the approximating procedure can be checked. Figure 1 shows the two dimensional reconstructed time series while figure 2 show the convex hull. Figures 3 and 4 show the image of the boundary for (4) and its approximation. These figures are similar enough to claim that the folding process which gives rise to the strange Hénon attractor can be detected using the above procedure. Figures 5 and 6 show second applications of the exact and approximate maps which confirm this claim.

The points at which the approximation is not very accurate are those which are not close to any data points. Since we do not have sample data in these regions this is not suprising and is a good reason for keeping transient data when reconstructing the dynamics. If this cannot be done then the problem might be avoided by using the pruning procedure suggested by Mees [3] and removing those edges of the triangulation whose length is greater than some arbitrary cutoff. In this case only the boundary of this subset of the triangulation would be mapped.

A Biological System

The above procedure can be applied to biological population data and this has been done for the well known Lynx data [6]. Figure 7 shows the data which appears to be a regular sampling of a continuous aperiodic trajectory. This suggests that the reconstructed dynamics can be described be a diffeomorphism. The aim here is to decide whether the underlying dynamical system is two dimensional by finding a chaotic folding process. The reconstruction is shown in figure 8. Figure 9 shows the convex hull and its image and there does not appear to be any obvious folding. In addition the inequality (2b) appears to be strongly violated. This suggests that either there is : not enough data to make a reasonable approximation; the data is very noisy; a deterministic description of the system requires more than two degrees of freedom. The latter is probably the case and correlation dimension results suggest that at least a three dimensional embedding is required.

REFERENCES

[1] Takens F., Detecting Strange Attractor in Turbulence., *Dynamical Systems and Turbulence. Lecture Notes in Mathematics. Vol. 898*,(1980), Rand D.A. and Young L.S., eds, pp. 365-381. Springer-Verlag, New York.

[2] Schuster H.G.,*Deterministic Chaos: An Introduction*, 1984, VCH, New York, pp. 117-139.

[3] Mees A.I., Modelling Complex Systems, these proceedings.

[4] Judd K., From Fixed Points to Folds, these proceedings.

[5] Hénon M., Comm.Math.Phys., 1976, **50**, pp. 690.

[6] Cambell M.J., and Walker, A.M., A Survey of Statistical Work on the Mackenzie River Series of Annual Canadian Lynx Trappings for the Years 1821-1934 and a New Analysis ,1977, *J.Roy.Stat.Soc A*, **140**, part 4, pp. 411-431.

PARTICIPANT'S COMMENTS

The determination of diffeomorphic invariants (such as dimension of attractors) using a time series of a single scalar observable has obvious application in biosystem modelling. James offers one method for reconstructing an approximation to the original map which seems to work quite well with the Hénon attractor. However when applied to the well known Lynx data, we are left in doubt as to its applicability. As suggested, the problem could indeed be with the data. Even if a strange attractor exists for this system, we cannot be certain that the system was on or near the attractor when the data was taken. Never the less, as suggested by James, it would be of interest to examine a three (or higher) dimensional embedding. A positive result would be most noteworthy.

Tom Vincent

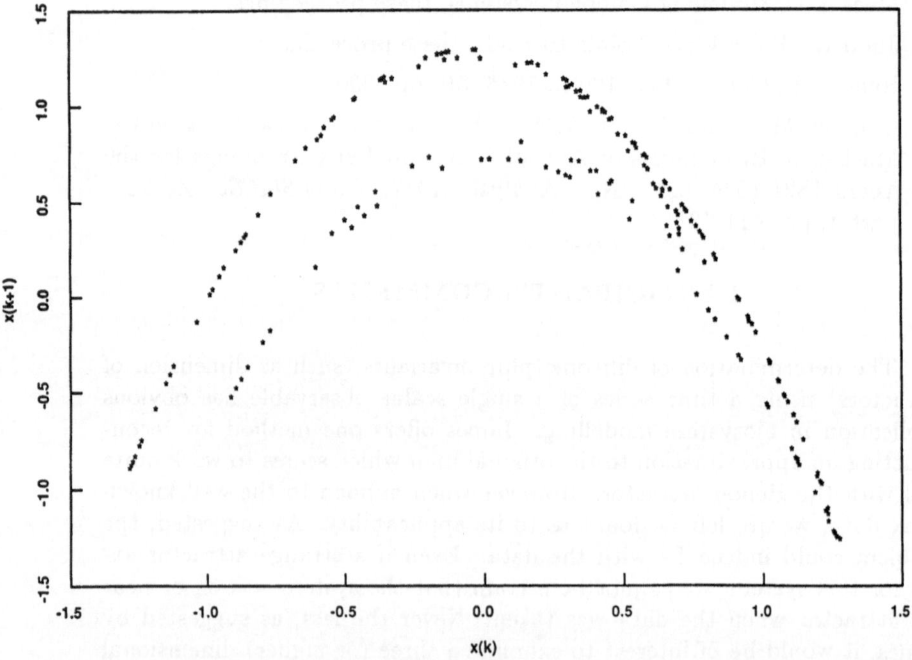

Figure 1. Reconstructed Hénon attractor.

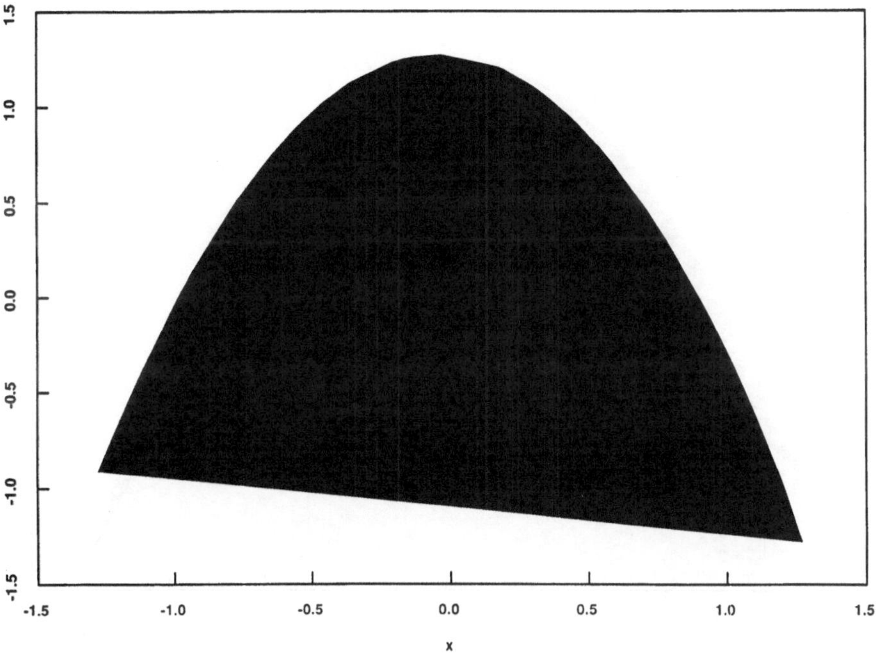

Figure 2. Convex hull of Hénon attractor.

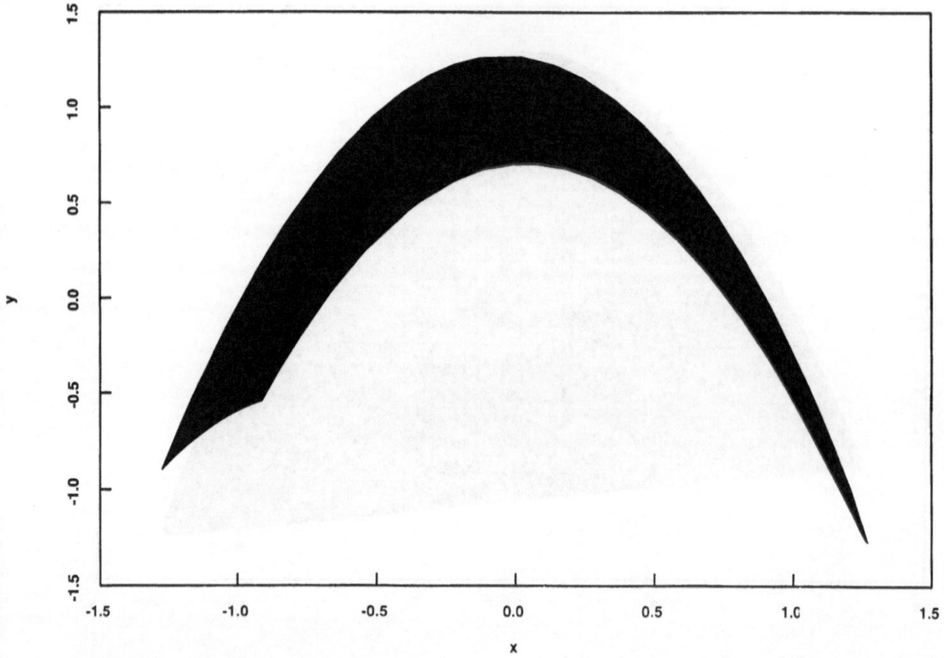

Figure 3. Exact map of convex hull.

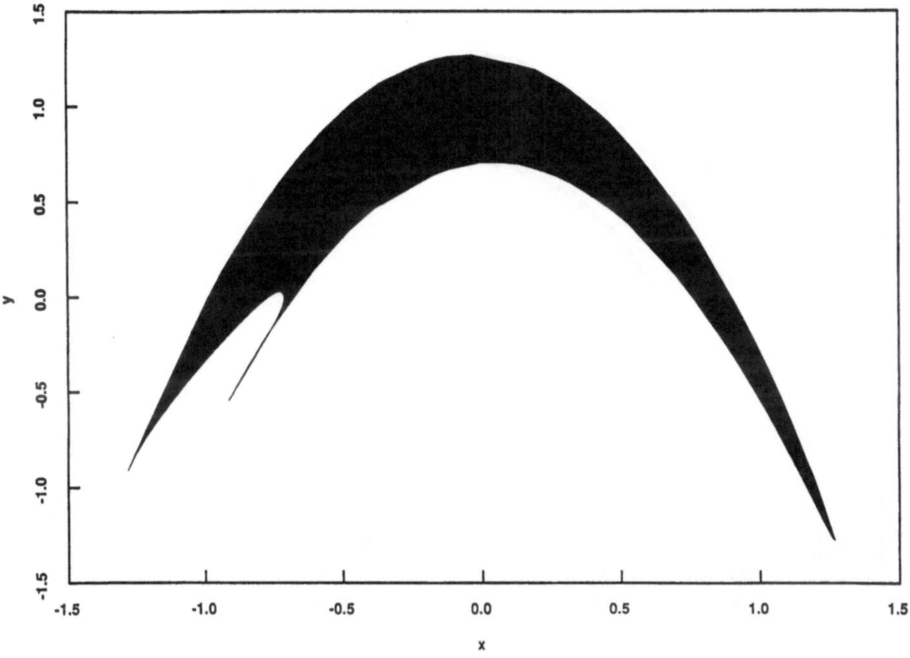

Figure 4. Map of convex hull.

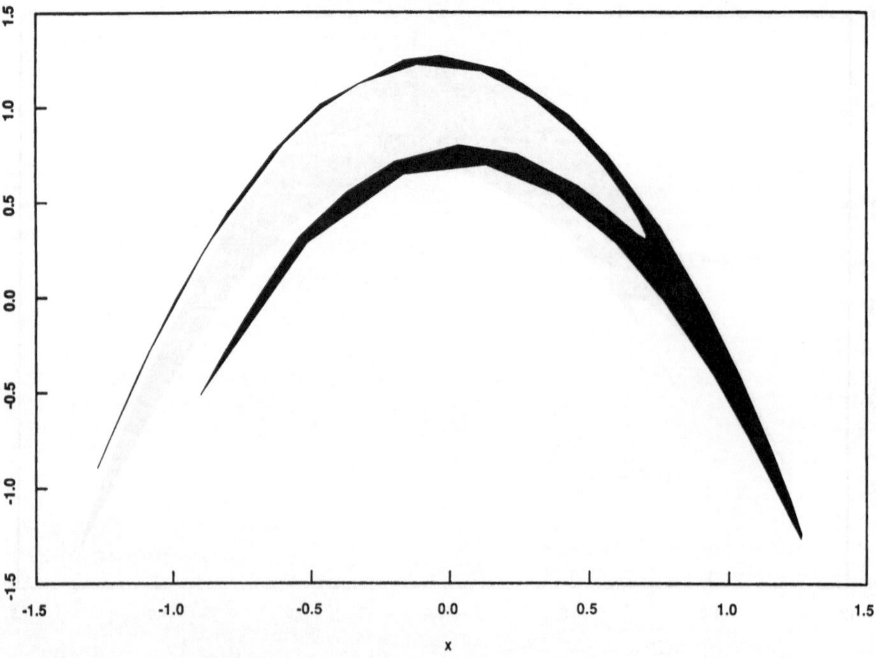

Figure 5. Exact second map of convex hull.

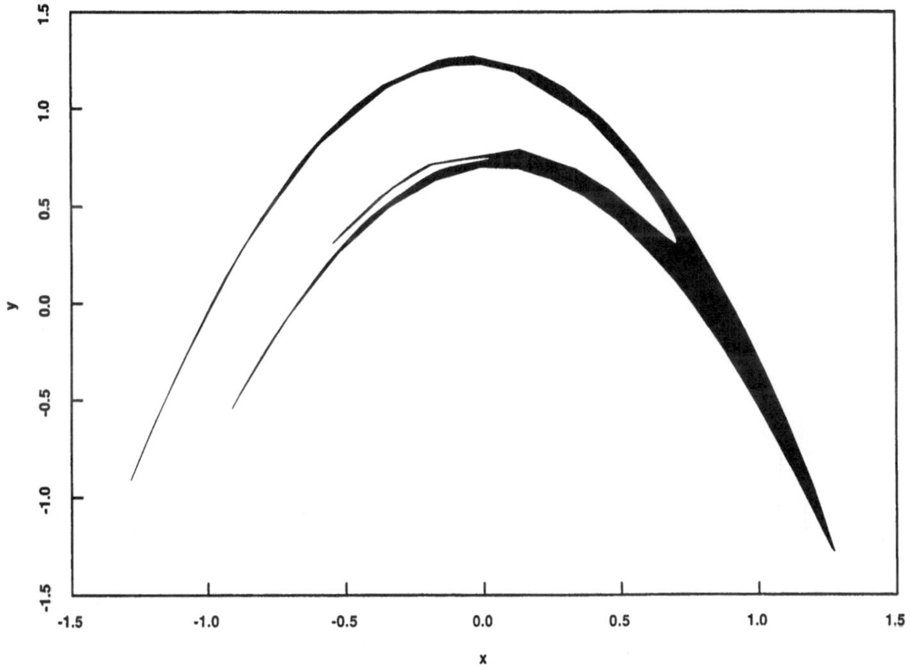

Figure 6. Second map of convex hull.

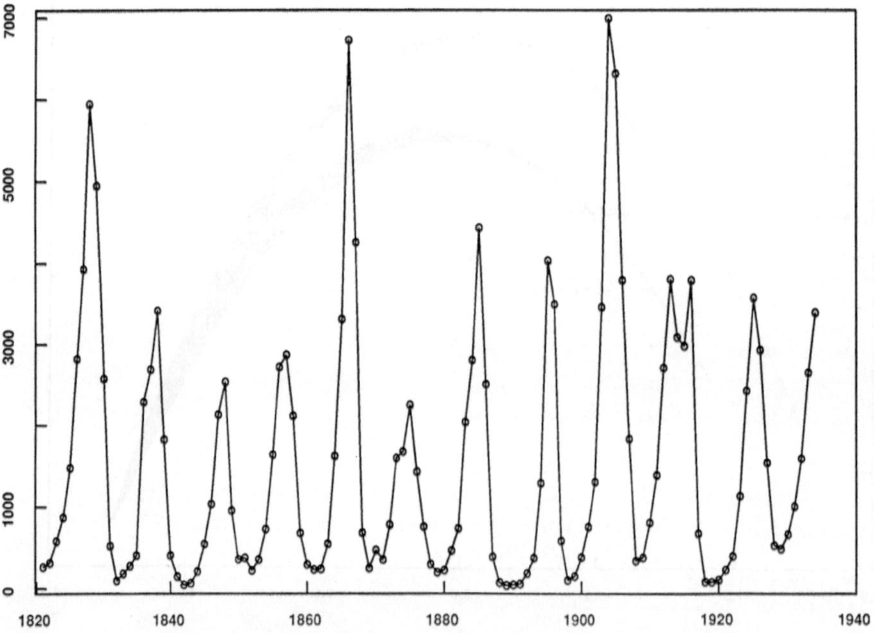

Figure 7. Lynx population data.

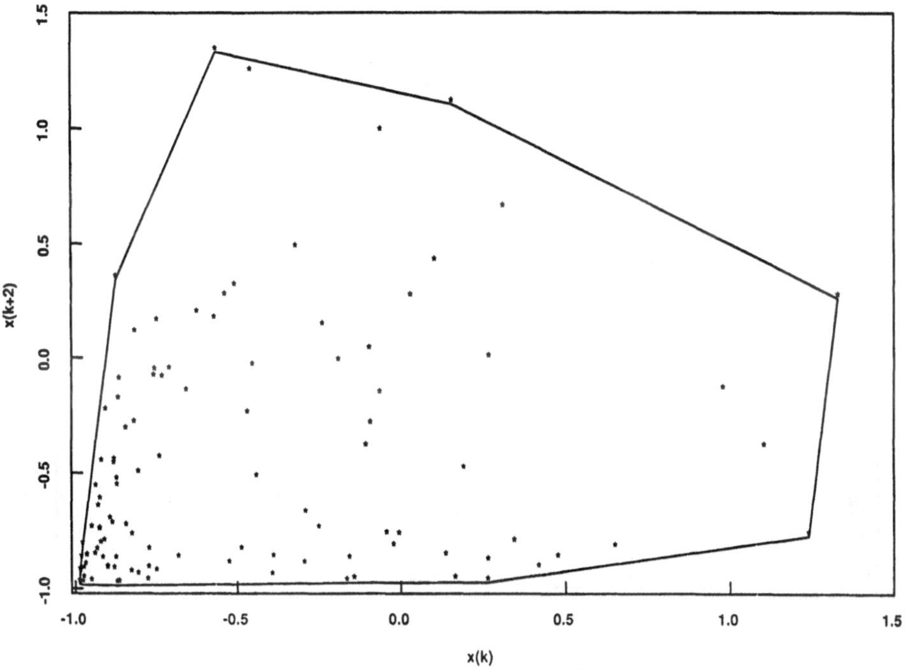

Figure 8. Reconstructed lynx data and convex hull.

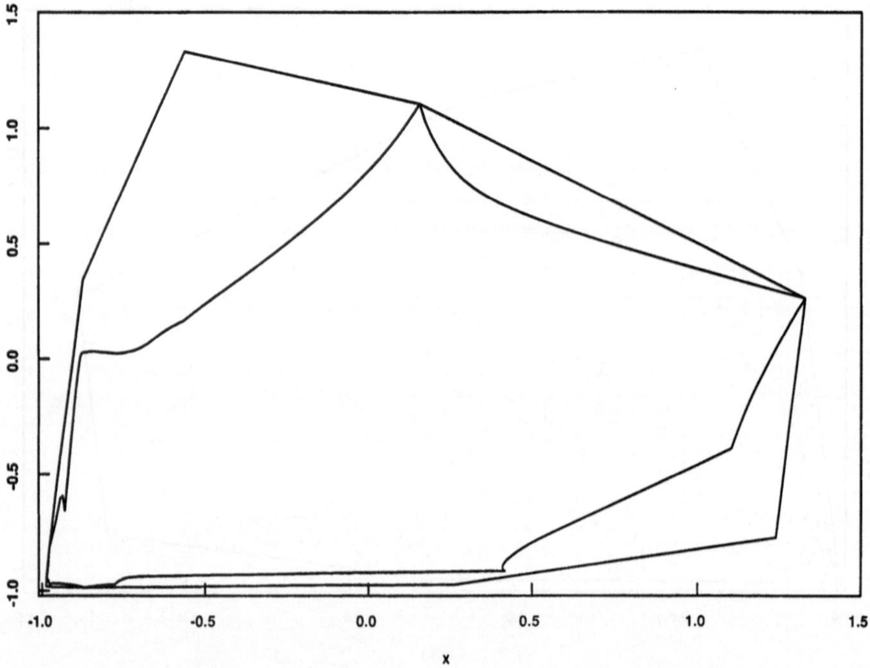

Figure 9. Map of convex hull.

Chaos in Complex Systems

KEVIN JUDD

Abstract

A principal new direction in the study of complex systems is through qualitative descriptions of the behaviour. Typically a system is modelled by a state space and a flow. The flow tells us how the state of the system changes with time. A lot can be said about the qualitative behaviour of the system by the geometry of the flow, in particular the geometry of the attractors in the flow. There are just a few elementary attractors: equilibria, periodic and quasi-periodic, which have the simple geometry of points, lines and surfaces. However there are also chaotic behaviours arising from strange attractors which have a fractal geometry. The behaviour of a system of interconnected elementary and chaotic subsystems will be described, in particular the change of the fractal dimension, a measure of how chaotic the system is.

1. Complex Systems

In modern dynamical systems theory there is now a large body of results which can be used to give qualitative descriptions of the behaviour of dynamical systems. This paper is intended to introduce these ideas to researchers not familiar with dynamical systems theory. The fundamental results will be stated without excessive explanation (details can be found in numerous texts) and will be presented in a way which illustrates how they can be used to understand the behaviour of some complex systems. Of particular interest is how chaotic behaviour arises in dynamical systems. Some time will be devoted to describing how chaos is typically produced in dynamical systems by folded or branched manifolds, and also how chaotic behaviour affects complex systems.

1.1 Division into Subsystems

A complex system, like an ecology or a brain, has a phase space far too large to be modelled entirely, but it may be possible, perhaps only under special circumstances, to view it as composed of a number of interconnected subsystems; see figure 1. This is analogous to an electronic system which might consist of various oscillators, filters, amplifiers, comparators etc. connected together and interacting by signals passing between them. The signals are a passage of energy between the subsystems which is more importantly a passage of information, where one system tells another something about its current state. In a biological system such as an ecology the signals, or currents of energy and materials, are the grazing or preying of populations on others; in a brain they are nerve impulses.

Each of the subsystems might have a phase space M, say, of relatively low dimension. The phase space of a system is a space in which a single point completely describes the system, that is the entire history and future can be determined. The entire history and future of a system is called its trajectory in phase space and the set of all trajectories defines the flow, or dynamics, on phase space (Hirsch M. and Smale P., 1974). The dynamics of a subsystem is determined not only by its state $x \in M$, but also by the states y of the other subsystems to which it is connected. In a continuous time model the dynamics of a subsystem might be given by

$$\dot{x} = f(x) + h(x, y), \tag{1}$$

where f determines the dynamics of the subsystem if it were 'isolated' from the whole system, and h is the interaction with the other subsystems.

It is possible, indeed likely, that the state x affects the state y of the connected subsystems, forming a feedback loop. Suppose that the subsystems can be chosen so that there are no significant feedback loops, that is information flows between the subsystems in one direction as in the complex system of figure 1. Physically this might be achieved when a connection to another subsystem will either passively supply or drain energy and materials but not actively amplify or suppress any of the subsystems' behaviour.

This assumption of passive connections of subsystems means that in the dynamic equations (1) the state y of the connected subsystems can be treated as 'unknown' functions of time, that is the equations are equivalent to

$$\dot{x} = f(x) + g(x, t)$$

where $g(x,t) = h(x, y(t))$. This description of the dynamics of a subsystem is similar to problems in perturbation theory, namely equations of the form,

$$\dot{x} = f(x) + \epsilon g(x,t), \tag{2}$$

where ϵ is a small parameter.

An elementary analysis of complex systems could begin by considering those which can be decomposed into low dimensional subsystems weakly coupled together, that is, the connections between subsystems are passive and the perturbative effects on each other are small so that each subsystem can be described by equations of the form (2) where ϵ is small.

This paper briefly describes what kinds of qualitative behaviour can arise in such systems, and in particular what kind of chaotic behaviour can arise.

Under these assumptions an individual subsystem is like a 'black box' with a phase space state $x \in M$ and in isolation a behaviour described by $\dot{x} = f(x)$. This black box has an input and an output; signals are fed into the input and read off from the output. The output signal x read off for a given input $y(t)$ satisfies the equation

$$\dot{x} = f(x) + \epsilon h(x, y(t)).$$

Although the output signal need not be x, a result of Takens' theorem (Takens F., 1980) is that generically for a nonlinear system any output signal contains enough information to reconstruct the dynamics of the black box (Mees A.I. this volume), so it can be assumed that the state x of the system is known and the output signal is x.

1.2 Classification of Subsystems

Generally there are four qualitative asymptotic behaviours a system can have: equilibrium, periodic, quasi-periodic or chaotic behaviour. Therefore there are four classes of output signals that can be expected: approaching steady, periodic or quasi-periodic and chaotic signals.

The question asked in this paper is, when a particular system is fed an input signal from one of these four classes what is the class of the output and what are its qualities: for example is it a chaotic signal and if so how does its complexity vary? Before attempting to answer this question an aside on the nature of chaos is useful.

2. Chaos from Folding and Branching

2.1 Measures of Complexity

By reconstructing a dynamical system from a signal by one of the many methods suggested by Takens' theorem, one can find the dimension of the attractor of the system. This dimension does not depend on the reconstruction and so can be called the dimension of the signal. The dimension of a signal is an important measure of its complexity. The dimension of a steady signal is zero, that of a periodic signal is one and that of a quasi-periodic signal of n incommeasurate frequencies is n, since a reconstruction of such signals give a point, a circle and an n-torus. The spectrum of a signal is a well established and important measure of the nature of a signal. Periodic and quasi-periodic signals have discrete spectra with the number of incommensurate frequency components equal to the dimension of the signal.

A random or noise signal has an infinite dimension because if one tries to reconstruct an attractor from such a signal in any finite dimensional space it will always densely fill an open set according to some distribution. Also the spectrum of such a signal is broad-band indicating no recurrent behaviour. Chaotic signals are somewhere between quasi-periodic signals and noise. Their spectrum is broad band but generally has sharp peaks indicating recurrent behaviour, but most importantly the reconstructed strange attractor of the signal has a complex fractal geometry and a non-integer dimension.

Why this is so needs further explanation but simply stated it is because chaotic signals and chaotic transients are typically generated by simple processes; they are foldings and branchings in the phase space by the dynamics of the system. Excellent examples of these processes are seen in the Duffing and Lorenz systems described later, but nearly every chaotic system encounted results from these simple processes.

2.2 Fractal Geometries

Fractal geometries and dimensions are nicely illustrated by the process of making croissants. In fact this is precisely the process referred to as a folding. Imagine a region of the phase space of a system is a block of dough which is to be made into croissants or pastries. The dynamics of phase space is the stretching out and folding over of the dough. Each particle in the dough represents a different initial configuration and the motion under the folding is the evolution of the system. The croissant that results from this process is the strange attractor of the system, at least within the bounds of the analogy. The croissant would be better described as an attracting set which produces the chaotic transients. The croissant-like strange attractor has a complex fractal geometry; any small piece is composed of uncountably many separate layers, not touching each other, but next to any layer there is another arbitarily close.

The fractal dimension of such an object measures how many layers are close to another, that is how complex the fractal set is. Figure 2 shows two foldings: (a) is a very fat fold producing a heavy croissant with a dimension close to 2, since it almost fills the plane; where as (b) is a very thin fold producing a light airy pastry, with dimension close to 1. These two folds would be expected to produce very different strange attractors and chaotic signals and it is in this way that the dimension measures the complexity of strange attractors and chaotic signals.

2.3 The Duffing System

The Duffing system is an example of chaos produced by a fold. Detailed descriptions of this system can be found in many texts (Guckenheimer J. and Holmes P., 1986). The system models a rigid pendulum with a steel bob. Either side of the usual equilibrium are magnets strong enough to attract the bob and produce two new stable equilibria near the magnets and a metastable equilibrium between them. The Duffing system models the pendulum's motion when it is periodically forced. The equations of motion are,

$$\begin{aligned} \dot{\theta} &= -\omega \\ \dot{\omega} &= \theta - \theta^3 + \gamma \cos \alpha t, \end{aligned}$$

where θ is the angular postion of the pendulum and ω is the angular momentum. Because the forcing is periodic it is natural to consider this system as the Poincaré map on a time cross section (Guckenheimer J. and Holmes P., 1986), that is to take stroboscopic pictures at the same frequency as the forcing of the trajectories of the system in (θ, ω)-space, the phase of the system. This will define a unique map of (θ, ω)-space, called the Poincaré map which can be used to study the system. In the Duffing system this map is a 'fold' map like the croissant mapping described before. It has a range of different folds depending on the frequency of the forcing. It will be discussed later how the chaotic dynamics arise from a Mel'nikov bifurcation, a homoclinic bifurcation which in the Poincaré map appears when the stable and unstable manifolds of a fixed point intersect transversally. Figure 3 shows the foldings that occur in the chaotic Duffing system's Poincaré map for high, moderate and low forcing frequencies. The folds on a particular region of the phase plane vary from producing small dimples in the region which form thin cavities that burrow deeper and deeper into the region with repeated folds, through basic croissant-like folds at moderate forcing frquencies, to folds that stretch out and roll up the region at low frequencies.

2.4 The Lorenz System

The Lorenz system is a much cited example of a chaotic system, it was first introduced to model convection processes (Sparrow C.,1982). Its chaotic behaviour results from a folding process with a branching and produces a butterfly shaped strange attractor; see figure 4(a). Trajectories in the flow move out horizontally from the central region, rotate out around the butterfly wings several times before entering the central region again where the trajectories cross over to the other wing. The chaos results from the unpredictable crossings of a trajectory between the wings. The fold of the system can be understood by considering a rectangular region of phase space in the central region. The flow of the chaotic Lorenz system stretches out this region horizontally. The ends are folded up and around on the rotating wings and are flattened in the process. The ends are then tucked back down into the space originally occupied by the region; see figure 4(b). This folding results in the butterfly pastry formed of layers crossing from one wing to the other.

There are many examples of different foldings: the Hénon map, the Rössler system, the Sil'nikov bifurcation and the solenoid map. There are others which include a branching; the Rikitake system combines both types of foldings.

3. The Interaction of Subsystems

We now return to the question posed earlier. Given a particular system what output signal results from an input signal from each of the four classes of signal, steady, periodic, quasi-periodic and chaotic? These questions are given incomplete answers in the following. Firstly linear and equilibrium systems are considered, then periodic and quasi-periodic. Special attention is given to some delicate systems, those with near homoclinic cycles, which can have complex changes in output signals. Finally a chaotic system's response to input signals is considered.

3.1 Linear Systems and Stable Equilibria

The simplest system to consider in the black box is a linear system with a linear perturbation. There is a well documented theory on how input and output signals behave with such systems (Mees A.I.,1981). A linear system will usually have several resonances which amplify parts of a signal's spectrum. The dimension of the signal, however, is unchanged; in a sense the system only distorts the reconstruction of the signal.

The simplest nonlinear systems are those which are neither chaotic nor have homoclinic or near homoclinic cycles, which will be dicussed later. These systems are those with stable equilibia, periodic or quasi-periodic attractors. A system with a stable equilibrium behaves essentially like a linear system with an equilibrium, which is to be expected because locally a stable equilibrium is like a linear system. The proof of this is elementary (Mees A.I.,1981). Thus a subsystem with a stable equilibrium behaves in response to input signals the same way as a linear system.

3.2 Periodic and Quasi-periodic Systems

A nonlinear system with a periodic attractor can significantly change the character of a signal through an effect called entrainment (Guckenheimer J. and Holmes P., 1986). A periodic signal with a subharmonic frequency of the natural frequency of the periodic attractor will resonate, as in a linear system, but a signal near a subharmonic of the system will 'lock in' on the subharmonic, sometimes at quite high orders. The signal in this case remains a one dimensional periodic signal but with the peaks in the spectrum shifted. If the signal does not 'lock in' the output signal has two independent frequency components, and is quasi-periodic. The signal has a dimension of two and a spectrum with new frequency components. When presented with a quasi-periodic signal the system may entrain some frequency components, none of them or all of them. If the quasi-periodic signal has dimension n the output signal may have a dimension between 1 and $n+1$, with a correspondingly different spectrum, with some peaks shifted or vanishing.

The question of what happens to a chaotic signal is difficult. There may be a tendency for the system to entrain the recurrent nearly periodic part of the signal but being chaotic the signal always departs from its recurrent behaviour and can never be entrained. This partial entrainment effect may be seen in the spectrum of the output signal by damping, shifting or spreading peaks. If entrainment is not a significant effect the chaotic signal acts as a perturbation of the periodic behaviour of the system. The output spectrum would have new component peaks and the signal's dimension would increase by one at most. The dimension could be less if there are entrainment effects.

If the nonlinear system in the black box were to have a quasi-periodic attractor with m incommeasurate frequencies then entrainment could occur with any, none or all of these components. The effects on signals are similar to the descriptions of the last paragraph. For example if the input were a quasi-periodic signal of dimension n the output signal could have a dimension between 1 and $m+n$, with a corresponding spectrum depending on how the frequency components are entrained. Chaotic signals would be affected similarly.

3.3 Systems with Homoclinic Cycles

Some of the most interesting behaviour arises from systems with homoclinic or near homoclinic cycles, such as the Duffing system. A system has a homoclinic cycle when a number of saddle points are connected by trajectories to form a loop with the flow in the same direction, see figure 5. A near homoclinic cycle is like a perturbation of a homoclinic cycle such that some or all of the saddle connections are broken but the separatrices of the saddle points still pass close to each other as illustrated in figure 5(b). When such systems are periodically perturbed they can become chaotic.

Mel'nikov theory (Guckenheimer J. and Holmes P., 1986) gives the conditions under which bifurcations to chaos can occur for small perturbations of such systems. Let $q_j(t)$, $j = 1, \ldots, n$ be trajectories of a two dimensional system $\dot{x} = f(x)$ which form a homoclinic cycle and $\epsilon g(x,t)$ be a periodic perturbation of the system. Then define n Mel'nikov functions by

$$m_j(\phi) = \int_{-\infty}^{\infty} f^{\perp}(q_j(t)) \cdot g(q_j(t), t+\phi)\, dt,$$

where $f^{\perp}$ is the perpendicular vector to f on the right hand side. Mel'nikov's theorem states that if all n Mel'nikov functions have simple zeroes, then the system is chaotic for sufficiently small ϵ. Typically for a periodic perturbation the system has a bifurcation diagram like figure 6 in the frequency amplitude domain of the periodic perturbation.

Mel'nikov theory says that the output of a homoclinic or near homoclinic system can change from a single periodic component one dimensional signal to an output with a chaotic signal with a broad complex spectrum and a noninteger dimension. The dimension of such a chaotic signal can be calculated (Judd K., 1989).

Mel'nikov's theorem applies to any recurrent signal and has also been extended to higher dimensional spaces (Salam F *et al* , 1983). With a recurrent signal a system is chaotic if for all t_0 there exit suitable simple zeroes for $t > t_0$ of each of the Mel'nikov functions $m_j(t)$. In particular a quasi-periodic signal of dimension n is the sum of n component signals of incommeasurate frequencies, but because a Mel'nikov function is formed with an integral it is the sum of the Mel'nikov functions of each component signals. The Mel'nikov functions of the components which have simple zeroes are the components which contribute to the chaotic behaviour, each changing the spectrum and the dimension of the output signal.

Essentially this is the same as plotting each component of the quasi-periodic signal on the bifurcation diagram to see which lie in the chaos region and contribute a major change to the signal. For example, if only one component of n causes the chaos, then the dimension d of a output signal resulting from an input signal of just that one component can be calculated (Judd K., 1989), and the dimension of the entire output signal would be less than $n + d - 1$. If m components contributed to the chaotic behaviour, where each by itself would give an output signal of dimension d_i, $i = 1, \ldots, m$, then the dimension of the entire output signal would be less than $n - m + d_1 + \ldots + d_m$. If none of the components cause chaotic behaviour the system typically will approach an equilibrium, periodic or quasi-periodic attractor and the output signal behaves asymptotically as described previously.

The method of the previous paragraphs extends to considering the effect of chaotic signals on systems with homoclinic or near homoclinic cycles. Mel'nikov's theorem will decide whether a chaotic signal will produce new chaotic behaviour. Essentially there are two possibilities, either the Mel'nikov functions do not have suitable simple zeroes, in which case the system behaves as one of the systems already described in response to the chaotic signal, the dimension and spectrum changing only a little, or the the system responds chaotically to the signal changing substantially its spectrum and dimension.

As described previously with a quasi-periodic signal only those components which lie in the chaos region of the bifurcation diagram had a significant effect; similarly if the spectrum of the chaotic signal is plotted on the bifurcation diagram only those peaks or parts of the spectrum which lie in the chaos region contribute to a major change in the chaotic nature of the signal. If none of the spectrum lies in the chaos region there is no new chaotic change, but if only one or a few peaks lie in this region then the change produced is much like that of a quasi-periodic signal with components the same as the peaks. Such knowledge can be used to predict the likely spectrum and dimension of the output signal.

4. Chaotic Systems

Analysis of the behaviour that results from a chaotic system in the black box is very incomplete. Following the earlier description of chaos as the result of foldings of state space a signal can be viewed as a modification of the folding for these systems. Behaviours like those already described for chaotic signals fed into other systems can be expected. For example, periodic signals could resonate and partially entrain a periodic folding or alternatively alter the folds in such a way as to change the nature of the chaotic signal, changing its spectrum and dimension. This research is still in progress.

REFERENCES

[1] Guckenheimer J. and Holmes P., 1986, *Nonlinear Oscillations, Dynamical Systems and Bifurcations of Vector Fields*, Springer-Verlag, New York.

[2] Hirsch M. and Smale P., 1974, *Differential Equations, Dynamical Systems and Linear Algebra*, Academic Press, New York.

[3] Judd K., 1989, *The Structure and Dimension of Strange Attractors*, Ph.D. Thesis, The University of Western Australia.

[4] Mees A.I.,1981, *Dynamics of Nonlinear Feedback Systems* Wiley, Cambridge.

[5] Salam F *et al* , 1983, Chaos and Arnold Diffusion in Dynamical Systems, *IEEE Transactions on Circuits and Systems*, **CAS-30**, No 9.

[6] Sparrow C.,1982, The Lorenz Equations: Bifurcations, Chaos and Strange Attractors, *Appl. Math. Sci.* **41.** Springer-Verlag, New York.

[7] Takens F., 1980, Detecting Strange Attractor in Turbulence., *Dynamical Systems and Turbulence. Lecture Notes in Mathematics. Vol. 898*, Rand D.A. and Young L.S., eds, pp. 365-381. Springer-Verlag, new York.

PARTICIPANT'S COMMENTS

The paper presents an introduction to some basic ideas about chaos, aimed at researchers unfamiliar with dynamical systems theory, and applies the ideas to complex interconnected systems. The introduction provided by the paper serves as a link between several other papers in the proceedings

Walter Grantham

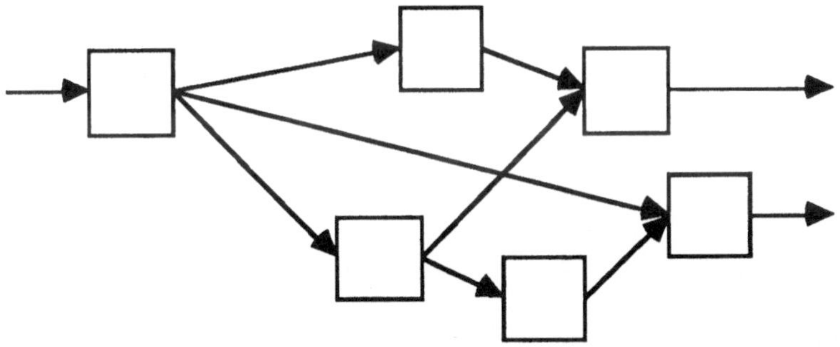

Figure 1. A complex dynamical system can in some circumstances be viewed as a number of simpler subsystems coupled together and interacting with signals passing between them.

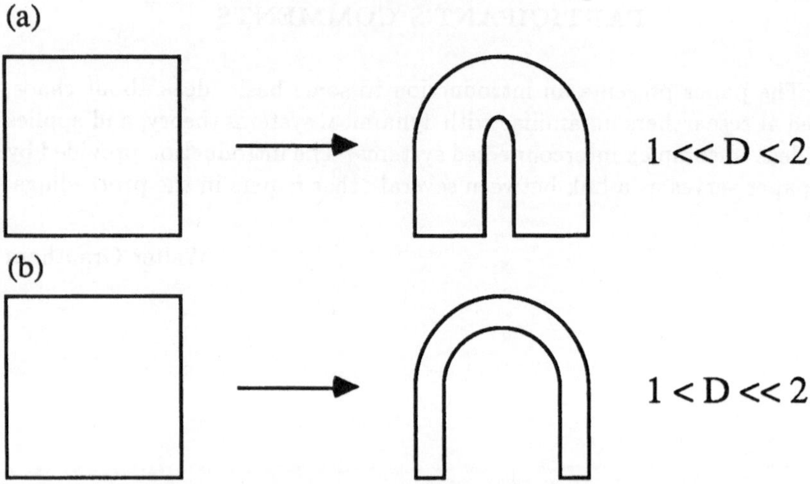

Figure 2. Two different basic folds: (a) is a thick fold, the dimension D of the attracting set is almost two since it very nearly fills the plane. (b) is a thin fold with a dimension D very nearly one because it is close to becoming a line.

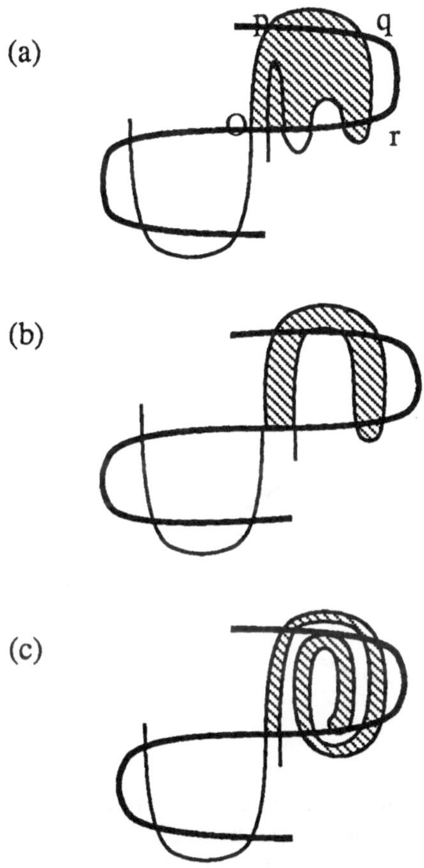

Figure 3. The fold in the Poincaré map of the Duffing equation for different forcing frequencies. The shaded area is the image of the region Opqr which is bounded by the stable and unstable manifolds of the fixed point (thick and thin lines). A high frequency forcing gives the fold (a) which when repeated forms thin cavities that burrow deeper and deeper into the region, seen here as the loops of the unstable manifold. At a moderate frequency basic croissant-like folds (b) are produced and at low forcing frequencies folds like (c) which stretch out and roll up the region Opqr.

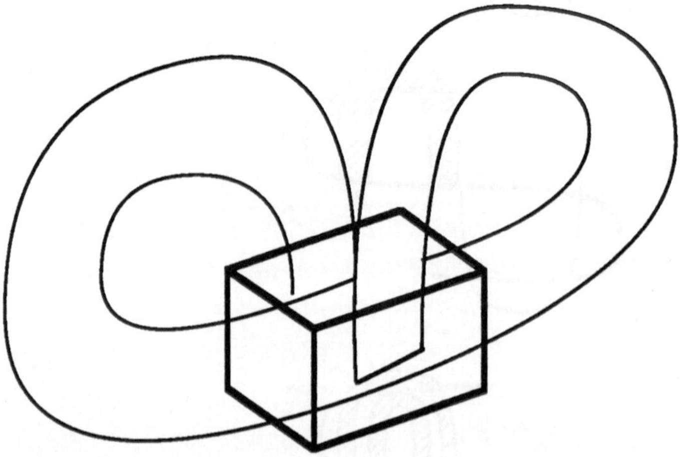

Figure 4. A trajectory on the attractor of the Lorenz system is show in (a) The fold of the Lorenz system can be viewed as deforming a rectangular region, the solid box in (b). The region is stretched out horizontally and the ends are folded up and around along the rotating wings and are flattened in the process; the ends are tucked back into the space originally occupied the the region. This continuous folding results in a butterfly pastry formed of layers crossing from one wing to the other.

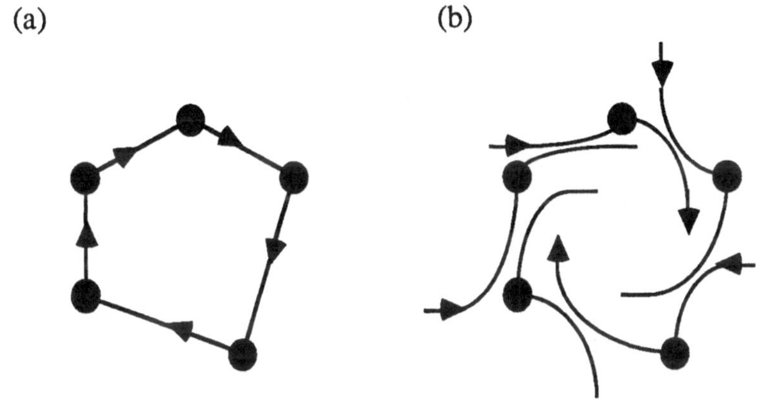

Figure 5. A system has a homoclinic cycle (a) when a number of saddle points are connected together to form a loop with the flow in the same direction. A near homoclinic (b) occurs if the stable and unstable manifolds of a number of saddle points pass close to each other like a broken homoclinic cycle.

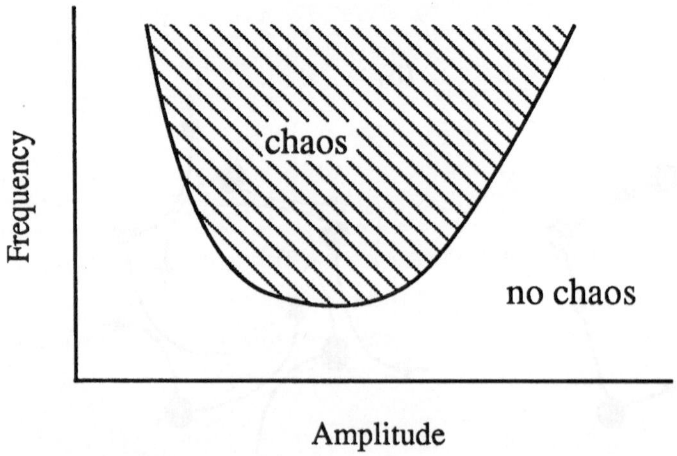

Figure 6. A typical bifurcation diagram of a (near) homoclinic cycle in response to a periodic perturbation as a function of the amplitude a frequency of the perturbation.

A Chaotic System: Discretization and Control

WALTER J. GRANTHAM AND AMIT M. ATHALYE

Abstract

This paper examines chaos in some population models for biological processes. In particular, the paper is concerned with chaos produced by discretization of continuous-time population models, and with the question of whether or not this chaotic behavior can occur when a feedback control management strategy is applied to continuous-time population models. Using two classical population models, it is shown that both exponential discretization and discretization associated with numerical simulation can produce chaos in the resulting discrete-time system. For example, a continuous-time Lotka-Volterra system exhibits periodic trajectories, but a particular discretization procedure is shown to yield discrete-time trajectories that all converge to a very complicated chaotic strange attractor, even if the discretization time steps are small. The results in the paper also show that this chaotic behavior can occur even if a feedback control management strategy is applied to the continuous-time system to stabilize the equilibrium point.

1. Introduction

The recent explosion of interest in the study of chaos and chaotic dynamical systems stems from the realization that even simple nonlinear deterministic systems can behave in an apparently unpredictable and chaotic

This research was supported in part by National Science Foundation and Australian Department of Science and Technology travel grants and by NASA-Ames Research Center under Interchange No. NCA 2-219.

manner. The existence of chaotic behavior can be found in a wide spectrum of scientific fields, such as economics, engineering, physics, chemistry, and life sciences [Mees and Sparrow, 1981; Peitgen and Richter, 1982]. Convection of a fluid heated from below [Lorenz, 1963], stirred chemical reactor systems [Roux, *et al* , 1981], and prey-predator population dynamics [May, 1974; Inoue and Kamifukumoto, 1984] are but a few examples of systems which can exhibit chaotic behavior.

A considerable body of knowledge about chaotic dynamical systems has developed in recent years, as indicated by several excellent texts dealing with chaos [Devaney, 1988; Guckenheimer and Holmes, 1986; Schuster, 1988; Hsu, 1987; Wiggens, 1988]. However, the rapid development of knowledge about chaos is by no means complete, as indicated by the fact that the basic reference texts [Devaney, 1988; Guckenheimer and Holmes, 1986; Schuster, 1988] have all been revised significantly at their second printings.

Because chaos is a very recent topic of research, many important areas remain to be studied. One of these is the problem of developing feedback controllers for chaotic systems. Research on the control of chaotic systems is in a stage of infancy, with results confined mostly to the occurrence of chaos in discrete-time control systems. Sufficient conditions for chaos in a class of linear sampled-data control systems with nonlinear controllers was investigated in [Ushio and Hirai, 1983]. They showed that chaotic motion occurs for a sufficiently large sampling time interval. Chaotic behavior in feedback systems was also reported in [Cook 1985, 1986]. Chaotic motion in an adaptive control system was investigated in [Rubio, *et al* , 1985] using power spectra.

In this paper we examine one particular aspect of the problem of controlling chaotic systems, in terms of two classical population dynamics models. Specifically, we will be concerned with chaos caused solely by transforming a continuous time model to a simpler discrete-time model. We will consider two discretization techniques. One is a common method employed by population biologists and the other method is, in some sense, typical of numerical integration techniques. Both techniques can produce chaotic behavior in the discretized models. From a control systems viewpoint, we will be concerned with whether or not this chaotic behavior can still occur, in the discrete-time model, even after a stabilizing feedback controller has been added to the original system.

The paper is structured as follows. In Section 2 we discuss continuous-time and discrete-time system models and present two types of approaches for obtaining discrete-time models from a continuous-time system. In Section 3 we discuss some topics related to chaotic dynamical systems. In

Section 4 we develop an "exponential discretization" model for a Lotka-Volterra two-species competition system The discrete-time model is the same one considered in [May, 1974], who showed that it is chaotic for certain parameter values. We verify this result via a state-space simulation of both the nonchaotic continuous-time and the chaotic discrete-time competition models. In Section 5 we consider a continuous-time Lotka-Volterra prey-predator system and present a "pseudo-Euler" numerical integration discrete-time model developed in [Peitgen and Richter, 1986]. Again we compare simulation results for the nonchaotic continuous time model and the chaotic discrete-time model. In Section 6 we develop a saturating state feedback controller for the continuous-time prey-predator system. We present simulation results for the feedback controller applied to both the continuous and the discrete-time prey-predator models. In Section 7 we summarize the results presented in the paper.

2. System Modelling

In the modelling, analysis, and control of biological population systems and other resource management systems, two different types of dynamical system models are typically employed: continuous-time models or discrete-time models. In this paper we will consider the effects of two different discretization approaches applied to classical Lotka-Volterra continuous-time models. It has been shown previously [Peitgen and Richter, 1986; May, 1974] that both discretization approaches can produce discrete time trajectories that are drastically different than those of the corresponding continuous time system. In particular, both approaches may yield chaotic motion and strange attractors in the discrete-time models, even though these phenomena do not occur in the continuous-time model. We show that the chaotic results for the discrete-time system can continue to occur even if a stabilizing feedback controller is applied to the continuous-time model.

A continuous-time model is a system of n differential equations of the form

$$\dot{x}(t) = f[x(t), u(t)], \tag{1}$$

where $x(t) \in R^n$ is the state of the system at time t, $u(t) \in \Omega \subset R^m$ is the control, Ω is a specified control constraint set, and $(\dot{\ })$ denotes the time derivative $d()/dt$.

A discrete-time model is a system of n algebraic equations of the form

$$x(k+1) = F[x(k), u(k)], \tag{2}$$

where $x(k)$ and $u(k)$ denote the state and control vectors at time $t_k = \Delta t$, Δt is some specified time step (not necessarily small), and $k = 0, \pm 1, \pm 2, \ldots$ is the discrete-time index.

For several reasons, population biologists may employ a discrete–time model instead of a continuous–time model. One reason is that the solution $x(k)$ to a system of algebraic equations of the form (2) is much easier to generate than the solution $x(t)$ to a system of nonlinear differential equations of the form (1). Another reason cited by population biologists is a belief that many biological systems are in fact discrete-event systems. For example, a species may reproduce only once per year. Another practical reason for using discrete–time models is that a field biologist may not be able to observe a species on a continuous basis. For example, it may be difficult for the biologist to be in the field on a daily or even monthly basis, or the population species may migrate annually and only be available for observation once a year.

For all of these reasons, population biologists often model dynamical population systems using discrete-time models, which yield a prediction of the population at the next cycle based on the population at the previous cycle. Although the length of a discretization cycle may vary from system to system, the cycle length is often very large compared to an approximate continuous-time period, such as yearly versus daily.

To illustrate one common approach for developing a discrete-time model from a continuous-time model, consider the scalar system

$$\dot{x} = ax \tag{3}$$

The solution to this differential equation is given by

$$x(t) = x(0)e^{at},$$

where $x(0)$ is the initial value of x. Thus the continuous-time system (3) is equivalent to the discrete-time system

$$x(k+1) = x(k)e^{a\Delta t}, \tag{4}$$

where Δt is the discrete time step. Note that both of the models (3) and (4) have the property that x decays (or grows) exponentially.

In a more general setting consider the scalar system

$$\dot{x} = xg(x). \tag{5}$$

Following the same procedure used to integrate (3), we can re-write (5) and then integrate the result as

$$\int \frac{dx}{x} = \int g[x(t)]\, dt. \tag{6}$$

Now, if we approximate $g[x(t)]$ as a constant from time $t_k = k\Delta t$ to time $t_{k+1} = (k+1)\Delta t$, integration yields the discrete-time model

$$x(k+1) = x(k)\exp\{g[x(k)]\Delta t\}. \tag{7}$$

This type of exponential discrete-time model has been employed previously for biological systems [May, 1974; Fisher and Grantham, 1985; Grantham and Fisher, 1987] and it has two properties that appeal to population biologists: 1) the population varies exponentially with time, which agrees with many observed population systems, and 2) the population always remains positive, which agrees with the original continuous-time model. In addition, note that the equilibrium points (constant state solutions) for the continuous-time system (5), corresponding to $x(t) \equiv 0$ or $g[x(t)] \equiv 0$, remain as fixed points for the discrete-time model (7).

The main disadvantage of the discrete-time model given by (7) is that the nonequilibrium solutions for (7) are not the same as those for the continuous-time model (5). In general the function $g(x)$ is not constant over a discrete-time cycle. Reproduction may occur only once during a cycle, but deaths occur throughout the cycle. This difference can be important for large Δt (e.g., one year versus one day) and, as we shall see later, the discrete-time model can produce fundamentally different predictions of population size than those associated with the continuous-time model.

Even if a continuous-time model is employed, the process of generating a solution $x(t)$ to a system of nonlinear differential equations generally requires the use of some numerical integration algorithm. This is equivalent to converting the continuous-time model to a corresponding discrete-time model. The simplest numerical integration scheme is Euler's method. Applied to (1), Euler's method yields the corresponding discrete-time system

$$x(k+1) = F[x(k), u(k)] = x(k) + f[x(k), u(k)]\Delta t. \tag{8}$$

For example, applying Euler's Method to the continuous-time system (5) yields the discrete-time model

$$x(k+1) = x(k)\{1 + g[x(k)]\Delta t\}. \tag{9}$$

Note that, unlike the exponential discrete-time model (7), the result in (9) does not insure that the population will remain positive. In fact if $g[x(k)]$ is sufficiently negative the predicted population will become negative!

3. Chaotic Systems

To discuss some properties of chaotic systems [May, 1976; Grebogi, *et al*, 1985; Li and Yorke, 1975], consider a system of autonomous differential equations of the form

$$\dot{x} = f(x), \tag{10}$$

where $f(\cdot)$ is continuous and has continuous first-order partial derivatives. From the theory of ordinary differential equations [Coddington and Levinson, 1955], a unique solution $x(t) = \xi(t, x_0)$ exists for each initial condition $x(0) = x_0$. Furthermore, the solution $x(t)$ at any time t is a continuous and continuously differentiable function of the initial state x_0. Theoretically, the state $x(t)$ at any time t is absolutely predictable, either forward or backward in time and even over very long time intervals. Furthermore, small (differential) changes in initial conditions yield small (differential) changes in the state at other times.

One property of chaotic systems is that they possess solutions that are very sensitive to changes in initial conditions [Schuster, 1988]. Even though the solutions to (10) are theoretically continuous and differentiable functions of the initial state, they may be so sensitive to changes in initial conditions that the initial state would have to be known exactly in order to predict future (or past) states over any large time interval. Since any real measurement contains noise, the initial state can not be known exactly. Thus in a chaotic system it is not possible to predict the future state, particularly over long time intervals.

Another property of chaotic systems is related to certain special types of solutions to (10). Constant-state solutions $x(t) \equiv \hat{x}$ are called equilibrium points and they can be either stable or unstable. If an equilibrium point $\hat{x}$ is asymptotically stable, then neighboring trajectories $x(t)$ approach $\hat{x}$ as $t \to \infty$ and the equilibrium point is called an attractor. A periodic solution, of the form $x(t) = x(t+T)$ with period T, is one for which the solution repeats itself. A periodic solution that has no periodic neighbors is called a limit cycle. Like equilibrium points, a limit cycle may also be an asymptotically stable attractor.

Normally, the only types of bounded solutions associated with (10) are equilibrium points, periodic orbits, and limit cycles. Chaotic systems however also possess strange attractors [Grebogi, *et al*, 1987]. These are bounded attracting trajectories that are not equilibrium points and are not periodic. They are called strange attractors partly because they have the unusual property that they are confined to a bounded region but they never repeat themselves. This in itself is not enough to warrant the label "strange". For example a quasiperiodic solution [Grebogi, *et al* , 1985] may exist which wraps itself around the surface of a torus and only repeats itself after an infinite time.

The truly strange property of strange attractors is that they are fractals [Mandelbrot, 1982]. That is, they are geometric objects that have a fractional dimension. In the usual notion of dimension, a point has dimension zero, a curve has dimension one, a plane has dimension two, and so forth. Thus an equilibrium point for (10) is zero-dimensional and any nonequilibrium solution $x(t)$ is theoretically a one-dimensional curve since it only takes one parameter value t to specify points along a solution. However, in a chaotic system a solution corresponding to a strange attractor wanders around erratically in such a way that it covers more than a one-dimensional region. For example, a nonchaotic system of three differential equations may generate a quasiperiodic solution that covers a torus and is thus a two-dimensional object. As a parameter in the differential equations is changed the solution may become a chaotic strange attractor which smears out and covers a region whose "dimension" is a fraction, greater than two. The existence of such objects requires an extension of the notion of dimension and many such extensions have been proposed. For a comparison of various dimension concepts see [Farmer, *et al* , 1983].

4. Lotka-Volterra Competition Models

Consider a Lotka-Volterra model for competition between two species, given by

$$\begin{aligned} \dot{x}_1 &= x_1\{r_1[c_1 - \alpha_{11}x_1 - \alpha_{12}x_2]/c_1\} \\ \dot{x}_2 &= x_2\{r_2[c_2 - \alpha_{21}x_1 - \alpha_{22}x_2]/c_2\}, \end{aligned} \tag{11}$$

where the population levels x_i and the parameters r_i, c_i, and α_{ij} are all positive. Note that if

$$\begin{aligned} d &= \alpha_{11}\alpha_{22} - \alpha_{21}\alpha_{22} > 0 \\ \hat{x}_1 &= (\alpha_{22}c_1 - \alpha_{12}c_2)/d > 0 \\ \hat{x}_2 &= (\alpha_{21}c_1 - \alpha_{11}c_2)/d > 0 \end{aligned} \tag{12}$$

then the equilibrium point $\hat{x}$ is asymptotically stable [May, 1974].

Using the procedure outlined in Section 2 we develop an exponential discretization model by first dividing each of the equations in (11) by the corresponding x_i and then treating the new right hand sides as constants while integrating over one time cycle. For $\Delta t = 1$ this yields

$$\begin{aligned} x_1(k+1) &= x_1(k)\exp\{r_1[c_1 - \alpha_{11}x_1(k) - \alpha_{12}x_2(k)]/c_1\} \\ x_2(k+1) &= x_2(k)\exp\{r_2[c_2 - \alpha_{21}x_1(k) - \alpha_{22}x_2(k)]/c_2\} \end{aligned} \tag{13}$$

This discrete-time model is the same as one considered in [May, 1974]. Note that the fixed points for the discrete-time model are the same as the equilibrium points for the continuous-time model. In [May, 1974], for $r_1 = r$, $r_2 = 2r$, $c_1 = c_2 = c$, $\alpha_{11} = \alpha_{22} = 1$, and $\alpha_{12} = \alpha_{21} = 0.5$, time plots of x_1/c show that the attractor for (13) becomes chaotic as r is increased in steps from $r = 1.1$ to $r = 4.0$.

Figure 1 shows a state-space plot of $x(k)$ for the $r = 2.5$, $c = 1$ case considered in [May, 1974], starting from $x_1(0) = x_2(0) = 0.5$. The discrete-time trajectory shown in Figure 1 wanders around chaotically and never repeats itself. Clearly this result does not agree with the results for the continuous-time system (11), which predict a trajectory that asymptotically converges to the equilibrium point at $\hat{x}_1 = \hat{x}_2 = 2/3$. The continuous-time trajectory is shown in Figure 2. It was constructed using fourth-order Runge-Kutta with a time step of $\Delta t = 0.1$.

5. Lotke-Volterra Prey-Predator Models

In this section we investigate chaos caused solely by a numerical integration procedure. In the application that we consider, a hint of chaos occurs even for a fourth-order Runge-Kutta integration algorithm with a fairly small time step. However, for simplicity, we will study an integration procedure akin to second-order Runge-Kutta integration, in which chaos can occur even for very small time steps.

For illustrative purposes consider a Lotka-Volterra prey-predator system given by

$$\begin{aligned} \dot{x}_1 &= x_1(\alpha - \beta x_2) \\ \dot{x}_2 &= x_2(-\gamma + \delta x_1), \end{aligned} \tag{14}$$

where x_1 and x_2 are the prey and predator population levels, respectively, and the parameters α,β, γ, and δ are all positive. In vector notation the differential equations (14) are in the form

$$\dot{x} = f(x). \tag{15}$$

Note that the equilibrium point at

$$\begin{aligned}\hat{x}_1 &= \gamma/\delta \\ \hat{x}_2 &= \alpha/\beta\end{aligned} \tag{16}$$

is stable but not asymptotically stable.

Figure 3 shows a state-space plot of numerical solutions $x(t_k) \approx x(k\Delta t)$ to (14) from various initial conditions for the case where $\alpha = \beta = \gamma = \delta = 1$. The solutions were generated using the standard fourth-order Runge-Kutta algorithm with a time step of $\Delta t = 0.1$. Figure 3 shows the individual solution points $x(k\Delta t)$ generated by the numerical integration procedure.

We purposely have not "connected the dots" in Figure 3, in order to show the cumulative errors that can occur even though the per-step error is of the order 10^{-5}. In particular, note that at the top of the outer trajectory in Figure 3 the solution points do not lay on a smooth curve, as they would for an exact integration of (14). In fact it is possible that this numerical trajectory, which approximates an exact solution that repeats itself once per revolution, is not periodic and is in fact chaotic. Evidently, this possibly chaotic behavior disappears if we take a small enough step size, say $\Delta t = 0.01$.

Our purpose here is to investigate chaotic motion caused by numerical integration. However, instead of using fourth-order Runge-Kutta, which can be made fairly accurate, consider the following "pseudo-Euler" numerical integration algorithm:

$$\begin{aligned}y(k) &= x(k) + f[x(k)]\Delta\tau \\ x(k+1) &= x(k) + \frac{\Delta t}{2}\{f[x(k)] + f[y(k)]\}\end{aligned} \tag{17}$$

where $\Delta\tau$ and Δt are time-step parameters. Note that the equilibrium points of (15) are fixed points for (17). It should also be noted that the algorithm reduces to the standard first-order Euler method for $\Delta\tau = 0$.

The name "pseudo-Euler" that we give to the algorithm (17) stems from the fact the algorithm has the same structure as a second-order Runge-Kutta procedure: evaluate the velocity vector $f(\cdot)$ at the current state $x(k)$ and at a test point $y(k)$ along the current velocity vector and then use a linear combination of the two velocity vectors in the move that updates the state to $x(k+1)$. When the two time steps Δt and $\Delta\tau$ are equal the algorithm is a true second-order Runge-Kutta procedure, known as the modified Euler method. However when $\Delta t \neq \Delta\tau$ the algorithm is not a second-order Runge-Kutta procedure. It has the proper structure but the solution $x(k+1)$ that it generates only agrees with the second-order Taylor series expansion of the actual solution $x(t_k + \Delta t)$ to (15) when $\Delta t = \Delta\tau$.

Peitgen and Richter [1986] introduced the algorithm (17) and used it to study chaos in the resulting discrete-time system corresponding to the prey-predator system (14). They showed that chaos occurs in roughly an expanding parabolic band of points in $(\Delta t, \Delta\tau)$ parameter space. For Δt values on the order of $0.01 \leq \Delta t \leq 0.8$ the chaotic region lies near but at values of $\Delta\tau$ below the $\Delta\tau = \Delta t$ second-order Runge-Kutta line. In the range $0.8 \leq \Delta t \leq 1$ chaos occurs for $\Delta\tau$ values above, below, and on the $\Delta\tau = \Delta t$ line.

Figure 4 shows the trajectory $x(k)$ generated from (14) by pseudo-Euler integration with $\Delta t = 0.7$ and $\Delta\tau = 0.63$ for the case where $\alpha = \beta = \gamma = \delta = 1$, with $x_1(0) = x_2(0) = 0.5$. This trajectory converges to a chaotic attractor that bears little resemblance to the periodic orbits in Figure 3 that are associated with the continuous-time system (14).

6. Prey-Predator Feedback Control

To investigate the effects of a feedback control management strategy applied to the continuous-time Lotka-Volterra prey-predator system (14), we add effort harvesting terms to each of equations (14), yielding the system

$$\begin{aligned} \dot{x}_1 &= x_1[\alpha - \beta x_2 - u_1] \\ \dot{x}_2 &= x_2[-\gamma + \delta x_1 - u_2]. \end{aligned} \tag{18}$$

The control vector $u = (u_1, u_2)$ is to be chosen subject to maximum effort and no stocking constraints of the form

$$\begin{aligned} 0 &\leq u_1 \leq u_{1,max} \\ 0 &\leq u_2 \leq u_{2,max}. \end{aligned} \tag{19}$$

We will design a feedback controller to stabilize the system (18) about the equilibrium point given by (16). In a neighborhood of this equilibrium point the linearized equations of motion can be stabilized by using a linear state feedback controller of the form

$$u = K(x - \hat{x}) \qquad K = \begin{bmatrix} k_{11} & k_{12} \\ k_{21} & k_{22} \end{bmatrix} \tag{20}$$

For simplicity, we consider the case where $\alpha = \beta = \gamma = \delta = 1$ and choose

$$k_{11} = k_{22} = \mu > 0, \qquad k_{12} = k_{21} = 0, \tag{21}$$

yielding the linear feedback controller

$$u_i = \mu(x_i - \hat{x}_i), \quad i = 1, 2. \tag{22}$$

For the nonlinear system (18) with the control constraints (19) we employ the linear state feedback controller (22), but we make the control values saturate when they reach their upper or lower bounds. The saturating state feedback controller for (18) is given explicitly by

$$u_i(x_i) = \begin{cases} u_{i,max} & \text{if } \mu(x_i - \hat{x}_i) > u_{i,max} \\ \mu(x_i - \hat{x}_i) & \text{if } 0 \leq \mu(x_i - \hat{x}_i) \leq u_{i,max} \\ 0 & \text{if } \mu(x_i - \hat{x}_i) < 0, \end{cases} \tag{23}$$

where $i = 1, 2$, μ is a parameter (the controller strength or gain), and $\hat{x}$ is defined by (16).

For the feedback controller given by (23) the prey-predator system (18) becomes a system of ordinary differential equations of the form

$$\dot{x} = f(x)$$

given by

$$\begin{aligned} \dot{x}_1 &= x_1[\alpha - \beta x_2 - u_1(x_1)] \\ \dot{x}_2 &= x_2[-\gamma + \delta x_1 - u_2(x_2)]. \end{aligned} \tag{24}$$

For any value of $\mu > 0$, the controller in (23) yields an asymptotically stable equilibrium at the point $\hat{x}$ a given by (16) and every initial state in the positive quadrant converges asymptotically to the equilibrium point.

Figure 5 shows various state-space trajectories for the continuous-time prey-predator feedback control system (23)-(24) for $\alpha = \beta = \delta = \gamma = \mu = u_{1,max} = u_{2,max} = 1$. These trajectories were generated with the standard fourth-order Runge-Kutta algorithm with a time step size of $\Delta t = 0.1$.

Figures 6-10 show pseudo-Euler state-space trajectories of (23)-(24) for $\mu = 1.0, 0.5, 0.1, 0.05$, and 0.01, for the case $\alpha = \beta = \gamma = \delta = u_{1,max} = u_{2,max} = 1$, with $x_1(0) = x_2(0) = 0.5$. The pseudo-Euler time step sizes were $\Delta t = 0.7$ and $\Delta\tau = 0.63$. In Figure 6 $\mu = 1.0$ produces a trajectory that asymptotically approaches the fixed point at $x_1 = x_2 = 1$. Similar results occur in Figure 7 for $\mu = 0.5$.

As shown in Figure 8, if we reduce the controller strength to $\mu = 0.1$ the fixed point becomes unstable and the trajectory converges to a closed curve. For clarity in Figure 8 we have shown only the actual states $x(k)$ generated by the pseudo-Euler integrator and have not connected the dots. As in Figures 6 and 7, the points $x(k)$ in Figure 8 move in a counter-clockwise direction, with approximately nine points per revolution. For a continuous-time system the attractor shown in Figure 8 would be called a limit cycle, with a period of about $9\Delta t = 6.3$ time units. However, for the discrete-time pseudo-Euler system the attractor is actually quasiperiodic. The state cycles around on the attractor with approximately nine points per revolution, but after each revolution the state is slightly beyond its position at the previous revolution and the process does not repeat except in the limit as $t_k \to \infty$.

Figure 9 shows the attractor for the pseudo-Euler prey-predator system (23)-(24) for a controller strength of $\mu = 0.05$. As in Figure 8 the attractor is a closed quasiperiodic curve, but it is beginning to have an erratic shape. Figure 10 shows the attractor for the case $\mu = 0.01$. For this case the attractor is nonperiodic and chaotic.

7. Summary

In this paper we have presented simulation results for two different chaotic discretization schemes applied to nonchaotic continuous-time systems. An exponential discretization scheme was applied to an asymptotically stable Lotka-Volterra competition model and it produced a chaotic trajectory, whereas the continuous-time system produced a corresponding trajectory that converged asymptotically to an equilibrium point. The second discretization method was a pseudo-Euler integration procedure. When it was applied to a nonchaotic continuous-time Lotka-Volterra prey-predator system with periodic solutions, the pseudo-Euler integrator produced a chaotic attractor instead of periodic orbits. When a saturating state feedback controller was added to the continuous-time system, the pseudo-Euler integrator yielded trajectories that approach the equilibrium point asymptotically, but only if the strength (or gain) of the controller was sufficiently high. As the controller strength was decreased, the equilibrium point became unstable under pseudo-Eulerre eegration and the attractor became a quasiperiodic curve. As the controller strength was decreased further, the attractor became chaotic, even though the continuous-time system still had an asymptotically stable equilibrium point as its attractor.

There are two main conclusions that can be drawn from the results of this paper. One is that discretization schemes, such as the exponential discretization models often employed by population biologists or even certain numerical integration schemes, can produce drastically different results than those associated with continuous-time ordinary differential equation models. Indeed, the discrete-time system may be chaotic even though the continuous-time system is not. The second conclusion is that, whether it occurs naturally or as a result of applying a feedback controller, asymptotic stability in a continuous-time system is not sufficient to eliminate chaotic motion in a corresponding discrete-time model of the system.

REFERENCES

[1] Coddington, E.A. and N. Levinson. 1955. *Theory of Ordinary Differential Equations*, McGraw-Hill, New York.

[2] Cook, P.A. 1985. Simple Feedback Systems with Chaotic Behaviour, *Systems and Control Letters,* **6**, pp. 223-227.

[3] Cook, P.A. 1986. Chaotic Behaviour in Feedback Systems, *Proc. 25th Conf. on Decision and Control*, Athens, Greece, pp. 1151-1154.

[4] Devaney, R.L. 1988. *An Introduction to Chaotic Dynamical Systems*, Revised 2nd Printing, Benjamin/Cummings Pub. Co., Menlo Park, CA.

[5] Farmer, J.D., E. Ott, and J.A. Yorke. 1983. The Dimension of Chaotic Attractors, *Physica* **7D**, pp. 53-180.

[6] Fisher, M.E. and W.J. Grantham. 1985. Estimating the Effect of Continual Disturbances on Discrete-Time Population Models, *J. of Math. Biology,* **22**, pp. 199-207.

[7] Grantham, W.J. and M.E. Fisher. 1987. Generating Reachable Set Boundaries for Discrete-Time Systems, in *Modeling and Management of Resources Under Uncertainty*, Vincent, T.L., *et al* , eds, Springer-Verlag, New York.

[8] Grebogi, C., E. Ott, and J.A. Yorke. 1985. Attractors on an N-Torus: Quasiperiodicity Versus Chaos, *Physica* **15D**, pp. 354-373.

[9] Grebogi, C., E. Ott, and J.A. Yorke. 1987. Chaos, Strange Attractors, and Fractal Basin Boundaries in Nonlinear Dynamics, *Science,* **238**, pp. 632-638.

[10] Guckenheimer, J. and P. Holmes. 1986. Nonlinear Oscillations, Dynamical Systems, and Bifurcations of Vector Fields, Revised 2nd Printing, *Applied Mathematical Sciences series, Vol 42.*, Springer-Verlag, New York.

[11] Hsu, C.S. 1987. Cell-to-Cell Mapping, *Applied Mathematical Sciences series, Vol 64.*, Springer-Verlag, New York.

[12] Inoue, M. and H. Kamifukumoto. 1984. Scenarios Leading to Chaos in a Forced Lotka Volterra Model, *Progress in Theoretical Physics, Vol 71, No. 5*, pp. 930-937.

[13] Li, T. and J.A. Yorke. 1975. Period Three Implies Chaos, *Amer. Math. Monthly,* **82**, pp. 985.

[14] Lorenz, E.N. 1963. Deterministic Nonperiodic Flows, *J. Atmos. Sci.,* **20**, pp. 130.

[15] Mandelbrot, B.B. 1982. *The Fractal Geometry of Nature*, Freeman Pub., San an ancisco.

[16] May, R.M. 1974. Biological Populations with Nonoverlapping Generations: Stable Points, Stable Cycles, and Chaos, *Science,* **186**, pp. 645-647.

[17] May, R.M. 1976. Simple Mathematical Models with Very Complicated Dynamics, *Nature,* **261**, pp. 459-467.

[18] Mees, A.I. and C.T. Sparrow. 1981. Chaos, *IEE Proc. Part D,* **128**, pp. 201-205.

[19] Peitgen, H.-O. and P.H. Richter. 1986. *The Beauty of Fractals*, Springer-Verlag, New York.

[20] Roux, J.C., A. Rossi, S. Bachelart, and C. Vidal. 1981. Experimental Observation of Complex Behaviour During a Chemical Reaction, *Physica* **2D**, pp. 395.

[21] Rubio, F.R., J. Aracil, and E.F. Camacho. 1985. Chaotic Motion in an Adaptive Control System, *Int. J. Control*, **42**, No. 2, pp. 353-360.

[22] Schuster, H.G. 1988. *Deterministic Chaos*, 2nd Revised Ed., VCH Pub., New York.

[23] Ushio, T. and K. Hirai. 1983. Chaos in Nonlinear Sampled-Data Control Systems, *Int. J. Control*, **38**, No. 5, pp. 1023-1033.

[24] Wiggens, S. 1988. Global Bifurcations and Chaos, *Applied Mathematical Sciences series, Vol. 73,* **Springer-Verlag**, pp. New York.

PARTICIPANT'S COMMENTS

This paper investigates the question of whether or not the solutions obtained by difference approximations accurately represent the solutions to the original differential equations. The authors demonstrate that not only may the difference approximation solutions be inaccurate, but their qualitative behaviour may be totally different from that of the solutions of the original differential equation. This has important ramifications in using certain numerical integration schemes to solve differential equations. They show that some numerical integration schemes can exhibit chaotic behaviour where none existed in the original differential equations. This chaotic behaviour may still be present in the numerical integration even when a stabilizing feedback controller has been added to the original system.

As the authors state, population biologists, for a variety of reasons, often use discrete-time models to model dynamical systems. One approach to obtaining a discrete-time model of a population system is to take a continuous-time model and discretize it. This approach has led, for example, to the so-called discrete analogues of the Logistic equation and the Lotka–Volterra equations. These discrete-time analogues are demonstrated to have a much richer and more complicated dynamical behaviour than the original systems. In modelling biological systems, however, one should be concerned with capturing the essence of the birth and death processes in the model. A discrete-time model should be used if these processes can be more effectively modelled as discrete processes. The resulting discrete-time model may, or may not, then correspond to one of the standard discretizations of a continuous-time model.

Mike Fisher

This paper considers two continuous-time population models for biological processes. It is assumed that a feedback control management strategy is applied to stabilize these continuous-time models about their equilibrium points. The authors then show that both exponential discretization and discretization associated with numerical simulation can produce chaos in the resulting discrete-time systems. This phenomenon is interesting, and carries an important message: extreme care must be taken when a continuous-time system is to be approximated by a discrete-time system.

Kok Lay Teo

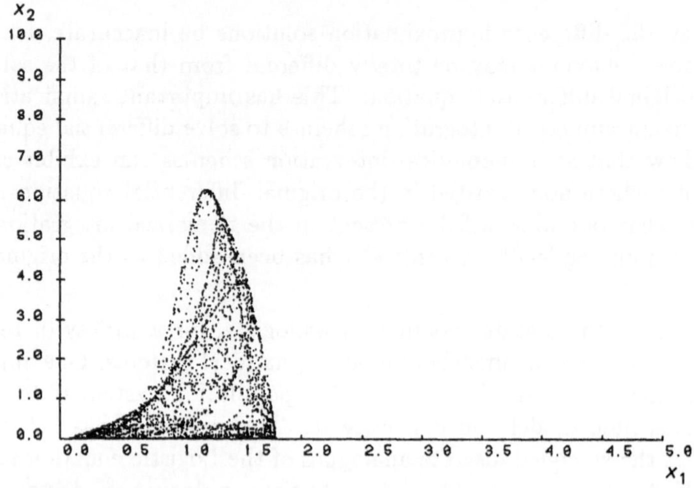

Figure 1. Chaotic motion in the exponential discrete-time competition model.

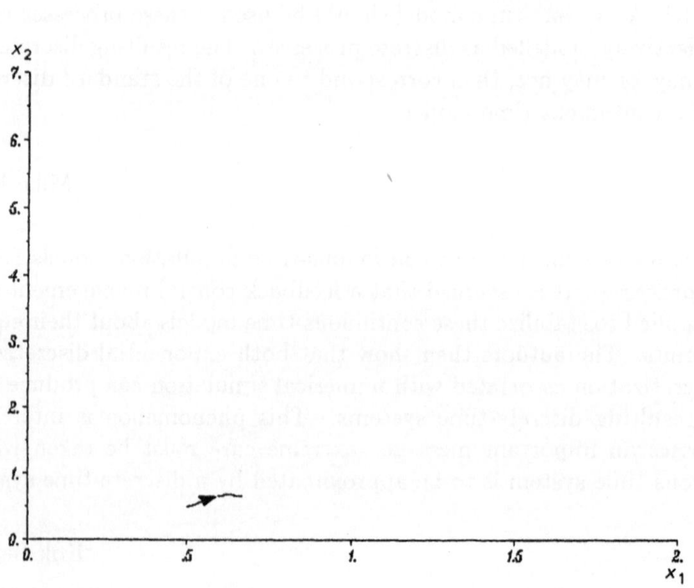

Figure 2. Asymptotically stable motion in the continuous-time competition model.

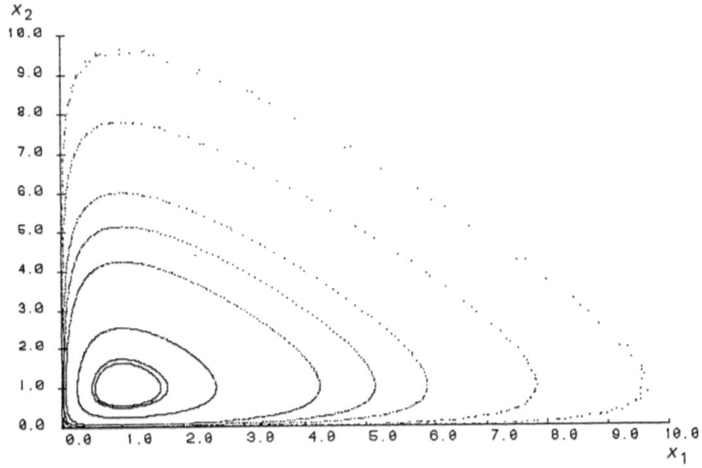

Figure 3. Periodic motion in the continuous-time prey-predator model.

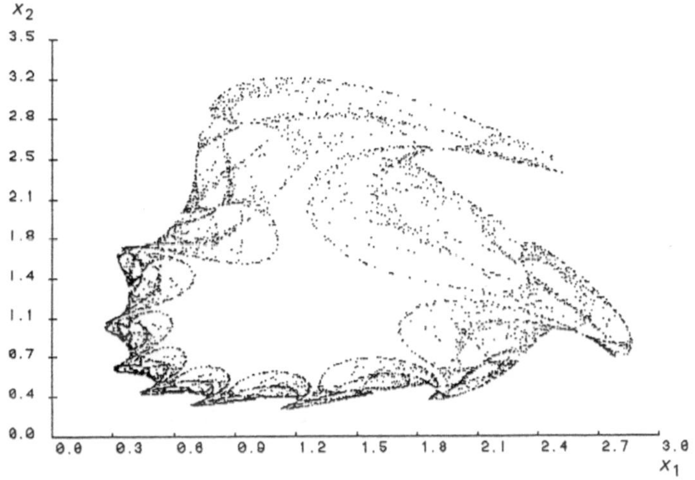

Figure 4. Chaotic discrete-time prey-predator motion ($\Delta t = 0.7, \Delta\tau = 0.63, \alpha = \beta = \gamma = \delta = 1$).

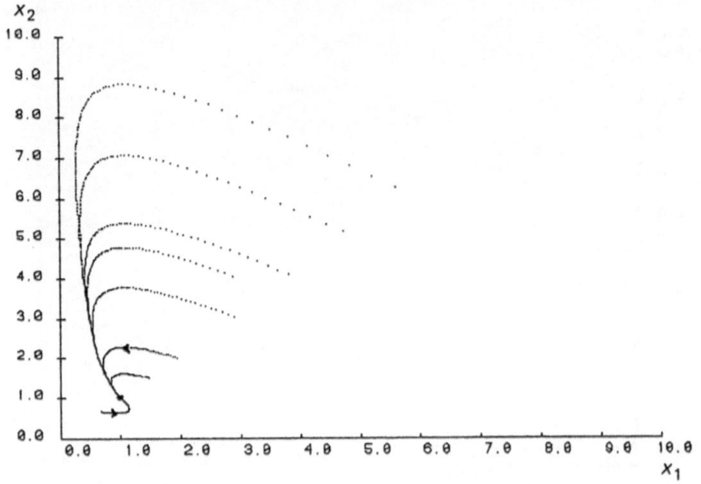

Figure 5. Asymptotic stability in continuous-time prey-predator control system ($\mu = 1$).

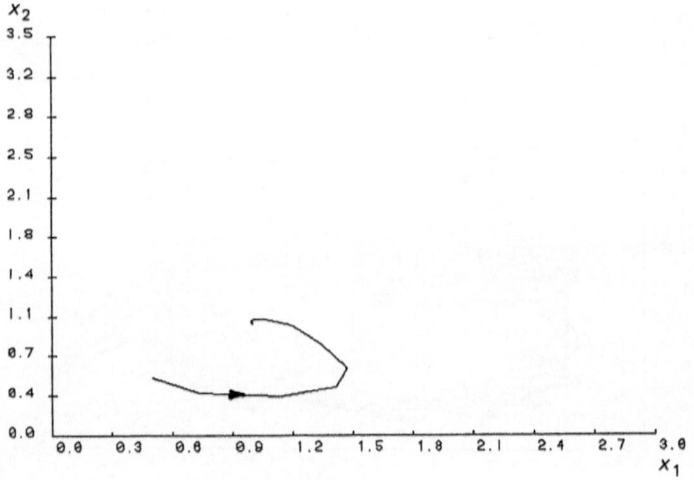

Figure 6. Asymptotic stability in discrete-time prey-predator control system ($\mu = 1$).

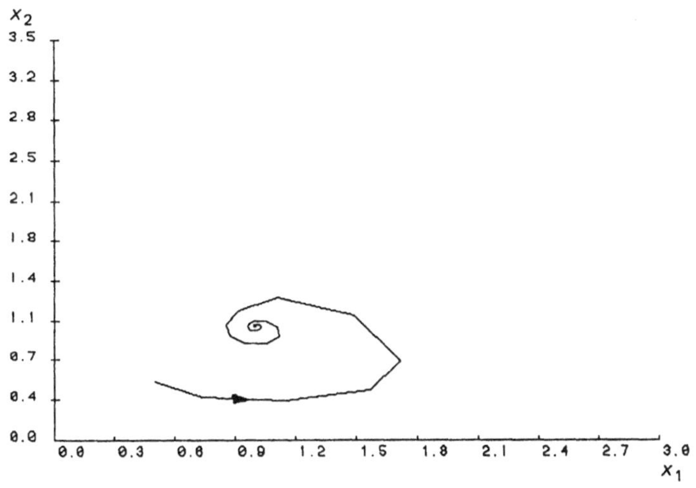

Figure 7. Asymptotic stability in discrete-time prey-predator control system ($\mu = 0.5$).

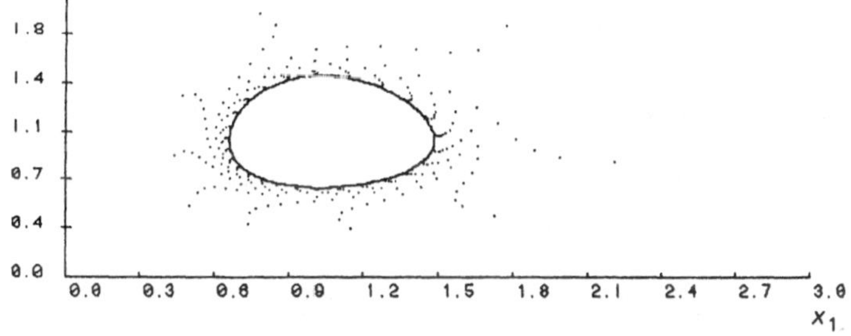

Figure 8. Quasiperiodic attractor for discrete-time prey-predator control system ($\mu = 0.1$).

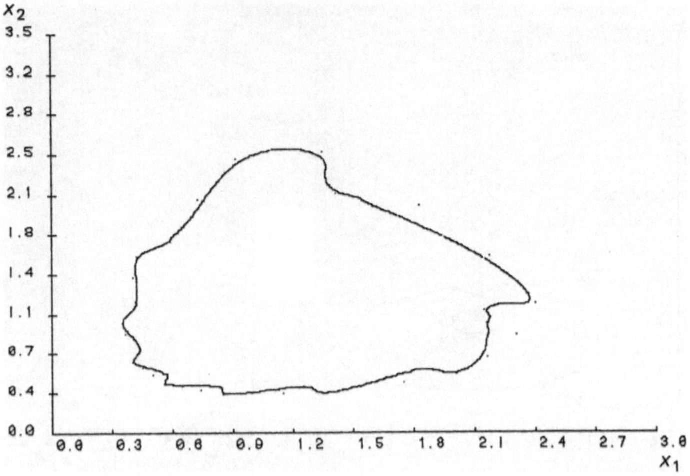

Figure 9. Quasiperiodic attractor for discrete-time prey-predator control system ($\mu = 0.05$).

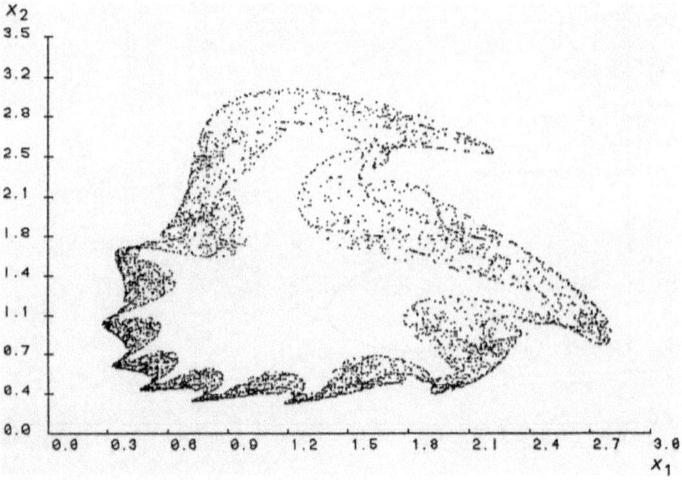

Figure 10. Chaotic motion in discrete-time prey-predator control system ($\mu = 0.01$).

Impulsive Evolution Equations and Population Models

PHIL DIAMOND

Abstract

Biological systems can be subjected to short-term fluctuations in the environment or in the character of the system. This paper models such dynamics by evolution equations with impulses. An appropriate setting is a space of Banach space valued functions of bounded variation. Some general properties, such as existence and uniqueness, are considered, along with some dynamical system concepts of impulsive evolution equations. The study is motivated by, and refers to an example of a continuously age-distributed population subjected to impulses from time to time.

Keywords: Impulses, evolution equations, measure operator equations, age-dependent population.

1. Introduction

Recall the abstract, semilinear initial value problem on a Banach space X:

$$dx/dt + Ax = f(t,x) \quad , \quad t \in (0,\tau) \quad , \quad x(0) = x_0 \tag{1.1}$$

It is assumed that $-A$ is the infinitesimal generator of a C_0-semigroup $T(t)$, $t \geq 0$, on X and that $f : I \times X \to X$ is continuous in $t \in I = [0,\tau]$ and uniformly Lipschitz continuous on X. Then $\|T(t)\| \leq Me^{\omega t}$, $M \geq 1$, $\omega \geq 0$, $t > 0$ and the conditions on A, together with the Lipschitzness of f, guarantee the existence of a *mild* solution to (1). That is, a continuous X-valued function $x(t)$ satisfying

$$x(t) = T(t)x_0 + \int_0^t T(t-s)f(s,x(s))\,ds \quad , \quad t \in (0,\tau] \quad . \tag{1.2}$$

Moreover the mapping $x_0 \to x$, defined by (2), is Lipschitz continuous from X into $C(I, X)$. For more complete details see, for example, Pazy [11], and for the equivalence with weak solutions Ball [3].

A slightly more complicated situation, which is the underlying model for much of our subsequent analysis, arises as follows: consider a herbivore with an age-dependent population density $x(t, a)$, grazing on a plant resource $R(t)$. The birthrate is assumed to be dependent on R which, in its turn, depends on some input $y(t)$. For example, a fish population in the ocean grazes on a phytoplankton/zooplankton resource, or perhaps even weed, which is itself controlled by an annual cycle of nutrient upwelling from the ocean depths. A typical model is the Gurtin-MacCamey system ([6],[13]) in the form of a hybrid lumped-distributed system [9]

$$\begin{aligned} \partial x/\partial t + \partial x/\partial a + \mu(a, x, R)x &= 0 \\ x(t, 0) &= \int_\alpha^{\alpha+\gamma} b(a, R)\, x(t, a)\, da \\ dR/dt &= g(t, x, R, y) \\ x(0, a) &= x_0(a) \end{aligned} \tag{1.3}$$

We wish to incorporate the idea of recurring, short-term fluctuations into systems (1),(3). To describe this impulsive behaviour, suppose $I = [0, \infty)$ and $U : I \to \mathbf{R}$ is a function of bounded variation on compact subintervals of I. This will induce a Lebesgue- Stieltjes measure $dU(t)$ and (1) is replaced by a *measure differential equation*

$$dx + (Ax - f(t, x))\, dt - g(t, x)\, dU(t) = 0 \quad , \quad x(0) = x_0 \quad . \tag{1.1'}$$

To see how impulses of this sort can be incorporated into a typical biological model, let $\{t_k\}$ be a positive increasing sequence of real numbers, which will be considered as consecutive times at which an impulse occurs. Suppose that the herbivore component of the system (3) is restocked at times $t = t_k$, at some rate $h(t, a, x, R)$. Harvesting can be considered as negative restocking. The impulses induced by instantaneous restocking can be modelled as

$$dx + (\partial x/\partial a + \mu(a, x, R)x)\, dt - h(t, a, x, R)\Big(\sum_{k=1}^{\infty} \delta(t_k)\, dt\Big) = 0 \quad , \tag{1.3'}$$

where $\delta(t_k)\, dt$ is the Dirac measure concentrated at $t = t_k$.

In equations of this sort, a solution could not be differentiable at the impulse times and even continuity is an artificial restriction. This suggests that some other form of regularity, such as bounded variation and one-sided continuity, should be sought as properties of the solutions. The integral associated with measure differential equations may be interpreted as a Bochner-Stieltjes integral defined piecewise on intervals between impulses, as was done for $X = \mathbf{R}^n$ by Pandit and Deo [10]. Here, a different approach will be taken and the Kurzweil integral used throughout. In addition, the definition of solutions is framed in the language of semigroups of operators. Although these tools give a slightly more abstract formulation, an ultimately (formally) clearer and natural description is obtained. In particular, it seems appropriate to define C_0-semigroups of bounded variation so as to accomodate the effect of impulses, and the treatment is general enough to allow for the set of impulse times to have points of accumulation. Even though most biological systems probably don't need this last device, there is little cost for including it as a possible modelling option.

This general approach to the problem of impulses in systems allows quite a lot to be said about a model of the form (1.3′). The remainder of this paper is concerned with defining just what is meant by a solution to such systems and what conditions are required for uniqueness. Moreover, the idea of an equilibrium solution, linearisation around an equilibrium and the relation between the stability of the linearised and original systems are all studied in the framework of dynamical systems and the limiting equations of the system. All these ideas are applied to a simple example of the type (1.3′).

The next section introduces BV-spaces and explores some of their properties. Section 3 defines measure evolution equations and defines the notion of a BC_0-semigroup, which seems appropriate for impulsive perturbations of semilinear evolution equations. Subsequent sections consider existence and uniqueness of solutions and explore some semidynamical system ideas for measure differential equations. In this context, equilibrium solutions and their stability are discussed. Finally, we return to the impulsive age-dependent equations (3′) and justify a formal analysis which displays a broad repertoire of complex behaviour. Two appendices are attached: the first sketches some steps necessary to show that measure differential equations induce semidynamical systems on an appropriate space, while the second contains some bibliographic notes and comments, featuring the Kurzweil integral.

2. The Spaces $BV(I,X)$

Let $I = [0,\tau]$ if $0 < \tau < \infty$, $I = [0,\infty)$ if $\tau = \infty$. Let X be a Banach space with norm $||\cdot||$ and write $B(X)$ for the space of bounded linear operators on X. Suppose that $f : I \to X$ and consider all possible partitions P : $a = s_0 < s_1 < \cdots < s_N = b$ of a compact subinterval $[a,b] \subset I$. Define the total variation of f on $[a,b]$, $V_X(f;[a,b]) = \sup_P\{\sum_{i=1}^N ||f(s_i) - f(s_{i-1})||\}$. The function f is said to be of bounded variation on I if $V_X(f;J) < \infty$ for all compact subintervals $J \subset I$, and the set $\{V_X(f;J) : J \subset I\}$ is bounded. Then $V_X(f;I) = \sup_{J\subset I} V_X(f;J)$ and the space of all such functions is denoted by $BV(I,X)$. With norm $||f||_{BV} = V_X(f;I) + ||f(0+)||$, $BV(I,X)$ is a Banach space. Observe that $\sup_{t\in I} ||x(t)|| := ||x||_\infty \leq ||x||_{BV}$ for all $x \in BV(I,X)$. For $U \in BV(I,\mathbf{R})$, let $dU(t)$, $d|U|(t)$ respectively denote Lebesgue-Stieltjes measure and total variation measure generated by U on I. Write $|U|$ for the total variation function of U.

Definition. 2.1 The space consisting of $x \in BV(I,X)$, with $x(\cdot)$ right continuous, is denoted by $BV^+(I,X)$.

Proposition 2.2. *$BV^+(I,X)$ is a Banach space with norm $||\cdot||_{BV}$.*

Proof: Let $\{x_n\}$ be a Cauchy sequence in $BV^+(I,X)$ converging to x. Now $x \in BV(I,X)$ and $||x - x_n|| \leq ||x - x_n||_{BV}$. Thus $\{x_n\}$ converges uniformly to x, $x(\cdot)$ is consequently right continuous and so $BV^+(I,X)$ is complete.

These X-valued functions of bounded variation share many properties with their real-valued counterparts. The major differences arise from the fact that a representation as a difference of monotonic functions has no meaning in a general space. However, the properties we require from $BV(I,X)$ do not depend upon this property.

Lemma 2.3. *Let $f \in BV(I,X)$. Then, at every interior point $t_0 \in I$, there exist left and right hand limits $f(t_0-)$, $f(t_0+)$.*

Proof: We show that $f(t_0-)$ exists, the proof for the other being very similar. If this limit does not exist, there is an increasing sequence $\{s_i\}$ in I, converging to t_0, such that $\{f(s_i)\}$ does not converge. That is, $\sum_{i=1}^\infty ||f(s_{i+1} - f(s_i)||$ diverges. But then f is not of bounded variation on $[s_1,t_0] \subset I$, contradicting $f \in BV(I,X)$.

Proposition 2.4. *If $f \in BV(I,X)$, then f has at most a countable number of discontinuities and these are the same as those of the function $V_X(f;[0,t])$.*

Proof: Follows the usual proof for $BV(I, R)$, see Graves [5], Theorem 10.24.

Theorem 2.5. *Let K be a compact subset of X. Then closed, bounded (in $\|\cdot\|_{BV}$) subsets of $BV(I,K)$ are compact subsets of $BV(I,X)$.*

Proof: Let $\{f_i\}$ be a sequence in a closed bounded subset of $BV(I,K)$. Let $E = \{t_k\}$ be a countable dense subset of I. Then $\{f_i(t_1)\}$ is a sequence of points in K, and so has a convergent subsequence $\{f_{i(1)}(t_1)\}$ with limit $g(t_1) \in K$. The usual diagonal construction gives a subsequence $\{f_{i(i)}\}$ such that $\lim_{i\to\infty} f_{i(i)}(t_k) = g(t_k)$ for $k = 1, 2, \cdots$. At each point $t \in I$, $g(t-), g(t+)$ are defined, by Lemma 2.3. If these limits coincide, set $g(t)$ equal to the common value, and g is at worst undefined on a denumerable set. The remainder of the proof now proceeds as in [5], and we observe that for any partition $P : a = s_0 < \cdots < s_N = b$ of $[a,b] \subset I$,

$$\sum_{l=1}^{N} \|g(s_l) - g(s_{l-1})\| = \lim_{i\to\infty} \sum_{l=1}^{N} \|f_{i(i)}(s_l) - f_{i(i)}(s_{l-1})\|$$
$$\leq \liminf_{i\to\infty} \; V_X(f_{i(i)}; [a,b]) \quad .$$

3. Measure Evolution Equations

We briefly recall some ideas and notations before proceeding to those definitions required later. Denote by $C_0^\infty(I,X)$ the space of infinitely differentiable functions $\phi : I \to X$, with compact support in I, and let $L^1_{loc}(I,X)$ be the space of all locally integrable functions on I. Any $f \in L^1_{loc}(I,X)$ defines a distribution in the dual $C_0^\infty(I,X)'$, by $S_f(\phi) = \int f(t)\phi(t)\,dt$, $\phi \in C_0^\infty(I,X)$, and we identify f with S_f. As usual, the distributional derivative is given by $Df = -\int f(t)\phi'(t)\,dt$. If $U \in BV^+(I,\mathbf{R})$, DU is the Lebesgue-Stieltjes measure $dU(t)$. Recall the definition of a C_0-semigroup:

Definition. 3.1 [11] A C_0-semigroup of bounded linear operators on a Banach space X is a family $\{T(t)\}_{t\geq 0} \subseteq B(X)$ such that

(i) $T(0) = E$, the identity operator on X;

(ii) $T(s)T(t) = T(s+t)$ for all $s, t \geq 0$;

(iii) for each fixed $x \in X$, $T(t)x \to x$ as $t \to 0+$.

It may be shown that there exist fixed real numbers $M \geq 1$, $\omega \geq 0$ such that $\|T(t)\| \leq Me^{\omega t}$, $t \geq 0$, and in this case the semigroup is said to belong to the class $C_0(M,\omega)$. The linear operator defined by $D(A) = \{x \in X : \lim_{t\to 0+}(T(t)x - x)/t \;\text{ exists}\}$ and $Ax = \lim_{t\to 0+}(T(t)x - x)/t$, $x \in D(A)$, is called the infinitesimal generator of the semigroup $T(t)$.

Definition. 3.2 Let X be a Banach space and $A : D(A) \subset X \to X$ a linear operator. Suppose that $U \in BV^+(I, \mathbf{R})$ and $f, g : I \times X \to X$. The initial value problem

$$dx(t) + (Ax - f(t,x))\,dt - g(t,x)\,dU(t) = 0\ ,\quad x(0) = x_0\ , \tag{3.1}$$

is called a measure differential equation. If $-A$ is the infinitesimal generator of a C_0-semigroup $T(t)$, a function $x \in BV^+(I, X)$ which satisfies, for all $t \in I$,

$$x(t) = T(t)x_0 + \int_0^t T(t-s)f(s, x(s))\,ds + \int_0^t T(t-s)g(s, x(s))\,dU(s)\ , \tag{3.2}$$

is said to be a mild solution of the initial value problem (3.1) on I.

Analysis in $BV(I, X)$ requires more than the continuity condition of Definition 3.1(iii) :

Definition. 3.3 Let $T(t)$ be a C_0-semigroup of bounded linear operators on X. Let J be a compact subinterval of I and $P : s_0 < s_1 < \cdots < s_N$ a partition of J. Suppose that, for each $x \in X$, there is a uniform constant v_J such that

$$\sup_P \sum_{i=1}^{N} \|T(s_i - s_{i-1})x - x\| \le v_J \|x\|\ .$$

Then $T(t)$ is said to be a BC_0- semigroup.

Proposition 3.4. *Let $T(t)$, $t \ge 0$, be a BC_0- semigroup of bounded linear operators on X. Let $x_0 \in X$ be fixed and suppose $x(\cdot) \in BV^+(I, X)$. Then, for every compact subinterval $J = [0, \tau] \subset I$,*

(i) $T(\cdot)x_0 \in BV^+(J, X)$ for every $x_0 \in X$;

(ii) $T(\cdot)x(\cdot) \in BV^+(J, X)$ for every $x(\cdot) \in BV^+(I, X)$;

Proof: Let $P : s_0 < s_1 < \cdots < s_N$ be any partition of J and write $M_J = \sup_{t \in J} \|T(t)\|$.

(i) Then,

$$\begin{aligned}\sum_{i=1}^{N} \|T(s_i)x_0 - T(s_{i-1})x_0\| &\le \sum_{i=1}^{N} \|T(s_{i-1})\|\, \|T(s_i - s_{i-1})x_0 - x_0\| \\ &\le M_J v_J \|x_0\|\ ,\end{aligned}$$

and $V_X(T(\cdot)x_0\,;\,J) \leq M_J v_J \|x_0\|$.

(ii) Again,

$$\begin{aligned}
&\sum_{i=1}^{N} \|T(s_i)x(s_i) - T(s_{i-1})x(s_{i-1})\| \\
&\qquad \leq M_J \sum_{i=1}^{N} \|T(s_i - s_{i-1})x(s_i) - x(s_{i-1})\| \\
&\qquad \leq M_J \sum_{i=1}^{N} \{\|T(s_i - s_{i-1})x_0 - x_0\| \\
&\qquad\quad + \|T(s_i - s_{i-1})(x(s_i) - x_0)\| + \|x_0 - x(s_{i-1})\| \\
&\qquad \leq M_J(V_X(T(\cdot)x_0\,;\,J) + (M_J+1)V_X(x(\cdot)\,;\,J)) \,.
\end{aligned}$$

Example Let $X = C^1(\overline{\mathbf{R}^+}\,, \mathbf{R})$. The translation semigroup $T(t)$, defined on X by $T(t)x(s) = x(s+t)$, $s,t \geq 0$, is a BC_0-semigroup, with infinitesimal generator $-A$, defined by $(-Ax)(t) = x'(t)$.

4. Existence and Uniqueness

Let $J = [0, \tau_0] \subset I$, $\tau_0 < \tau$. The measure differential equation initial value problem (2.1) is shown below to have a unique mild solution in $BV^+(J,X)$, provided $-A$ is the infinitesimal generator of a BC_0- semigroup and f, g satisfy certain variation conditions and f is Lipschitz on compact subsets of X.

Some intuitive explanation should be given here concerning the conditions on g , U in Definition 4.2 below. These are closely linked to ideas and technical requirements relating to the existence of a Bochner-Kurzweil (or generalised Perron) integral for the term $\int_{t_0}^{t} g(s,x(s))\,dU(s)$ appearing in (1.1'). Instead of the usual Riemann partition $s_0 < s_1 < \cdots < s_N$ of the interval $[t_0,t]$, define a Kurzweil partition to be a collection $\mathcal{P}$: $\sigma_1, \cdots, \sigma_N\,;\, s_0, \cdots, s_N$, where $\sigma_i \in [t_0,t]$ and $\{s_i\}$ is a Riemann partition of the interval. A guage on $[t_0,t]$ is a positive function $\delta : [t_0,t] \to (0,\infty)$. The partition $\mathcal{P}$ is said to be δ-fine if $[s_{i-1},s_i] \subset [\sigma_i - \delta(\sigma_i), \sigma_i + \delta(\sigma_i)]$ for $i = 1,2,\cdots,N$. For every partition $\mathcal{P}$, there is an associated Kurzweil-Stieltjes sum

$$S_U(g,\mathcal{P}) = \sum_{i=1}^{N} [g(s_i, x(\sigma_i))U(s_i) - g(s_{i-1}, x(\sigma_i))U(s_{i-1}) \,.$$

The function $g(\cdot, x(\cdot))$ will be called dU-integrable on $[t_0, t]$ if there exists an element I of the Banach space X such that for every $\gamma > 0$ there exists a gauge $\delta : [t_0, t] \to (0, \infty)$ such that for every δ-fine partition $\mathcal{P}$ of $[t_0, t]$ then

$$\|S_U(g, \mathcal{P}) - I\| < \gamma .$$

The element I will be called the Bochner-Kurzweil integral $\int_{t_0}^{t} g(s, x(s))\, dU(s)$ and $S_U(g, \mathcal{P})$ will be termed an approximating sum. A clear exposition, for the case $X = \mathbf{R}^n$, under stronger conditions than those of Definition 4.2, may be found in Artstein [1].

Definition. 4.1 For every compact set $K \subset X$, the class $\mathcal{F}_J$ consists of all continuous functions $f : J \times K \to X$ which are uniformly Lipschitz continuous on K.

Definition. 4.2 Let $U \in BV(J, \mathbf{R})$. For every compact set $K \subset X$, let $\mu_K(\epsilon)$ be a positive function $\mu_K : (0, \infty) \to (0, \infty)$. The class $\mathcal{G}_{J,U}$ consists of all continuous functions $g : I \times K \to X$ such that

(i) $g(\cdot, x(\cdot))$ is dU-integrable for each $x(\cdot) \in BV^+(J, K)$;

(ii) if $t_0 = s_0 \le s_1 \le \cdots \le s_N = t \le \tau_0$, with $t - t_0 < \mu_K(\epsilon)$, and if $x_1, x_2, \cdots, x_N \in K$, then

$$\sum_{i=1}^{N} \| g(s_i, x_i)|U|(s_i) - g(s_{i-1}, x_i)|U|(s_{i-1}) \| < \epsilon \tag{4.1}$$

(iii) for every compact $K \subset X$ there exists a positive nondecreasing continuous function $h = h_{g,K}$ such that if $x, y \in K$, then

$$\begin{aligned} \| (g(s, x) - g(s, y))|U|(s) - (g(s', x) - g(s', y))|U|(s') \| \\ \le \|x - y\| \, |h(s) - h(s')| \end{aligned} \tag{4.2}$$

Theorem 4.3. *Let $-A$ be the infinitesimal generator of a BC_0-semigroup $T(t)$, $t \ge 0$, on X and let $U \in BV^+(J, \mathbf{R})$. For each $f \in \mathcal{F}_J$, $g \in \mathcal{G}_{J,U}$ and $(t_0, x_0) \in J \times X$, there exists a unique mild solution $x \in BV^+(J, X)$ to the initial value problem (3.1) in $[t_0, t]$.*

Proof: For a given $x_0 \in X$, let $K_\eta = \{x \in X : ||x - x_0|| \le \eta\}$ and define a mapping G on $BV^+([t_0, \tau_0], K_\eta)$ by

$$(Gx)(t) = (Fx)(t) + (Hx)(t) \ , \ t \in (t_0, \tau_0) \ ,$$

where $(Fx)(t) = T(t-t_0)x_0 + \int_{t_0}^t T(t-s)f(s, x(s))\, ds$, $(Hx)(t) = \int_{t_0}^t T(t-s)g(s, x(s))\, dU(s)$.

(1) First, $Gx \in BV^+([t_0, \tau_0], K_\eta)$. For, $T(\cdot)x_0$ lies in this space by Proposition 3.4, while for any partition $t_0 = s_0 < s_1 < \cdots < s_N = t$ of $[t_0, t]$:

$$\begin{aligned}
&\sum_{i=1}^N || \int_{t_0}^{s_i} T(s_i - s)f(s, x(s))\, ds - \int_{t_0}^{s_{i-1}} T(s_{i-1} - s)f(s, x(s))\, ds|| \\
&\le \sum_{i=1}^N \{|| \int_{s_{i-1}}^{s_i} T(s_i - s)f(s, x(s))\, ds|| + || \int_{t_0}^{s_{i-1}} T(s_{i-1} - s)(T(s_i - s_{i-1}) \\
&\qquad - E)f(s, x(s))\, ds||\} \\
&\le M_J(\int_{t_0}^t ||f(s, x(s))||\, ds + \int_{t_0}^t v_J ||f(s, x(s))||\, ds) \ .
\end{aligned}$$

From continuity, $f_{K_\eta} := \sup_{(s,x)\in J\times K_\eta} ||f(s, x(s))|| < \infty$. Thus, $Fx \in BV^+([t_0, \tau_0], K_\eta)$. Further,

$$\begin{aligned}
\sum_{i=1}^N ||(Hx)(s_i) - (Hx)(s_{i-1})|| &\le \sum_i \{\int_{s_{i-1}}^{s_i} ||T(t-s)g(s, x(s))||\, d|U|(s) \\
&\quad + M_J \int_{t_0}^t ||(T(s_i - s_{i-1}) - E)g(s, x(s))||\, d|U|(s)\} \\
&\le M_J(1 + v_J) \int_{t_0}^t ||g(s, x(s))||\, d|U|(s) \ .
\end{aligned}$$

The integral $\int_{t_0}^t ||g(s, x(s))||\, d|U|(s)$ exists and is approximated by a Kurzweil-Stieltjes sum

$$\begin{aligned}
&|\sum_i ||g(s_i, x(\sigma_i))||\, |U|(s_i) - ||g(s_{i-1}, x(\sigma_i))||\, |U|(s_i - 1)| \\
&\qquad \le \sum_i ||g(s_i, x(\sigma_i))|U|(s_i) - g(s_{i-1}, x(\sigma_i))|U|(s_{i-1})||
\end{aligned}$$

where $\sigma_1, \cdots, \sigma_N \in [t_0, t]$. From (4.1) and the supposition that $x(\sigma_i) \in K_\eta$, $i = 1, \cdots, N$, $\sup_P \sum_{i=1}^N ||(Hx)(s_i) - (Hx)(s_{i-1})|| \leq \infty$, and so $(Hx)(\cdot) \in BV^+([t_0, \tau_0), K_\eta)$, and so the same is true for Gx.

(2) Let $\eta = M_J(1 + ||x_0||)$. Then $(Gx)(t) \in K_\eta$ for $|t - t_0|$ sufficiently small. To see this, note that if $\tau_0 < t_0 + 1/2f_{K_\eta}$,

$$\begin{aligned}||(Fx)(t)|| &\leq ||T(t)x_0|| + \int_{t_0}^{t} ||T(t-s)|| \, ||f(s, x(s))|| \, ds \\ &\leq M_J(||x_0|| + (\tau_0 - t_0) f_{K_\eta}) \ ,\end{aligned}$$

and $(Fx)(t) \in K_\eta$ for all $t \in J = [t_0, \tau_0]$. Furthermore, $||(Hx)(t)|| \leq M_J \int_{t_0}^t ||g(s, x(s))|| \, d|U|(s)$, and this integral is approximated by a sum as above. By (4.1), this sum is $< 1/2f_{K_\eta}$ for $t - t_0 < \mu_{K_\eta}(1/2f_{K_\eta})$. Hence, G maps $BV^+(J, K_\eta)$ to itself.

(3) Finally, G is a contraction on $BV^+(J, K_\eta)$. For, given $x, y \in BV^+(J, K_\eta)$, first

$$\begin{aligned}&\{||(Fx)(s_i) - (Fy)(s_i)|| - ||(Fx)(s_{i-1}) - (Fy)(s_{i-1})||\}| \\ &\quad \leq ||(Fx)(s_i) - (Fy)(s_i) - ((Fx)(s_{i-1}) - (Fy)(s_{i-1}))|| \\ &\quad \leq || \int_{s_{i-1}}^{s_i} T(s_i - s)[f(s, x(s)) - f(s, y(s))] \, ds|| \\ &\quad + || \int_{t_0}^{si-1} (T(s_i - s) - T(s_{i-1} - s))[f(s, x(s)) - f(s, y(s))] \, ds \\ &\quad \leq M_J L \int_{s_{i-1}}^{s_i} ||x - y||_{BV} \\ &\quad + M_J \int_{t_0}^{s_{i-1}} ||T(s_i - s_{i-1}) - E)[f(s, x(s)) - f(s, y(s))]|| \, ds \\ &\quad \leq M_J L \, ||x - y||_{BV}(s_i - s_{i-1}) \\ &\quad + M_J \int_{t_0}^{t} ||T(s_i - s_{i-1}) - E)[f(s, x(s)) - f(s, y(s))]|| \, ds \ .\end{aligned}$$

Summing from 1 to N and taking the supremum over all partitions P of $[t_0, t]$ gives :

$$\begin{aligned}||Fx - Fy||_{BV} &\leq M_J L ||x - y||_{BV}(t - t_0) \\ &\quad + M_J \int_{t_0}^{t} v_J ||f(s, x(s)) - f(s, y(s))|| \, ds \qquad (4.3) \\ &\leq M_J L \, (1 + v_J) ||x - y||_{BV}(t - t_0) \ .\end{aligned}$$

Now set $(\tau_0 - t_0)M_J L(1 + v_J) \leq \rho$, $0 < \rho < 1/2$. Performing a similar calculation with H,

$$\sum_i ||(Hx - Hy)(s_i) - (Hx - Hy)(s_{i-1})|| \\ \leq M_J(1 + v_J) \int_{t_0}^{t_1} ||g(s, x(s)) - g(s, y(s))|| \, d|U|(s) . \tag{4.4}$$

The integral here can be approximated by

$$|\sum_i \{||g(s_i, x(\sigma_i)) - g(s_i, y(\xi_i))|| \, |U|(s_i) \\ - ||g(s_{i-1}, x(\sigma_i)) - g(s_{i-1}, y(\xi_i))|| \, |U|(s_{i-1})\}| \\ \leq \sum_i ||x(\sigma_i) - y(\xi_i)|| \, (h(s_i) - h(s_{i-1})) \\ \leq (h(t) - h(t_0))||x - y||_{BV} \tag{4.5}$$

for appropriate μ_{K_η}-fine partitions $\{(s_i, \sigma_i)\}$, $\{(s_i, \xi_i)\}$. From (4.4),(4.5) and taking the supremum over μ_{K_η}-fine partitions, for $t_0 \leq t \leq \tau_0$

$$||(Hx)(\cdot) - (Hy)(\cdot)||_{BV} \leq M_J(1 + v_J)(h(t) - h(t_0))||x - y||_{BV} . \tag{4.6}$$

Since h is continuous, $\tau_0 - t_0$ may be taken so small that $||Hx - Hy||_{BV} \leq \rho||x - y||_{BV}$, and consequently $||Gx - Gy||_{BV} \leq 2\rho||x - y||_{BV}$, thus showing that G is a contraction on $BV^+([t_0, \tau_0], K_\eta)$.

Proposition 4.4 . *Let $f \in \mathcal{F}_{\mathbf{R}+}$, $g \in \mathcal{G}_{\mathbf{R}+,U}$ and $(t_0, x_0) \in \mathbf{R}^+ \times X$. Then there exists a unique maximal mild solution $x(t)$ of (3.1), defined on an open interval (t_0, ω). Either $\omega = \infty$ or $x(t)$ converges to the boundary of a compact set $K \ni x_0$ as $t \to \omega-$.*

Proof: Essentially follows the standard proof for ordinary differential equations, involving $x_0 \in K$, except that the interval of existence must remain open on the right to prevent "jumps" in the dU-integral from destroying the compactness argument.

Remark In Definition 4.2, the function $h(\cdot)$ may be taken as nondecreasing and right-continuous only, without losing existence and uniqueness. The first is proved using a Peano-like construction, similar to [7], and involves the Compactness Theorem 2.5. Uniqueness is proved by a "Kurzweil-Gronwall Lemma" akin to inequality (3.4) of [7]. These will be discussed elsewhere. Using the Gronwall-type estimate it is not difficult to show that the mild solutions are Lipschitz continuous with respect to initial condition x_0.

5. Limiting Equations and Equilibrium Solutions

The central idea behind limiting equations is to relate the asymptotic behaviour of the functional form of an ordinary or measure differential equation to the asymptotic behaviour of solutions to the equation. For example, an equation of the form $dx/dt = a(x) + b(t,x)$ may be ***asymptotically autonomous*** in the sense that it is indistinguishable (in a certain sense) from the autonomous equation $dx/dt = a(x)$ as $t \to \infty$ and, moreover, the asymptotic behaviour of the autonomous equation determines that of the time-dependent equation. That is to say, in terms of an appropriate form of convergence, the translates $b^{\tau}(t,x) = b(t+\tau,x) \to 0$ as $\tau \to \infty$. In such a case the limiting equation is unique (see [2] and references therein for more complete treatment of these ideas).

This section considers limiting equations for measure evolution equations. In general, a limiting equation takes the form of an operator equation of convolution type, which is essentially the translation limit points of the equation expressing mild solutions. The principal result is that when there is only one limiting equation, then it is an autonomous limiting equation. This is the analogue of asymptotically autonomous and allows a definition of mild equilibrium solutions.

Let $\mu : [0,\infty) \to [0,\infty]$ be a nondecreasing, right-continuous function with $\mu(0) = 0$. Say that the function $\gamma : I \to X$ admits μ as a modulus of variation if $V_X(\gamma; [s,t]) \le \mu(|t-s|)$ for every $s,t \in I$.

Definition. 5.1 A mapping $\Phi : I \times BV(I,X) \to BV(I,X)$, written as $\Phi_a x$ for $a \in I, x \in BV(I,X)$, is said to be an $\mathcal{H}$-operator if

(i) $\Phi_a : BV([a,\tau],X) \to BV([a,I],X)$ is continuous in the norm $\|\cdot\|_{BV}$, for each $J_a := [a,\tau] \subseteq I$;

(ii) $(\Phi_a x)(t) = T(t-s)(\Phi_a x)(s) + (\Phi_s x)(t)$ for all $a,s,t \in I$, $a \le s \le t$;

(iii) for every compact subset $K \subset X$, and $x \in BV(J_a, K)$, there is a modulus of variation μ_K admitted by Φ_a.

Lemma 5.2. *Suppose that $T(t)$ is a BC_0-semigroup on X and let $f \in \mathcal{F}_J$, $g \in \mathcal{G}_{J,U}$. Then Φ, defined by $(\Phi_a x)(t) = \int_a^t T(t-s) f(s,x(s))\,ds + \int_a^t T(t-s) g(s,x(s)\,dU(s)$, is an $\mathcal{H}$-operator.*

Proof: For (i), write $J_a = [a, \tau]$, let $x, y \in BV(J_a, K)$, and let $P : a = s_0 < s_1 < \cdots < s_N = \tau$ be a partition of J_a. Then Φ_a is quite similar to F in Theorem 4.3, and so the estimates (4.3),(4.6) give

$$||\Phi_a x - \Phi_a y||_{BV} \leq M_{J_a}(1 + v_{J_a})L_K(\tau - a) + (h(\tau) - h(a))||x - y||_{BV} \,. \tag{5.1}$$

(ii) is trivial, while (iii) follows from estimate (5.1), with a linear function

$$\mu_K(\theta) = \text{diam}(K)\, M_{J_a}(1 + v_{J_a})(L_K \theta + h(\theta))$$

Let $f \in \mathcal{F}_{\mathbf{R}^+}$, $g \in \mathcal{G}_{\mathbf{R}^+, U}$ and consider the initial value problem (1.1), on $J_a = [a, \tau]$, in the case where $-A$ is the infinitesimal generator of a BC_0-semigroup. The $\mathcal{H}$-operators include not only the integrals which occur in mild solutions, but also limits of these integrals.

Definition. 5.3 For $\sigma \in \mathbf{R}^+$, define the translates f^σ, g^σ of f, g by $f^\sigma(t, x) = f(t + \sigma, x), g^\sigma(t, x) = g(t + \sigma, x)$. The $\mathcal{H}$-operator equation $x(t) = T(t - a)x_0 + (\Phi_a x)(t)$ is said to be a limiting equation as $\sigma \to \infty$ of $dx + (Ax - f(t, x))dt = g(t, x)dU(t)$, in the sense of mild solutions, if there is an unbounded increasing sequence $\{\sigma_k\}$ of positive real numbers such that, whenever $\{x_k\}$ is a convergent sequence in $BV(J_a, X)$, $\int_a^t T(t - s)f^{\sigma_k}(s, x_k(s))\, ds + \int_a^t T(t - s)g^{\sigma_k}(s, x_k(s)\, dU^{\sigma_k}(s)$ converges to $(\Phi_a x)(t)$. in the BV-norm. Observe that this convergence implies that, for $\sigma \geq 0$, the integral operators involving $f^{\sigma_k + \sigma}$, $g^{\sigma_k + \sigma}$ converge to a translate Φ_a^σ of Φ_a.

Proposition 5.4 . *Let $\{\sigma_k\}$ be an unbounded increasing sequence of positive real numbers. Suppose that whenever $x_k \in BV(J_a, X)$ converges, then*

$$\int_a^\tau T(t - s)f(s + \sigma_k, x_k(s))\, ds + \int_a^\tau T(t - s)g(s + \sigma_k, x_k(s))\, dU^{\sigma_k}(s)$$

is a Cauchy sequence in X. Then the operator Φ defined by

$$(\Phi_a x)(t) = \lim_{k \to \infty} \{\int_a^t T(t - s)f(s + \sigma_k, x(s))\, ds$$
$$+ \int_a^t T(t - s)g(s + \sigma_k, x(s))\, dU^{\sigma_k}(s)\} \,, \quad a \leq t \leq \tau \,,$$

is an $\mathcal{H}$-operator. Consequently, $x(t) = T(t - a)x_0 + (\Phi_a x)(t)$ is a limiting equation of $dx + (Ax - f(t, x))dt = g(t, x)dU(t)$, in the sense of mild solutions.

Proof: Let $x_k \to x$, and write $(\Phi_a^{(k)}x)(t) = \int_a^t T(t-s)f(s+\sigma_k, x_k(s))\,ds + \int_a^t T(t-s)g(s+\sigma_k, x_k(s))\,dU^{\sigma_k}(s)$. Now,

$$V_X(\Phi_a x;[a,t]) \le \|\Phi_a x - \Phi_a^{(k)}x_k\|_{BV} + \|\Phi_a^{(k)}x_k - \Phi_a^{(k)}x\|_{BV} + V_X(\Phi_a^{(k)}x;[a,t]) .$$

By Lemma 5.2, each $\Phi^{(k)}$ admits a modulus of variation μ_K on compact subsets K of X, and $V_X(\Phi_a^{(k)}x;[a,t]) \le \mu_K(t-a)$. By the Cauchy property of $\{\Phi_a^{(k)}\}$, the other two summands of the inequality are $\le \mu_K(t-a)$ for k sufficiently large, and so Φ admits $3\mu_K(\cdot)$ as a modulus of variation. That $\Phi_a : BV(J_a, X) \to BV(J_a; X)$ again from Lemma 5.2, since $f_K^\sigma = \sup_{I\times K}\|f(t+\sigma, x)\| \le f_K$.

Definition. 5.5 The class $\mathcal{F}_\infty$ consists of all functions $f \in \mathcal{F}_{\mathbf{R}^+}$ such that for every compact subset $K \subset X$ there exists a function $b_K \in BV^+(\mathbf{R}^+, \mathbf{R})$ with

$$\|f(s,x) - f(\sigma,x)\| \le |b_K(s) - b_K(\sigma)| .$$

Definition. 5.6 A function $g \in \mathcal{G}_{\mathbf{R}^+,U}$ is said to be U-diminishing if for every compact set $K \subset X$, there is a function $\nu_K; \mathbf{R}^+ \to \mathbf{R}^+$ so that $\lim_{\sigma\to\infty}\nu_K(\sigma) = 0$ and $|\int_\sigma^{\sigma+t} g(s,x)\,dU(s)| \le \nu_K(\sigma)$ for every $(t,x) \in [0,1]\times K$ and $\sigma \in \mathbf{R}^+$.

Theorem 5.7 . *Suppose that there is only one limiting equation, in the mild sense, of (3.1) as $\sigma \to \infty$, for $f \in \mathcal{F}_\infty$, $g \in \mathcal{G}_{\mathbf{R}^+,U}$. Then this limiting equation is an autonomous evolution equation, $dx/dt + Ax = f_0(x)$ if and only if $g\,dU(t) = g_0\,dt + g_1\,dU(t)$, where*

(i) g_0, $g_1 \in \mathcal{G}_{\mathbf{R}^+,U}$, and g_0 is autonomous;

(ii) g_1 is U-diminishing.

Proof: *Sufficiency :* the limiting equation is an $\mathcal{H}$-operator by the result above. From the condition on $\int_0^t g(s,x)\,dU(s)$, we may assume without loss of generality that $g_0 \equiv 0$ and so it remains only to show the existence of a continuous function $f_0 : X \to X$, uniformly Lipschitz on compact subsets $K \subset X$, such that $(\Phi_a x)(t) = \int_a^t T(t-s)f_0(x(s))\,ds$, $a \le t \le \tau$, for every $x \in BV(J_a, X)$. Consider the constant function $x(t) = x$ and define $e_x(t) = (\Phi_0 x)(t)$. Let $\{\sigma_k\}$ be an unbounded increasing sequence of positive real numbers. Then $\int_0^t T(t-s)f(s+\sigma_k, x)\,ds$ is Cauchy in X, uniformly for $0 \le t \le \tau$, because

$$\|\int_0^t T(t-s)(f(s+\sigma_k,x) - f(s+\sigma_l,x))\|\,ds \le M_{J_a}|b_K(\sigma_k) - b_K(\sigma_l)| ,$$

and since b_K is of bounded variation on $[0,\infty)$, $|b_K(\sigma_k) - b_K(\sigma_l)| \to 0$ as $\sigma_k, \sigma_l \to \infty$. Moreover, $\lim_{k\to\infty} \int_0^t T(t-s) g_1^{\sigma_k}(s,x)\, dU^{\sigma_k} = 0$. Since $\int_0^t T(t-s) f(s,x(s))\, ds$ is continuous, it follows that $e_x(t)$ is continuous in $[0,\infty)$. Uniqueness of the limiting $\mathcal{H}$-operator Φ implies that any translate $\Phi^\tau = \Phi$, so $e_x(2t) = 2e_x(t)$. Since $e_x(t)$ is continuous, a standard argument gives $e_x(t) = te_x(1)$. Put $f_0(x) = e_x(1)$. Then $(\Phi_0 x)(t) = tf_0(x)$. The continuity of Φ_0 on $BV(J_0, X)$ implies that f_0 is a continuous function on X, and $D(f_0) = X$. Define $f_1 : D(A) \to D(A)$ by $e_x(t) = \int_0^t T(t-s) f_1(x)\, ds$. Then

$$\begin{aligned} f_0(x) &= \lim_{t\to 0+} t^{-1} \int_0^t T(t-s) f_1(x)\, ds \\ &= T(0) f_1(x) + \lim_{t\to 0+} \int_0^t T(t-s) A f_1(x)\, ds \\ &= f_1(x) \,. \end{aligned}$$

Let $y(\cdot) \in BV(J_0, X)$ and suppose that $0 = t_0 < t_1 < \cdots < t_k = t$, $0 \le t \le \tau$, is a uniform partition of $[0,t]$. For $k = 1, 2, \cdots$, define $y_k(s) = y(t_j)$, $t_j \le s < t_{j+1}$, $j = 0, 1, \cdots, k$. Then $y_k \to y$ uniformly in $BV([0,t], X)$. Since y_k is piecewise constant,

$$\begin{aligned} (\Phi_0 y_k)(t) = \sum_{j=1}^{k} (\Phi_{t_{j-1}} y_k)(t_j) &= \sum_{j=1}^{k} \int_{t_{j-1}}^{t_j} T(t-s) f_0(y_k)\, ds \\ &= \sum_{j=1}^{k} \int_{t_{j-1}}^{t_j} T(t-s) f_0(y(t_{j-1}))\, ds \end{aligned}$$

Consequently, for k sufficiently large

$$\left\| \int_0^t T(t-s) f(s+\sigma_k, y_k(s))\, ds - \sum_{j=1}^{k} \int_{t_{j-1}}^{t_j} T(t-s) f_0(y(t_{j-1}))\, ds \right\| < \epsilon/4 \,.$$

But $\{\int_0^t T(t-s) f(s+\sigma_k, y_k(s))\, ds\}$ is a Cauchy sequence, since

$$\begin{aligned} &\left\| \int_0^t T(t-s) f(s+\sigma_k, y_k(s))\, ds - \int_0^t T(t-s) f(s+\sigma_l, y_l(s))\, ds \right\| \\ &\qquad \le \left\| \int_0^t T(t-s)[f(s+\sigma_k, y_k(s)) - f(s+\sigma_l, y_k(s))]\, ds \right\| \\ &\qquad + \left\| \int_0^t T(t-s)[f(s+\sigma_l, y_k(s)) - f(s+\sigma_l, y_l(s))\, ds \right\| \\ &\qquad \le M_J |b_K(\sigma_k) - b_K(\sigma_l)| + M_J L \|y_k - y_l\| \,, \end{aligned}$$

for k, l sufficiently large. Thus $\|\int_0^t T(t-s)f(s+\sigma_k, y(s)) - (\Phi_0 y)(t)\| \leq \epsilon/4$. Similar bounds hold for $\|\sum_{j=1}^{k}\int_{t_{j-1}}^{t_j} T(t-s)f_0(y(t_{j-1}))\,ds - \int_0^t T(t-s)f_0(y(s))\,ds\|$ and $\|\int_0^t T(t-s)f(s+\sigma_k, y_k(s))\,ds - \int_0^t T(t-s)f(s+\sigma_k, y(s))\,ds\|$. Hence $\|(\Phi_0 y)(t) - \int_0^t T(t-s)f_0(y(s))\,ds\| \leq \epsilon$.

Necessity : suppose that $f^\sigma\,dt + g^\sigma\,dU^\sigma(t) \to g_2\,dt$ as $\sigma \to \infty$. As $f \in \mathcal{F}_\infty$, as in the proof of sufficiency there exists a continuous function $f_0 : X \to X$ such that $\int_0^t T(t-s)(f(s+\sigma_k, x) - f_0(x))\,ds \to 0$ uniformly on compact subsets of $\mathbf{R}^+ \times X$. Set $g_0 = g_2 - f_0$. Then $\int_0^t g_1^{\sigma_k}(s,x)\,dU^{\sigma_k}(s) \to 0$ as $k \to \infty$. Put $\nu_K(\sigma) = \sup_{x\in K, t\in[0,1]} \|\int_0^t g_1^\sigma(s,x)\,dU^\sigma(s)\|$. Then $\nu_K(\sigma) \to 0$ as $\sigma \to \infty$, while $\int_\sigma^{t+\sigma} g_1(s,x)\,dU(s) = \int_0^t g_1^\sigma(s',x)\,dU^\sigma(s') \leq \nu_K(\sigma)$ for every $(t,x) \in [0,1]\times K$ and $\sigma \in \mathbf{R}^+$. Hence g_1 is U-diminishing.

Definition. 5.8 Let the measure evolution equation (3.1) have a unique limiting equation of the form $dx/dt + Ax = f_0(x)$. An element $\phi \in X$ is said to be an equilibrium of (3.1) if it is a solution of $Ax - f_0(x) = 0$.

Proposition 5.9. *Let $T(t)$ be a BC_0-semigroup satisfying $\|T(t)\| \leq Me^{-\rho t}$, $t > 0$, for constants $M \geq 1$, $\rho > 0$. Let $f \in \mathcal{F}_\infty$, $g \in \mathcal{G}_{\mathbf{R}^+,U}$. If for all $x \in K$, $\lim_{t\to\infty} f(t,x) = f_1$, $\lim_{t\to\infty} g(t,x) = 0$, then $x(t)$, the mild solution of (3.1), satisfies $\lim_{t\to\infty} x(t) = A^{-1}f_1$. Moreover, this is an equilibrium solution.*

Proof: Observe that the limiting equation is unique under these assumptions. Then the proof is standard (see [11], Theorem 4.4.4).

Note: The measure evolution equation induces a system analogous to a flow, and it is the concept of the resultant generalised semidynamical system which lies behind the results of this section. Definition of the flow requires the idea of a convergence structure on the spaces $\mathcal{F}_\infty$, $\mathcal{G}_{\mathbf{R}^+,U}$, and some of these matters are sketched in Appendix 1.

6. Stability of Zero Solution

It may be assumed that $\phi(t) = 0$ is a solution of (3.1), and that $f(t,0)$, $g(t,0) = 0$. For, if $\psi(t)$ is any solution of (3.1), redefine $f(t,x) := f(t, x+\psi(t)) - f(t,x)$, $g(t,x) := g(t, x+\psi(t)) - g(t,x)$. It is not difficult to see that the spaces $\mathcal{F}_\infty$, $\mathcal{G}_{\mathbf{R}^+,U}$ are invariant under such changes. Moreover, the zero solution is then a solution. In this section, we examine its stability. The following fixed point theorem is helpful:

Theorem 6.1 . *(Krasnoselskii) Let B be a closed bounded subset of a Banach space. Suppose that $P : B \to B$ is a contraction mapping and $S : B \to B$ is a compact map. Then $P + S$ has a fixed point in B.*

Our first result gives conditions under which the zero solution is stable, in the sense that at least one solution stays close to it. Write $B_\eta = \{x \in BV^+(\mathbf{R}^+, X) : ||x||_{BV} \leq \eta\}$.

Theorem 6.2 . *Let $T(t)$ be a BC_0-semigroup. Let $f \in \mathcal{F}_\infty$, $g \in \mathcal{G}_{\mathbf{R}^+,U}$ and let $f(t,0) = g(t,0) = 0$ for all $t \in [0,\tau]$. Suppose further that, for every $\epsilon > 0$ there is an $\eta > 0$ such that*

$$||g(t,x) - g(t,y)|| \leq \epsilon||x-y|| \ , \ ||f(t,x)|| \leq \epsilon||x||$$

whenever $||x||, ||y|| \leq \eta$. Then if $||z_0|| < \eta/(4M_J v_J)$, there is a mild solution $z(\cdot)$ of (3.1) such that $z(0) = z_0$ and $z(t) \in B_\eta$ for all $t \in (0,\tau]$.

Proof: Recall the form of mild solution,

$$z(t) = T(t)z_0 + \int_0^t T(t-s)f(s,z(s))\,ds + \int_0^t T(t-s)g(s,z(s))\,dU(s) \ .$$

Put $(Sz)(t) = T(t)z_0 + \int_0^t T(t-s)f(s,z(s))\,ds$, $(Pz)(t) = \int_0^t T(t-s)g(s,z(s))\,dU(s)$. First, $P+S$ maps B_η into itself for η sufficiently small, since if $0 = s_0 < s_1 < \cdots < s_N = t$ is a partition of $[0,t]$,

$$\begin{aligned}||(Sz)(s_i) - (Sz)(s_{i-1})|| &\leq ||T(s_{i-1})||\ ||T(s_i - s_{i-1}) - E||\ ||z_0|| \\ &+ \int_{s_{i-1}}^{s_i} ||T(s_i - s)||\ ||f(s,z(s))||\,ds \\ &+ \int_0^{s_i} ||T(s_{i-1} - s)||\ ||T(s_i - s_{i-1}) - E||\ ||f(s,z(s))||\,ds \ .\end{aligned}$$

Consequently, on $J = [0,t]$, for ϵ sufficiently small,

$$||Sz(\cdot)||_{BV} \leq M_J v_J ||z_0|| + tM_J\epsilon\eta + \int_0^t M_J v_J \epsilon\eta\,ds \ < \ \eta/2$$

$$||Pz(\cdot)||_{BV} \leq M_J\epsilon\eta V(U;J) + \int_0^t M_J v_J \epsilon\eta\,d|U|(s) \ < \ \eta/2 \ .$$

Now, P is a contraction on B_η, since

$$\sum_i ||(Pz)(s_i) - (Py)(s_i) - ((Pz)(s_{i-1}) - (Py)(s_{i-1}))||$$

$$\leq \sum_i [\int_{s_{i-1}}^{s_i} ||T(s_i - s)|| \, ||g(s,z(s)) - g(s,y(s))|| \, d|U|(s)$$

$$+ \int_0^{s_{i-1}} ||T(s_{i-1} - s)|| \, ||T(s_i - s_{i-1}) - E|| \, ||g(s,z(s))$$

$$- g(y(s))|| \, d|U|(s)]$$

$$\leq \int_0^t M_J \epsilon ||z - y||_{BV} \, d|U|(s) + \int_0^t M_J v_J \epsilon ||z - y||_{BV} \, d|U|(s)$$

$$\leq ||z - y||_{BV}/2$$

for ϵ small enough. Further, S is compact. To see this, let $\{z_k\}$ be a sequence in B_η. By Theorem 2.5, a subsequence $\{z_{k(i)}\}$ converges to z^*, and $||z^*||_{BV} \leq \liminf ||z_{k(i)}||_{BV} < \eta$. Thus B_η is compact. Since f is Lipschitz, S is continuous and $S(B_\eta)$ is compact. So, by Krasnoselkii's fixed point theorem, there exists $z \in B_\eta$ such that $z = Sz + Pz$, and the result is proved.

Theorem 6.3 . *Let the conditions of Theorem 6.2 hold and let $T(t)$ be a BC_0-semigroup satisfying $||T(t)|| \leq Me^{-\rho t}$, for constants $M \geq 1, \rho > 0$. Then the zero solution to (3.1) is uniformly asymptotically stable.*

Proof: Let

$$B_\eta^0 = \{x : x \text{ is a mild solution in } B_\eta \text{ on } J_0 = [t_0, t] \text{ with } x(t_0) = x_0 \, , \, ||x_0|| \leq \eta\} \, .$$

Note that $B_\eta^0 \subset B_\eta$ and is closed, and hence compact. Place $(P_1 x)(t) = \int_{t_0}^t T(t-s)g(s,x(s))\,dU(s)$, $(S_1 x)(t) = T(t-t_0)x_0 + \int_{t_0}^t T(t-s)f(s,x(s))\,ds$. The previous estimate for P is modified as

$$||P_1 z - P_1 y||_{BV} \leq Me^{-\rho(t-t_0)}\epsilon(1 + v_{J_0})V(|U|; J_0) \, ||z - y||_{BV} \, .$$

A straightforward calculation gives

$$||S_1 z - S_1 y||_{BV} \leq Me^{-\rho(t-t_0)}(v_{J_0} + (t-t_0)L + (t-t_0)Lv_{J_0}) \, ||z - y||_{BV} \, .$$

So $P_1 + S_1$ is an exponential contraction of B_η^0 as $t \to \infty$, and the result follows.

7. The Age–Dependent Equation

Consider the example given in equation (1.3′), together with the resource equation and boundary and initial conditions of (1.3). For simplicity of exposition, suppose that $t_k = k\sigma_0$, $k = 1, 2, \cdots$, where σ_0 is a constant and that the system has, in some sense, a natural solution. By this we mean that there is an equilibrium (or possibly a periodic solution), along the lines of [12],[13], for example. Assume that it is possible to linearise around this solution. This assumption is not trivial, and demands further conditions on the functions f, g since these in general may not be differentiable, or formal linearisation only densely defined in X. Some of these problems are more completely discussed in [13]. Suppose that the linearisation, expressed in terms of the deviations $z(t, a)$ from the grazer solution, and $r(t)$ from the resource solution, is of the form

$$dz + (\partial z/\partial a + \overline{\mu} z)\, dt + \overline{h}(t) \sum_{k=1}^{\infty} (\delta(t_k)\, dt) = 0 \tag{7.1a}$$

$$z(t, 0) = \overline{g} r(t) + \overline{b} \int_{\alpha}^{\alpha+\gamma} z(t, a)\, da \tag{7.1b}$$

$$dr/dt = -\, Fr(t) - Bn(t) + Cy(t) \tag{7.1c}$$

Here $n(t)$ is the total population as measured from the equilibrium solution, $n(t) = \int_0^{\infty} z(t, a)\, da$. The constants determined by the linearisation are $\overline{\mu}, \overline{g}, F, B, C$, the birthrate $b(a, R) = \overline{b}$ is assumed constant over the breeding age window $(\alpha, \alpha + \gamma)$, while $\overline{h}(t)$ is the form the impulse takes measured from the equilibrium level. For expliciteness and simplicity, suppose that $\overline{h}(t) = N$ is a constant function. Adapting an argument of Oster and Takahashi [9], the Laplace-Stieltjes transform of (7.1a) is $(\overline{\mu} + s)Z(s, a) = dZ/da + \sum_k Ne^{-sk\sigma_0}$. Solving from (7.1b,c), using capitals Z, R, Y to denote transforms of $z(\cdot, a), r(\cdot), y(\cdot)$,

$$Z(s, a) = aNe^{-s\sigma_0}[(s + \mu)(1 - e^{-s\sigma_0})]^{-1} + e^{-(s+\overline{mu})} H(s)R(s) \tag{7.2a}$$

$$H(s) = \overline{g}[1 - b(1 - e^{-(s+\overline{\mu})\gamma})e^{-(s+\overline{\mu})\alpha}/(s + \overline{\mu})]^{-1} \tag{7.2b}$$

$$R(s) = CY(s)[s + F + BH(s)/(s + \overline{\mu})]^{-1} . \tag{7.2c}$$

The characteristic equation of the linearised system is given by the denominator of (7.2c) as

$$s + F + BH(s)/(s + \overline{\mu}) = 0 .$$

There are simple roots $s = -A, -\overline{\mu}$, and a set of complex roots. The structure of this set for various ranges of parameters induces a rich and varied qualitative behaviour for the system. Different forms of bifurcation seem possible, as well as chaotic evolution. The simple case of $b(a) = b^*\delta(a-\alpha)$ (point birth) in nonimpulsive situations has been considered in [8], where a biological interpretation of bifurcation was used to explain the age-distributed "waves" seen in some laboratory predator-prey interactions.

Acknowledgements: This research was done while I was on leave, at the Mathematics Institute, Warwick University and Mathematics Department, Leicester University. Thanks are due to both for providing a stimulating environment, and generous provision of academic facilities. I am grateful to K.L. Teo for pointing out references in the control literature.

REFERENCES

[1] Z. Artstein, 1977. Topological dynamics of ordinary differential and Kurzweil euations, *J. Differential Equations* **23**, pp. 224–243.

[2] Z. Artstein, 1977. The limiting equations of nonautonomous ordinary differential equations, *J. Differential Equations* **25**, pp. 184–20.

[3] J.M. Ball, 1977. Strongly continuous semigroups, weak solutions and the variation of constants formula, *Proc. Amer. Math. Soc.* **63**, pp. 370–373.

[4] T.S. Chew, 1988. On Kurzweil generalized ordinary differential equations, *J. Diff. Eqs.* **76**, pp. 286–293.

[5] L.M. Graves, 1946. *The Theory of Functions of Real Variables*, McGraw-Hill New York.

[6] M.E. Gurtin and R.C. MacCamy, 1974. *Nonlinear age-dependent population dynamics*, *Arch. Rat. Mech. Anal.* **54**, pp. 281–300.

[7] J. Kurzweil, 1958. Generalized ordinary differential equations, *Czechoslovak Math. J.* **8**, pp. 360–389.

[8] G. Oster and J. Guckenheimer, 1975. Bifurcation Phenomena in population models, in J.E. Marsden and M. McCracken (Eds.), *The Hopf Bifurcation and its Applications*, Springer-Verlag, New York

[9] G. Oster and Y. Takahashi, 1974. Models for age-specific interactions in a periodic environment, *Ecol. Monogr.* **44**, pp. 483–501.

[10] S.G. Pandit and S.G. Deo, 1982. Differential Systems Involving Impulses, *Lecture Notes in Mathematics No. 954*, Springer-Verlag Berlin.

[11] A. Pazy, 1983. *Semigroups of Linear Operators and Applications to Partial Differential Equations*, Springer-Verlag New York.

[12] J. Pruss, 1983. On the qualitative behaviour of populations with age-specific interactions, *Comp. and Maths. with Appls.* **9**, pp. 327–339.

[13] G.F. Webb, 1985. *Theory of nonlinear age-dependent population dynamics*, Marcel Dekker, New York.

Appendix 1 : Impulsive Semidynamical Systems

Recall the definition of a semidynamical system ([A13] and its comprehensive bibliography):

Definition. A1.1 The pair $(\mathcal{Y}, \pi)$ is called a continuous semidynamical system if $\mathcal{Y}$ is a Hausdorff topological space and $\pi : \mathbf{R}^+ \times \mathcal{Y} \to \mathcal{Y}$ satisfies

(i) $\pi(0, y) = y$ for each $y \in \mathcal{Y}$;

(ii) $\pi(s, \pi(t, y)) = \pi(s + t, y)$ for each $y \in \mathcal{Y}$ and $t, s \in \mathbf{R}^+$;

(iii) π is continuous in the product topology of $\mathbf{R}^+ \times \mathcal{Y}$.

If (ii) is weakened so that only maximal intervals of existence (t_0, ω) , $\omega < \infty$, are used (see Proposition 5.9), then the pair $(\mathcal{Y}, \pi)$ is called a local semidynamical system. Without a great loss of generality ([A13], ch.4), such systems can always be considered as global, i.e. $\omega = \infty$, by reparametrising if necessary. Of much more interest in the context of this paper is weakening (iii). Because mild solutions are in $BV^+(\mathbf{R}^+, X)$, rather than in a space of continuous functions, continuity in t is inappropriate. LaSalle [A7] defined the more general notion of a D_1-system, using the notion of a convergence space (or $\mathcal{L}^*$ space, see Kuratowski [A5],Vol.1,pp.188 ff.), and we will use this.

Definition. A1.2 A set $\mathcal{Y}$ is said to be a convergence space if to certain sequences $\{y_n\}$ in $\mathcal{Y}$, called the convergent sequences, there corresponds a limit y such that

(i) $y_n \to y$ implies $y_{n(k)} \to y$ for each subsequence $\{y_{n(k)}\}$ of $\{y_n\}$;

(ii) if, for every n , $y_n = y$, then $y_n \to y$;

(iii) if $y_n \not\to y$, then it contains a subsequence $\{y_{n(k)}\}$ none of whose subsequences converges to y.

Definition. A1.3 $(\mathcal{Y}, \pi)$ is called a D_1-system if, in Definition A1.1, condition (iii) is replaced by

(iii)$_1$ $\mathcal{Y}$ is a convergence space and, if $y_n \to y$, then $\pi(t, y_n) \to \pi(t, y)$ for each $t \in \mathbf{R}^+$.

Uniform convergence of continuous functions, and the Helly-Bray theorem, respectively translate as :

Lemma A1.4 . *$\mathcal{F}_{\mathbf{R}^+}$ is a convergence space with "$\to$" defined by uniform convergence on compact subsets of $\mathbf{R}^+ \times X$.*

Lemma A1.5 . *$\mathcal{G}_{\mathbf{R}^+,U}$ is a convergence space with "$\to$" defined by $g_n \to g$ iff for each $(t, x) \in \mathbf{R}^+ \times X$*

$$\lim_{n\to\infty} \int_0^t T(t-s) g_n(s,x)\, dU(s) = \int_0^t T(t-s) g(s,x)\, dU(s) \ ,$$

for every BC_0-semigroup $T(t)$, $t \geq 0$.

Denote $\mathrm{tran}(U) = \{U^t : t \geq 0\}$. If $U \in BV^+$ $(\mathbf{R}^+, \mathbf{R})$, then $\mathrm{tran}(U)$ is precompact. The semidynamical system is set up on the space $\mathbf{R}^+ \times \mathcal{Y}$, where $\mathcal{Y} = X \times \mathcal{F}_\infty \times \mathcal{G}_{\mathbf{R}^+,U} \times \mathrm{tran}(U)$ as follows : let $x(t)$ be a mild solution of (3.1) and put

$$\pi(\tau, x, f, g, U) = (T(\tau)x, f^\tau, g^\tau, U^\tau) \ .$$

The semigroup property is obvious, and clearly $\pi(0, \cdot, \cdot, \cdot, \cdot)$ is the identity on $\mathcal{Y}$. It remains to check (iii)$_1$ holds.

Lemma A1.6 . *For each $t \in \mathbf{R}^+$, $x(t)$ is continuous in $X \times \mathcal{F}_\infty \times \mathcal{G}_{\mathbf{R}^+,U}$.*

Proof: Let the sequence of functions $z_n \to z_0$ equivariationally in BV^+ $([0,t], X)$, let $f_n \to f_0$ in $\mathcal{F}_\infty$ and $g_n \to g_0$ in $\mathcal{G}_{\mathbf{R}^+,U}$. If $z_0 \in K$, a compact subset of X, for n sufficiently large all values $z_n(\tau) \in K$, $0 \leq \tau \leq t$. Hence

$$\begin{aligned} &\left\| \int_0^t T(t-s)[f_n(s, z_n(s)) - f_0(s, z_0(s))]\, ds \right\| \\ &\quad \leq \left\| \int_0^t T(t-s)[f_n(s, z_n(s)) - f_0(s, z_n(s))]\, ds \right\| \\ &\quad + \left\| \int_0^t T(t-s)[f_0(s, z_n(s)) - f_0(s, z_0(s))]\, ds \right\| \ . \end{aligned}$$

The first integral on the right is $< \epsilon$ for n so large that $||f_n - f_0||_\infty < \epsilon/M_{[0,t]}$, while the second for $||z_n - z_0||_{BV} < \epsilon/(LM_{[0,t]})$. Then also, $||T(t)(z_n - z_0)||_{BV} < v_{[0,t]}\epsilon/L$. Moreover, let $y(s)$ be a piecewise X-valued function on a partition $0 = s_0 < s_1 < \cdots < s_N = t$ which is a uniform approximation to $z_0(\cdot)$

$$
\begin{aligned}
&|| \int_0^t T(t-s)[g_n(s, z_n(s)) - g_0(s, z_0(s))]\, dU(s)|| \\
&\qquad \leq || \int_0^t T(t-s)[g_n(s, z_n(s)) - g_n(s, y(s)]\, dU(s)|| \\
&\qquad\quad + || \int_0^t T(t-s)[g_n(s, y(s)) - g_0(s, y(s))]\, dU(s)|| \\
&\qquad\quad + || \int_0^t T(t-s)[g_0(s, y(s)) - g_0(s, z_0(s))]\, dU(s)\ .
\end{aligned}
$$

By condition (iii) of Definition 4.2, the first integral is $< h_{g_n,K}(t) \max_{0 \leq s \leq t} ||z_n(s) - y(s)||$ (compare with equation (4.5)), and is small if y is a good approximation and n is sufficiently large. Similarly, the third integral is small if y is close to z_0. If $y_i = y(s)$ is a constant K-valued function on the subinterval $[s_{i-1}, s_i]$, then the convergence structure in $\mathcal{G}_{\mathbf{R}^+,U}$ gives $\int_{s_{i-1}}^{s_i} T(t-s)[g_n(s, y_i) - g_0(s, y_i)]\, dU(s) \to 0$. Thus , if y is piecewise constant, the second integral is a finite sum of N such terms, each of which $\to 0$ as $n \to \infty$, because $g_n \to g_0$ in $\mathcal{G}_{\mathbf{R}^+,U}$.

These two lemmas, together with the semigroup property and identity at $t = 0$ give

Theorem A1.7 . *For each $t \in \mathbf{R}^+$, $\pi(t, \cdot) : \mathcal{Y} \to \mathcal{Y}$ is continuous, and $(\mathcal{Y}, \pi)$ is a D_1-system.*

Finally, precompactness of $\mathcal{Y}$ can be shown. This involves an easy proof for the compactness of $\mathcal{F}_\infty$, but the compactness of the $\mathcal{G}$-space requires a further equibounded growth assumption on the functions $h_{g,K}$ of (iii), Definition 4.2. The proof is relatively long and will not be discussed here. However, this then sets up all the prerequisites and machinery for an invariance principle and stability theory for abstract measure evolution equations.

Appendix 2 : Bibliographic Notes and Comments

Section 2. Spaces $BV(I,X)$ have been discussed previously in the 1930s by Clarkson [A4], Bochner and Taylor [A2]. Clarkson showed that ***uniform convexity*** of X was a sufficient condition for $f \in BV(I,X)$ to be Fréchet differentiable almost everywhere in I. For example, if $X = L^p([0,1])$, $p > 1$, or l^p, $p \geq 1$, these spaces are uniformly convex and BV-functions $I \to X$ have the differentiabilty a.e. property. But without uniform convexity, even if X be separable, the property need no longer hold, even in the weak sense (i.e. $(f(t+h) - f(t))/h$ does not converge weakly, as $h \to 0$). Consider the example ([A4], p.405) of $f : [0,1] \to L^1([0,1])$ given by

$$f(t) = \phi_t(s) = \begin{cases} 1\,, & s \leq t \\ 0\,, & s > t. \end{cases}$$

It is easy to see that $f \in BV([0,1], L^1([0,1]))$, but that $\lim_{h\to 0}(f(t+h) - f(t))/h$ does not exist for any $t \in (0,1)$.

Bochner and Taylor were concerned with what might be called the Bochner- Stieltjes integral and the relationship between integrals and absolutely continuous functions for spaces of X-valued functions, and studied spaces of type $BV(I,X)$ among others.

Because we seek mild, rather than strong (i.e. differentiable) solutions to measure evolution equations, our conditions are more relaxed than those of [A2],[A4]. However, the natural spaces for many important semigroups are uniformly convex. Consequently, in such cases mild solutions would be expected to have some differentiability properties, if only in the weak sense.

Section 3. Measure differential equations have been considered on $\mathbf{R}^n$, and details, together with many references, are given by Pandit and Deo [10]. Impulses in functional differential equations have been studied by Bainov and his colleagues (e.g.[A10]), but the inherent infinite dimensionality of the problem was not exploited.

Cesari [A3] has considered optimality problems and ordinary differential equations whose solution involves BV discontinuous functions. Recently, Ahmed [A1] studied control of evolution systems with measure-valued controls. The notion of a semigroup of bounded variation (BC_0-semigroup) appears to be new.

Section 4. The conditions on $f \in \mathcal{F}_J$ can be weakened to the Carathéodory conditions without noticeably weakening the conclusion of Theorem 4.3. We have not done so for reasons of simplicity and exposition. The conditions defining $\mathcal{G}_{J,U}$ are weaker than those of Artstein [7], but stronger than Kurzweil's [7] (see the remark at end of Section 4). The proof of existence and uniqueness is shortened as a result, but still uses elements of the Kurzweil integral, on Banach spaces. Further references to this elegant and general integral are Kurzweil's original paper [A6], the books by McLeod [A8] and McShane [A9], and the clear and succinct monograph by Schwabik [A12]. All these references relate to integrals of $\mathbf{R}^n$-valued functions, but the extension to Banach space values, via Kurzweil-Stieltjes sums $S_U(g, \mathcal{P})$, is straightforward and along similar lines to [A2].

Section 5. Even though these ideas are developed in a Banach space setting, many of the techniques are essentially those of Artstein [1],[2], see also the book by Saperstone [A12] in which a very complete bibliography can be found. U-diminishing functions are a natural extension of the diminishing functions which arise in the theory of eventually asymptotic systems, see references in [A12].

Section 6. The treatment involving Krasnoselskii's fixed point theorem extends $\mathbf{R}^n$ theory of [10] to the evolution in an abstract space. All these ideas, and those in previous sections, can be further generalised :

(1) The infinitesimal generator depends on t. Then $A(t)$ generates an *evolutionary system* $U(t,s)$, satisfying a translation property $U(t,r)U(r,s) = U(t,s)$, $0 \le s \le r \le t$, $U(s,s) = E$, with the continuity extending to the map $(t,s) \to U(t,s)$. See [9], ch.5.

(2) Nonlinear semigroup theory, using accretive operators, can be used. In the age-dependent system, the nonlinear generator is then $(Ax)(t,a) = \partial x/\partial a - (Gx)(t,a)$, where $Gx = -\mu(a,x,R)x$, $x \in L^1$, $a \ge 0$ [13]. Much of the analogy with transfer functions is lost in nonlinear semigroup theory, and even the variation of constants formula (appearing in the form of mild solution) needs be replaced by the notion of *integral solution*. Pavel [A11] gives a readable account of the general theory.

REFERENCES

[A1] N.U. Ahmed, 1983. Properties of relaxed trajectories for a class of nonlinear evolution equations on a Banach space, *SIAM J. Optim. Control* **21**, pp. 953–967.

[A2] S. Bochner and A.E. Taylor, 1983. Linear functionals on certain spaces of abstractly-valued functions, *Ann. Math.* **39**, pp. 913–944.

[A3] L. Cesari, 1988. Discontinuous optimal solutions and applications, in K.L. Teo et al (Eds), *International Conference on Optimization Techniques and Applications*, Singapore, 8-10 April, 1987, University of Singapore pp. 107–121.

[A4] J.A. Clarkson, 1936. Uniformly convex spaces, *Trans. Amer. Math. Soc.* **40**, pp. 396–414.

[A5] K. Kuratowski. 1966. *Topology* Academic Press New York.

[A6] J. Kurzweil, 1957. Generalized ordinary differential equations and continuous dependence on a parameter, *Czechoslovak Math. J.* **7**, pp. 418–446.

[A7] J.P. LaSalle, 1976. Stability theory and invariance principles, in L. Cesari, J.K. Hale and J.P. LaSalle (Eds), *Dynamical Systems*, Vol. 1, Academic Press, New York pp. 211–222.

[A8] R.M. McLeod, 1980. *The Generalized Riemann Integral*, The Mathematical Association of America, Carus Mathematical Monographs.

[A9] E.J. McShane. 1983. *Unified Integration* Academic Press New York.

[A10] S.D. Milusheva and D.D. Bainov, 1982. Justification of the averaging method for a class of functional differential equations with impulses, *J. Lond. Math. Soc.* **25**, pp. 309–331.

[A11] N.H. Pavel, 1987. Nonlinear Evolution Operators and Semigroups, *Lecture Notes in Mathematics 1260* Springer-Verlag, Berlin.

[A12] S. Schwabik, 1985. *Generalized Differential Equations* , Rozpravy Ceskoslovenske Akademie Ved, Prague.

[A13] S.H. Saperstone, 1981. *Semidynamical Systems in Infinite Dimensional Spaces*, Springer-Verlag, New York.

PARTICIPANT'S COMMENTS

Diamond's paper provides a very general theoretical framework for situations which are governed by an evolution equation with an impulsive driving term. His theory allows impulse instants to be countable in number and to form a dense subset of the time set. However, this level of generality seems to require a very abstract mathematical setting, and I wonder if it has not passed beyond the reach of population biologists. Indeed, in my naiveté I would imagine that impulsive disturbances of real biological systems will occur only at isolated instants. Consequently differential equations describing such systems should be sectionally solvable, i.e., by integrating from one impulse to the next with special consideration being given to how an impulse induces initial conditions for the subsequent time interval. Diamond's example in §7 is a linear system with evenly spaced impulses, and his use of the Laplace transform is a time–honoured and simple way of approaching such a problem.

Diamond's work may be interesting from a different direction. Over the last three decades there has been developed an L_2–theory of stochastic differential equations allowing square integrable semi-martingale driving terms. This theory admits certain classes of point processes as driving functions, as well as processes whose sample paths have finite locally bounded variation. In addition, if there is no local martingale component, then solutions of the differential equation can be pathwise constructed. Much of this theory carries over to a theory of random evolution equations. Diamond's paper could be construed as a theory of pathwise constructions for random evolution equations driven by processes of locally bounded variation, and hence there should be a strong connection with existing work, with some oppportunity for mutual reinforcement. This seems to be worth exploring.

A.G. Pakes

REPLY

I agree with Tony that it is more appealing and intuitive to think of impulsive disturbances as sequentially isolated events in time. However, this heuristic plausibility perhaps should not totally preclude the possibility of accumulation points in impulse times. Accumulations occur in parameter values for period doubling bifurcations in quite simple mathematical systems, such as

$$x_{n+1} = rx_n(1-x_n)\ ,\ n = 0,1,2,\cdots,\ 0 < r < 4\ .$$

It is not impossible to visualise a system where something of this sort is an interconnected component of a larger system, where the bifurcations produce impulses at times t_k, condensing to some sort of chaos at finite time t_∞.

Nevertheless, isolated impulses of an amenable form can be sectionally solved, as Tony suggests, to give relatively explicit results. As an example, let $t_1 < t_2 < \cdots < t_k < \cdots$ be isolated, increasing impulse times, $t_k \to \infty$ as $k \to \infty$, and suppose that $U(t) = t + \sum_{k=1}^{\infty} c_k H_{t_k}(t)$, where $H_p(t)$ denotes the unit step function, zero for $t < p$ and 1 if $t \geq p$. Then the corresponding measure is $dU(t) = dt[1 + \sum_{k=1}^{\infty} c_k \delta(t - t_k)]$ where $\delta(\cdot)$ is the Dirac delta function with unit weight at 0. Consider the measure equation

$$dx + Ax\, dU(t) = f(t,x)\, dt + g(t,x)\, dU(t) \ , \ x(0) = x_0 \ .$$

This is very similar to (1.1′), but there is an explicit impulse added to the linear infinitesimal generator term. Since the t_k are isolated, for each $t > 0$ there is a unique $k > 0$ such that $t_{k-1} \leq t < t_k$. Assume that for each k, $B_k = E - c_k A$ is invertible (i.e. $1/c_k$ is in the resolvent set $\rho(A)$ of A, for all k). Then it can be shown that

$$x(t) = T(t-t_k)x(t_k) + \int_{t_k}^{t} T(s-t_k) f(s,x(s))\, ds + \int_{t_k}^{t} T(s-t_k) g(s,x(s))\, dU(s)$$

for $t_k \leq t < t_{k+1}$, $k = 1, 2, \cdots$ and where

$$x(t_k) = [\prod_{i=1}^{k} B_{k-i}^{-1} T(t_{k-i} - t_{k-i-1})] x_0 + \sum_{i=1}^{k} c_i [\prod_{j=1}^{k-i} B_{k-j}^{-1} T(t_{k-j} - t_{k-j-1})](I_i + J_i)$$

and

$$I_i = \int_{t_{i-1}}^{t_i} f(s,x(s))\, ds \ , \ J_i = \int_{t_{i-1}}^{t_i} g(s,x(s))\, dU(s) \ .$$

There may indeed be some relation with, and possible extension to the theory of stochastic evolution equations, and investigation along the suggested lines could well be worthwhile. Of possible relevance here are :

[R1] N.U. Ahmed, 1987. Abstract stochastic evolution equations and related control and stability problems, *Lecture Notes in Control and Information Sciences* **97**, pp. 107–120. Springer–Verlag, New York.

[R2] Yu.A. Rozanov, 1988. Some boundary value problems for generalized PDE, Bielefeld–Bochum–Stochastic, preprint No.313

However, these papers may address the topic from a different point of view and seem only concerned with Wiener measure rather than general processes of locally bounded variation. Since I am not familiar with the latter theory, although further investigation does look interesting, I can offer no more comment at this stage.

Phil Diamond

Scaling as a Tool for the Analysis of Biological Models

D. L. S. MCELWAIN

Abstract

Many mathematical models of biological systems lead to differential equations. Although numerical methods are available to solve these equations, the introduction of dimensionless variables often simplifies the form of the equations. In addition, if scaled dimensionless variables are introduced then it is sometimes possible to obtain useful approximations to the solution using, for example, perturbation methods. A discussion of the use of these techniques is illustrated by examining a model developed by Volterra for population growth in a closed system.

Derivation of the Volterra Equation for a Closed System

We consider a population growing in a closed system. We assume that, *in the absence of any effects due to the fact that the system is closed*, the population density u is governed by the logistic equation

$$\frac{du}{dt} = au - bu^2 \tag{1}$$

where a and b are constants and t is the time. We now extend this model by assuming that the effect of the closed environment is to allow toxins to build up and that the organism whose population density is u has a per capita death rate which is proportional to the toxin concentration v. *For the closed system* , Equation (1) becomes

$$\frac{du}{dt} = au - bu^2 - \lambda uv \tag{2}$$

where λ is a constant. If we further assume that the toxin is produced by the organism itself at a rate proportional to the organism population density then

$$\frac{dv}{dt} = \mu u \tag{3}$$

where μ is a constant. We assume that the system is toxin–free at time $t = 0$, so that $v(0) = 0$.

If we integrate Equation (3) and substitute this into Equation (2) we obtain

$$\frac{du}{dt} = au - bu^2 - cu \int_0^t u(\tau)d\tau, \quad u(0) = u_0 \tag{4}$$

where we have denoted the initial value of u by u_0 and we have written $c = \lambda\mu$.

An alternative interpretation for Equation (4) is based on the idea of depletion of resources. We assume that the system has a resource which is essential to the organism and that the population density is given by

$$\frac{du}{dt} = \gamma u(R - \beta u) \tag{5}$$

where R is a measure of the abundance of the resource and γ and β are constants. If R is constant, this equation displays the same features as the logistic equation. However, here we assume that this resource is consumed by the organism at a rate proportional to its population density so that

$$\frac{dR}{dt} = -\alpha u \tag{6}$$

where α is a constant. If we denote the initial resource abundance in the closed system by R_0 then integrating equation (6) and substituting for R(t) in equation (5) gives

$$\frac{du}{dt} = \gamma u \left(R_0 - \alpha \int_0^t u(\tau)d\tau - \beta u \right)$$

or

$$\frac{du}{dt} = au - bu^2 - cu \int_0^t u(\tau)d\tau$$

where we have put $a = \gamma R_0$, $b = \beta\gamma$ and $c = \alpha\gamma$.

For convenience of presentation we will limit our discussion to the toxin build–up model.

Equation (4) has three parameters. The parameter a is the intrinsic growth rate, the parameter b is a measure of the effects of crowding on the population density whereas the parameter c measures the sensitivity of the death rate of the organism to the presence of toxins.

Scudo (1971) claims that Volterra used this equation as a model for the growth of an organism in a closed system.

Use of Dimensionless Variables

Equation (4) has no closed form solution, although Small(1983) points out that a "quasi-closed" form solution can be obtained by using the equivalent system (2) and (3) to eliminate t and to obtain an equation in u and v. However, this still only leads to an inverse integral form for v which is not very informative so that the behaviour of the solution of (4) for various sets of values of the parameters must be obtained by numerical methods.

A full numerical investigation would require exploring the solution for a range of values of all three parameters a, b and c together with a range of initial values u_0. To determine the dependence of the solution on a certain parameter, say a, a plausible procedure would be to fix the values of b and c and to examine the behaviour of the solution as the parameter a varies. Separate calculations would have to be performed to determine the dependence on b and c. It is obvious that a large catalogue would be needed to elucidate the solution's dependence on the three parameters and the initial value.

The use of dimensionless variables is an attempt to reduce the number of parameters.

In this problem, we introduce dimensionless variables

$$\hat{u} = \frac{u}{\bar{u}} \text{ and } \hat{t} = \frac{t}{\bar{t}} \tag{7}$$

where a quantity with a bar has the same dimensions as the corresponding physical variable.

In order to decide on appropriate values of $\bar{u}$ and $\bar{t}$ we investigate the dimensions of all the variables and parameters in Equation (4). We denote the dimensions of u and t by $[u]$ and T, respectively.

Then, by consideration of the individual terms of Equation (4) we can establish the following:

$$[a] = \frac{1}{T}$$

$$[b] = \frac{1}{[u]T} \tag{8}$$

and $$[c] = \frac{1}{[u]T^2} \; .$$

We see that one candidate for $\bar{t}$, a quantity which has the same dimensions as the time t, is $\frac{1}{a}$ but another would be $\frac{b}{c}$, for example. Similarly, candidates for $\bar{u}$ are $\frac{a}{b}$ and $\frac{a^2}{c}$. In fact, there are four different cases to consider (Small, 1983):

(A) If we choose $\bar{u} = \frac{a}{b}$ and $\bar{t} = \frac{1}{a}$, then in terms of the dimensionless variables $u_1 = \frac{b}{a}u$ and $t_1 = at$ Equation (4) becomes

$$\frac{du_1}{dt_1} = u_1 - u_1^2 - \epsilon u_1 \int_0^{t_1} u_1(\tau_1)d\tau_1 \tag{9}$$

where ϵ is a dimensionless parameter which is given by

$$\epsilon = \frac{c}{ab} \tag{10}$$

(B) With $\bar{u} = \frac{a}{b}$ and $\bar{t} = \frac{b}{c}$ then the dimensionless variables become $u_2 = \frac{b}{a}u$ and $t_2 = \frac{c}{b}t$ so that Equation (4) is transformed to

$$\epsilon\frac{du_2}{dt_2} = u_2 - u_2^2 - u_2 \int_0^{t_2} u_2(\tau_2)d\tau_2 \ ; \tag{11}$$

(C) With $\bar{u} = \frac{a^2}{c}$ and $\bar{t} = \frac{1}{a}$ we get the dimensionless variables $u_3 = \frac{c}{a^2}u$ and $t_3 = at$ and Equation (4) is transformed to

$$\frac{du_3}{dt_3} = u_3 - \frac{1}{\epsilon}u_3^2 - u_3 \int_0^{t_3} u_3(\tau_3)d\tau_3 \ ; \tag{12}$$

and finally

(D) With $\bar{u} = \frac{c}{b^2}$ and $\bar{t} = \frac{b}{c}$ we get the dimensionless variable $u_4 = \frac{b^2}{c}u$ and $t_4 = \frac{c}{b}t$ and Equation (4) is transformed to

$$\frac{du_4}{dt_4} = \frac{1}{\epsilon}u_4 - u_4^2 - u_4 \int_0^{t_4} u_4(\tau_4)d\tau_4 \tag{13}$$

Whereas the original equation had three parameters a, b and c we see that, whichever form of non–dimensionalization we choose, just the one parameter, ϵ, together with the initial value characterizes the solution. This demonstrates the fact that even if numerical methods are used to explore the solution, a good deal of simplification of the problem can be obtained by the the use of dimensionless variables. In addition, the dimensionless form of the equation is "unit free". In the original problem we had to give consideration to the dimensions of all the parameters and variables whereas the dimensionless version of the problem applies to any appropriate real situation provided a consistent set of units has been chosen for the parameters and variables.

The governing dimensionless parameter ϵ can be interpreted in terms of the original problem.

We denote a typical value of the population density by $\bar{u}$ and consider the ratio

$$\frac{\text{Death rate due to toxin if the toxin has built up for a time } \frac{1}{a}}{\text{Death rate due to crowding effects if } u = \bar{u}}$$

This can then be written as

$$\begin{aligned} & \frac{c\bar{u}\int_0^{\frac{1}{a}} \bar{u}\, d\tau}{b\bar{u}^2} \\ = & \frac{c\bar{u}\bar{u}\frac{1}{a}}{b\bar{u}^2} \\ = & \frac{c}{ab} \end{aligned}$$

which is the definition of ϵ. As we have seen above, and as we will argue below, the quantity $\frac{1}{a}$ is a measure of the time. Thus we see that ϵ is a measure of the sensitivity of the population to toxins compared with the sensitivity to crowding in the environment. Figure 1 shows the numerical solution of equation (14) for $u_1(0) = 0.1$ and various values of ϵ.

In order to illustrate an approximation method, *in the remainder of this paper we will restrict our consideration to the case in which the sensitivity to toxin is small*, that is, where the parameter ϵ is small. This means that the initial growth is dominated by crowding effects and toxins only come into play later on in the evolution of the system.

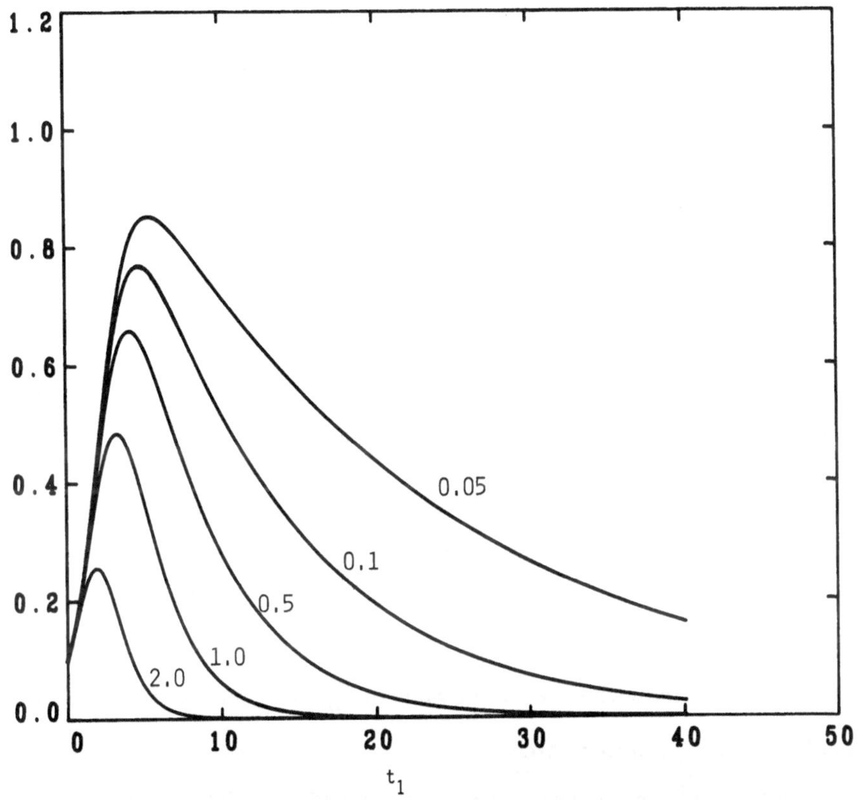

Figure 1. Numerical solution of the closed system equation (9) when $u_1(0) = 0.1$ for various values of $\epsilon = \dfrac{c}{ab}$.

Short–time Approximation

We have to decide which choice of dimensionless variables is appropriate in this case. We wish to make approximations by neglecting certain terms in the governing equation. Different terms, as we shall see, may be neglected in different parts of the time domain. In order to decide which term or terms may be dropped we have to use *scaled* dimensionless variables (Segel,1972; Segel,1984; Segel and Slemrod,1989). Following this methodology, the dependent variable, u in this case, should be scaled so that its maximum value does not exceed unity (in magnitude) and the independent variable, t in this case, should be scaled so that the dependent variable changes significantly over the characteristic time scale. This leads us to adopt the (A) scaling above for our initial investigation, namely

$$\frac{du_1}{dt_1} = u_1 - u_1^2 - \epsilon u_1 \int_0^{t_1} u_1(\tau_1) d\tau_1 \tag{14}$$

and we shall look at the case where u(0) is less than $\frac{a}{b}$ so that $u_1(0)$ is less than unity. This scaling can be justified using the following argument:

We want the population u scaled by what we expect to be its largest value, namely $\frac{a}{b}$ so that u_1 will be less than unity. To justify the time scale, we look at the method suggested by Segel (1984, P56) namely

$$\bar{t} = \frac{u_{max}}{\left(\frac{du}{dt}\right)_{max}} . \tag{15}$$

If the effect of toxins is small, we estimate $u_{max} = \frac{a}{b}$, approximately, since we expect the solution to follow the logistic equation. From the same equation we estimate the maximum value of $\frac{du}{dt}$ as $\frac{a^2}{4b}$ and thus from (15) we obtain $\bar{t} = \frac{1}{a}$, approximately.

We could also argue that since we are considering the situation where the sensitivity to toxins is small relative to the sensitivity to crowding effects we should choose a scaling in which the integral term is small. That is, one in which the parameter ϵ multiplies the integral term.

We now look at obtaining an approximate solution to equation (14). If ϵ is small then we expect the solution to be well approximated by the logistic equation

$$\frac{du_1}{dt_1} = u_1 - u_1^2$$

which has solution

$$u_1(t_1) = \frac{Be^{t_1}}{1 + Be^{t_1}} \tag{16}$$

where B is a constant given by $\dfrac{u_1(0)}{1 - u_1(0)}$. This solution, as we would expect, tends to 1 as $t_1 \rightarrow \infty$. We cannot expect it to remain a good approximation when the integral term in (14) becomes important. If we estimate the integral in the original equation (4) by putting $u = \frac{a}{b}$ then we would expect this term to become important at a time t^* such that

$$a.\frac{a}{b} \approx c.\frac{a}{b}.t^*.\frac{a}{b}$$

where we have compared the first and third terms on the right-hand side of (4). This gives $t^* \approx \frac{b}{c}$ and so for times of this order we require a new characteristic scale. The appropriate scaling for u is still $\frac{a}{b}$ since we expect this to be the maximum value. This is the time scale over which toxin effects are important. This is precisely the case(B) of the possible scalings discussed earlier, namely

$$\epsilon\frac{du_2}{dt_2} = u_2 - u_2^2 - u_2\int_0^{t_2} u_2(\tau_2)d\tau_2 \tag{17}$$

where we should also note that, as we would expect from the definition of ϵ, $t_2 = \epsilon t_1$.

Long–time approximation

Since we are assuming ϵ is small we solve Equation(17) with the term involving ϵ neglected, so that

$$0 = u_2 - u_2^2 - u_2\int_0^{t_2} u_2(\tau_2)d\tau_2 \tag{18}$$

The solution of this is

$$u_2(t_2) = e^{-t_2} . \tag{19}$$

where there is no arbitrary constant. Figure 2 shows the two approximations for $\epsilon = 0.05$ and $u_1(0) = 0.1$ together with the solution obtained by numerical integration.

As can be seen from Figure 2, although the short–time approximation is quite accurate the long–time exponential decay approximation appears to have the correct sort of behaviour but with a shift. This shift is related to the fact that the long–time approximation has been forced to satisfy the initial condition $u_2(0) = 1$, whereas the "exact" solution shows that the long-time approximation should be shifted to take account of the fact that the dimensionless solution does not start at unity. We can see that this shift must depend on the initial value for $u_1(0)$ and we have to consider two cases: B small, which corresponds to $u(0)$ small in comparison with $\frac{a}{b}$ and B large, which corresponds to $u(0)$ near, but still less than, $\frac{a}{b}$.

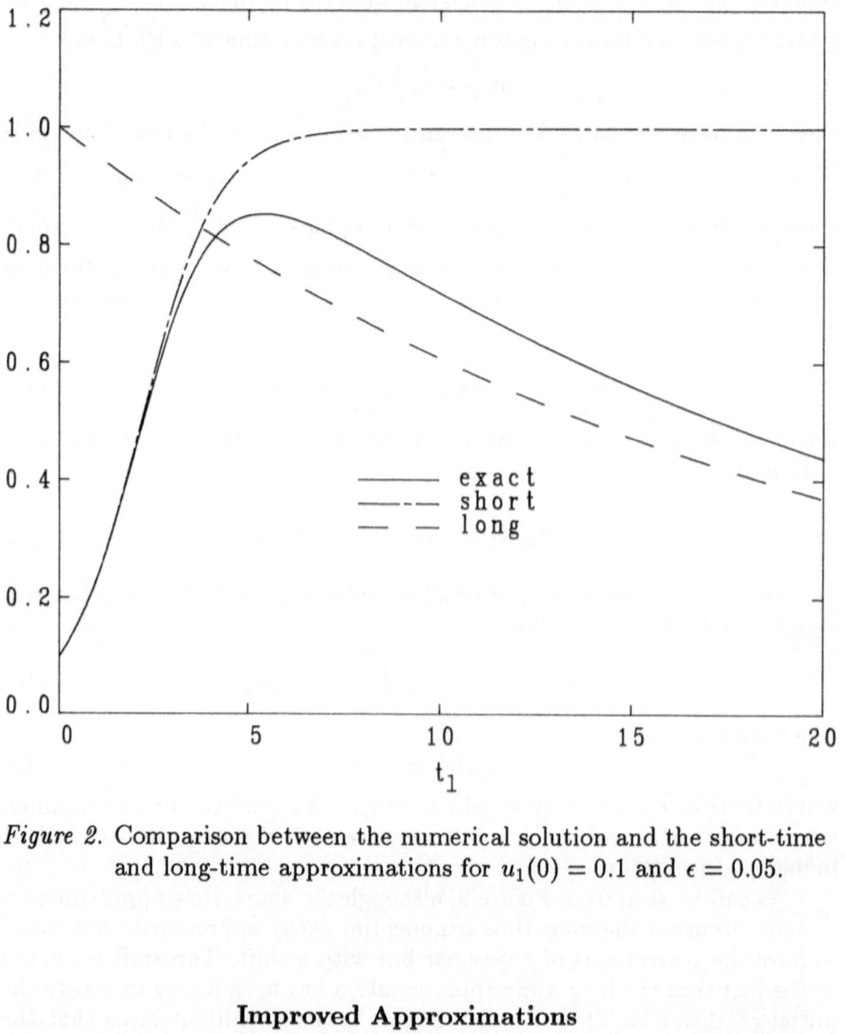

Figure 2. Comparison between the numerical solution and the short-time and long-time approximations for $u_1(0) = 0.1$ and $\epsilon = 0.05$.

Improved Approximations

Here we outline a method for improving the approximation using the method of matched asymptotic expansions (Van Dyke, 1975). The details are to be found in McElwain and Norbury (in preparation).

For small B we put

$$u_1 = \frac{Be^{t_1}}{1 + Be^{t_1}} + \epsilon v_1$$

for short times and (20)

$$u_2 = e^{-t_2} + \epsilon v_2$$

for long times where the v_i are perturbations to the zeroth order approximations found earlier. This leads to new approximations

$$u_1 = \frac{Be^{t_1}}{1 + Be^{t_1}} + \epsilon t_0 \left(1 - e^{-t_1}\right) - \epsilon t_1 \tag{21}$$

and

$$u_2 = e^{-\epsilon(t_1 - t_0)} \tag{22}$$

where $t_0 = ln(\frac{1}{B}) + 1$ and for ease of presentation we have written u_2 in terms of t_1. From this we can immediately see that the effect of the initial rise on the long-term solution is a shift which depends on the initial value. We are also able to obtain a uniformly valid approximation (Van Dyke, 1975)

$$u_v = \frac{Be^{t_1}}{1 + Be^{t_1}} + e^{-\epsilon(t_1 - t_0)} - 1 - \epsilon t_0 e^{-t_1} \tag{23}$$

Figure 3 shows a comparison of this approximation with the solution obtained by numerical methods.

When B is large, that is, $u(0)$ is near $\frac{a}{b}$, then we obtain different improved approximations, namely

$$u_1 = 1 - \left[\frac{1}{1 + B} + \epsilon\right] e^{-t_1} - \epsilon t_1 + \epsilon \tag{24}$$

for the short time scale and

$$u_2 = e^{-\epsilon(t_1 - 1)} \tag{25}$$

for the long time scale. The uniformly valid approximation obtained from these is

$$u_v = \left[\frac{1}{1 + B} + \epsilon\right] e^{-t_1} + e^{-\epsilon(t_1 - 1)} \tag{26}$$

Figure 4 compares this approximation with the solution obtained by numerical integration.

As can be seen from Figures 3 and 4 quite accurate approximations have been found by the use of scaled dimensionless variables and perturbation methods.

Finally, we should point out that the cases (C) and (D) of the scalings described earlier are appropriate when the sensitivity to toxins is large(Small,1983).

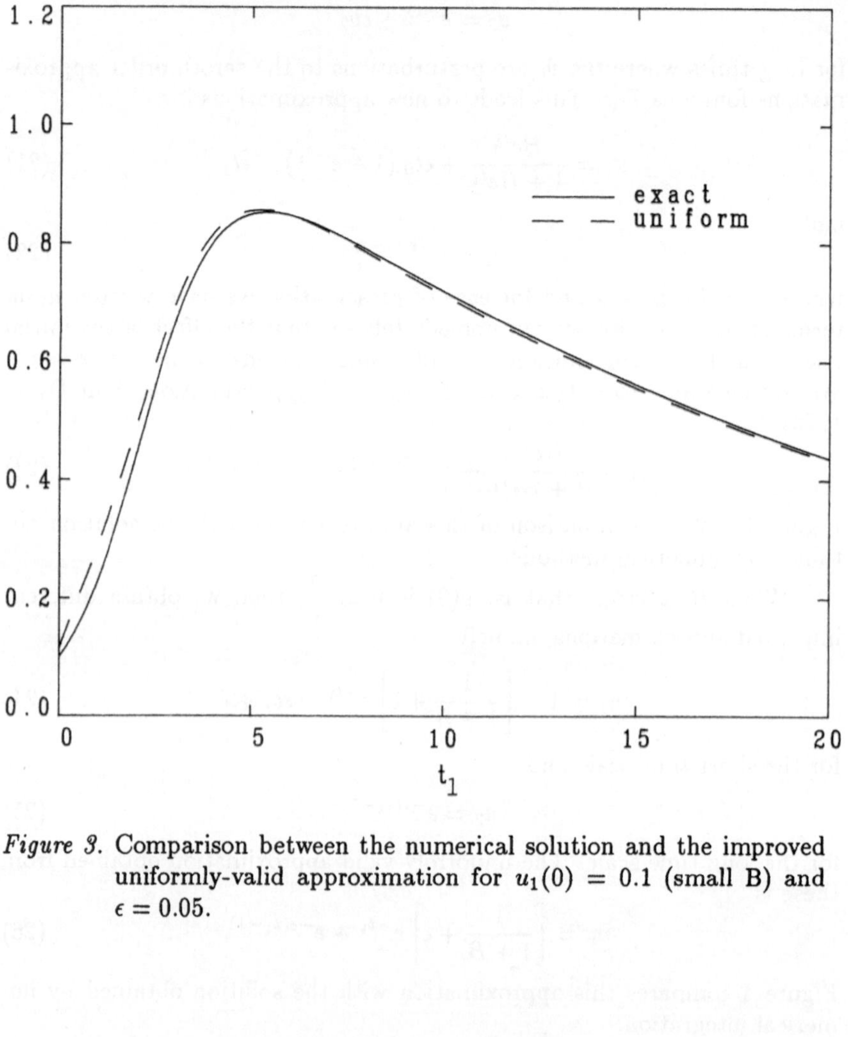

Figure 3. Comparison between the numerical solution and the improved uniformly-valid approximation for $u_1(0) = 0.1$ (small B) and $\epsilon = 0.05$.

Summary and Discussion

We have argued that one of the most important tools of biological modelling is the ability to non-dimensionalize a given model equation. This involves choosing characteristic quantities with which to scale the original variables. The choice is vitally important when we wish to obtain approximations to the solution.

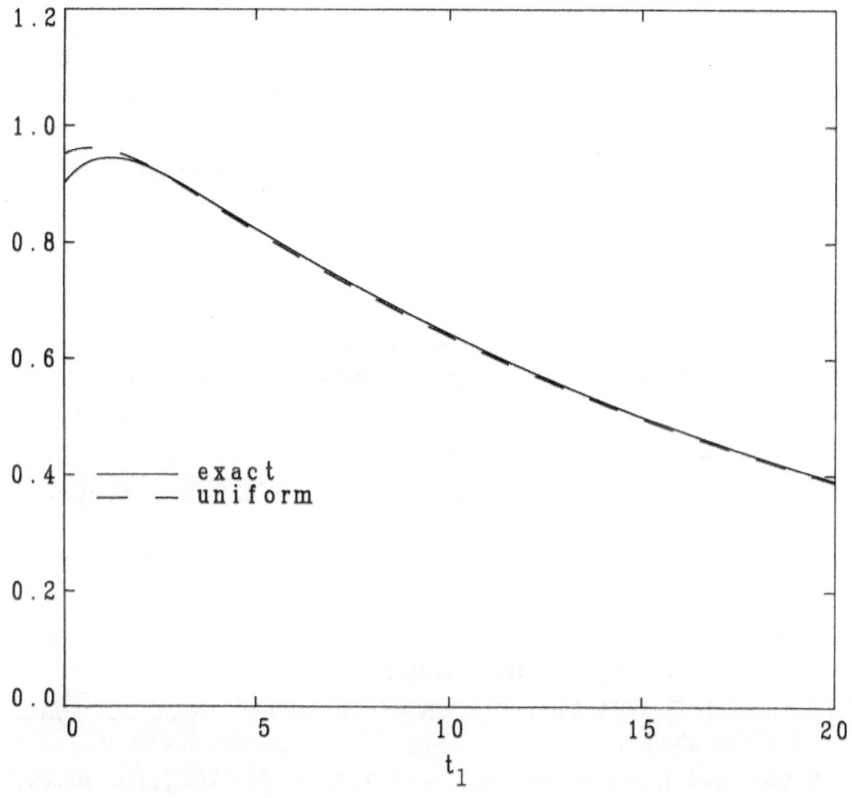

Figure 4. Comparison between the numerical solution and the improved uniformly-valid approximation for $u_1(0) = 0.9$ (large B) and $\epsilon = 0.05$.

The method has been illustrated by applying it to a problem which is said to have originated with Volterra. The use of scaled dimensionless variables and perturbation methods leads to a simple closed form approximation to the solution which is excellent over the whole time domain for low toxin sensitivity. In all cases when modelling biological phenomena, the equation should be made dimensionless. The advantages include: (a) a reduction in the number of parameters; (b) freedom from consideration of the units of the original parameters and variables; and (c) the possibility of developing good approximate solutions.

The choice of dimensionless variables should not be taken lightly, especially if approximations are sought. Quite often, as in this case, different scalings are required in different parts of the domains of the independent variables. Goodwin(1969) puts it well(the additions in brackets are mine):

"The discovery of appropriate (scaled dimensionless) variables for (mathematical) biology is itself an act of creation".

REFERENCES

[1] Goodwin, B.C. 1969. In C.H Waddington (ed.) *Towards a Theoretical Biology,* **2**, pp. 337. Edinburgh University Press: Edinburgh.

[2] Scudo,F.M. 1971. Vito Volterra and theoretical ecology, *Theoret. Pop. Biol.,* **2**, pp. 1-23.

[3] Segel, L.A. 1972. Simplification and scaling, *SIAM Rev.,* **14**, pp. 547-571.

[4] Segel, L.A. 1984. *Modeling Dynamic Phenomena in Molecular and Cellular Biology,* Cambridge University Press, Cambridge, 300 p.

[5] Segel, L.A. and M. Slemrod. 1989. *The quasi-steady state assumption: A case study in perturbation* (preprint).

[6] Small, R.D. 1983. Population growth in a closed system, *SIAM Rev.,* **25**, pp. 93-95.

[7] Van Dyke, M. 1975. *Perturbation Methods in Fluid Mechanics,* Parabolic Press, Stanford, California, 271 p.

PARTICIPANT'S COMMENTS

The author shows how a dynamical system with three parameters representing a birth rate, an over-crowding effect and a self poisoning effect, can be simplified to a canonical form with just one dimensionless parameter and dimensionless 'time'. Different values for this parameter determine the relative importance of over crowding or poisoning. The second part of the paper shows how the dynamics in time can be simplified for two different parameter ranges, each approximate system solved, and then matched across the time/parameter gap. The paper shows that much can be achieved in terms of classical analysis without having to resort to computer overkill on the original three parameter dynamical system.

Les Jennings

The moral of this paper could well be "look before you non–dimensionalise/leap". The example used to demonstrate this principle, an ordinary integro–differential equation for population growth in a closed system, is simple. The considerations on which non–dimensionalization is based can then be easily illustrated with minimal mathematical fuss.

Scaling and non–dimensionalization is an important tool in biological modelling. Most equations in this context are nonlinear. Analysis of such equations, as in this case, often requires approximation by taking one or more terms to be very small or very large. Further, model equations can be proposed in full and then simplified by neglecting terms preceded by a very small coefficient, in an analogous way to the derivation of Stokes equations for low Reynolds number fluid flow. The example given here nicely illustrates the point that it is important to use the correct scaling so that the variables in the domain of interest are of $\mathcal{O}(1)$ and so that approximations and simplifications become clear after suitable scaling.

M. R. Myerscough

A Numerical Algorithm for Constrained Optimal Control Problems with Applications to Harvesting

L.S. JENNINGS and K.L. TEO

1. Introduction

We are interested in optimal control problems with constraints involving the state and control variables for all values of the independent variable (time say) in some interval. These problems while substantially discussed in the literature from a theoretical viewpoint (see for example Cesari (1983)) and an algorithmic viewpoint (see Goh and Teo (1987), (1988), Miele (1975), Miele *et al* (1970), Miele *et al* (1986), Teo and Goh (1987), (1989), Wong *et al* (1986)) continue to be examined with the view to improving the computational algorithms in both efficiency and stability. We have chosen the control parameterization route (Goh and Teo (1988) and Teo and Goh (1989)) but we note that there is a family of gradient restoration algorithms due to Miele and his co-workers. A general purpose optimal control software package, MISER, (see Goh and Teo (1987)) has been developed based on the control parameterization technique and we use this development as the starting point to improve efficiency and stability. In the current version of MISER, a particular constraint transcription proposed by Teo and Goh (1987) is used to handle inequality continuous (in time) state constraints involving state and control variables. A disadvantage of this constraint transcription is that the usual constraint qualification is not satisfied as the gradient of the constraint is zero at all feasible solutions. Hence convergence of the numerical algorithm is not guaranteed and some oscillation can occur in computation close to the solution. Our experience indicates this is the case with the MISER software which uses the NLPQL algorithm (Schittkowski (1985)). We have noted failure to converge cleanly, that is the software is unaware it is close to the solution and the original constraints are violated by a small amount for some sub-intervals of time.

To overcome these two stability deficiencies, we propose a new constraint transcription. Numerical experience with a number of test examples demonstrates that the new algorithm is stable in the sense of converging cleanly and giving a feasible solution. We have had to put some extra conditions on the form of the constraints, in that the control functions cannot appear in the inequality continuous state constraint. Future theoretical work may overcome this disadvantage.

To show the effectiveness of the new algorithm, we compute the optimal controls for an example in Ryan (1987) using a range of parameter values which determine different management strategies of a fishery under different economic scenarios. Other examples will appear once the theoretical results are forthcoming.

2. Problem Statement

Consider a process described by the following system of differential equations on the fixed time interval $(0, T]$,

$$\dot{x}(t) = f(t, x(t), u(t)), \tag{1a}$$

where $x = [x_1, \ldots, x_n]^\mathsf{T} \in R^n$, $u = [u_1, \ldots, u_r]^\mathsf{T} \in R^r$ are, respectively, the state and control vectors; $f = [f_1, \ldots, f_n]^\mathsf{T} \in R^n$; and the superscript $^\mathsf{T}$ denotes transpose. The initial condition for the differential equation (1a) is

$$x(0) = x_0 \in R^n. \tag{1b}$$

Define

$$U \equiv \{u = [u_1, \ldots, u_r]^\mathsf{T} \in R^r \quad : \quad \alpha_i \le u_i \le \beta_i,\ i = 1, \ldots, r\} \tag{2}$$

where α_i, $i = 1, \ldots, r$, and β_i, $i = 1, \ldots, r$, are real numbers. Clearly U is compact and a convex set of R^r.

Any measureable function defined on $[0, T]$ with values in U is called an *admissible control.* Let $\mathcal{U}$ be the class of all such admissible controls.

Let L_∞^r denote the Banach space $L_\infty([0,T], R^r)$ of all essentially bounded measurable functions from $[0,T]$ into R^r. Its norm is

$$\|u\|_\infty = \operatorname*{ess\,sup}_{t\in[0,T]} \left\{ \sum_{i=1}^{r} (u_i(t))^2 \right\}^{\frac{1}{2}}.$$

For each $u \in L_\infty^r$, let $x(\cdot|u)$ be an absolutely continuous function defined on $[0,T]$ which satisfies the differential equation(1a) almost everywhere in $(0,T]$ and the initial condition (1b). This function is called the solution of the system (1) corresponding to the control $u \in L_\infty^r$.

The terminal state inequality constraints and continuous state inequality constraints are specified as follows:

$$\phi_i(x(T|u)) \leq 0, \quad i = 1, \ldots, N_T, \tag{3a}$$

where ϕ_i, $i = 1, \ldots, N_T$, are real valued functions defined on R^n, and

$$g_i(t, x(t|u)) \leq 0, \quad \forall t \in [0,T], \quad i = 1, \ldots, N, \tag{3b}$$

where g_i, $i = 1, \ldots, N$, are real valued functions defined on $[0,T] \times R^n$.

Let $\mathcal{F}$ be the set that consists of all those elements from $\mathcal{U}$ such that the constraints (3) are satisfied. *Elements from $\mathcal{F}$ are called feasible controls and $\mathcal{F}$ is called the class of feasible controls.*

We may now state the optimal control problem as follows:

Problem P. Given the system (1), find the control $u \in \mathcal{U}$ such that the cost functional

$$J(u) = \phi_0(x(T|u)) + \int_0^T \mathcal{L}_0(t, x(t|u), u(t))\, dt, \tag{4}$$

is minimised over $\mathcal{F}$, where ϕ_0 and $\mathcal{L}_0$ are given real valued functions.

We assume that the following conditions are satisfied:

(A1) $f : [0,T] \times R^n \times R^r \to R^n$ is piecewise continuous on $[0,T]$ for each $(x,u) \in R^n \times R^r$, and continuously differentiable with respect to each of the components of x and u for each $t \in [0,t]$; and furthermore, for any given compact set $\Omega \subset R^r$, there exists a constant $K > 0$ such that for all $(t,x,u) \in [0,T] \times R^n \times \Omega$,

$$|f(t,x,u)| \leq K(1 + |x|),$$

where $|\cdot|$ denotes the usual Euclidean norm;

(A2) For each $i = 1, \ldots, N_T$, $\phi_i : R^n \to R$ is continuously differentiable;

(A3) For each $i = 1, \ldots, N$, $g_i : [0,T] \times R^n \to R$ is continuously differentiable;

(A4) $\phi_0 : R^n \to R$ is continuously differentiable;

(A5) $\mathcal{L}_0 : [0,T] \times R^n \times R \to R$ is piecewise continuous on $[0,T]$ for each $(x,u) \in R^n \times R^r$, and continuously differentiable with respect to each of the components x and u for each $t \in [0,T]$.

Remark 2.1. From the theory of differential equations, we recall that the system (1) admits a unique solution, $x(\cdot|u)$, corresponding to each $u \in L_\infty^r$, and hence for each $u \in \mathcal{U}$.

3. Constraints Approximation.

For each $i = 1, \ldots, N$, the corresponding continuous state equality constraint in (3b) is equivalent to

$$G_i(u) = \int_0^T \max\{0, g_i(t, x(t|u))\}\, dt = 0, \tag{5}$$

which is, however, non-smooth in u. For convenience, let problem P with (3b) replaced by (5) be again denoted by P.

Let

$$\Theta = \{u \in \mathcal{U} \ : \ \phi_i(x(T|u)) \leq 0, \ \ i = 1, \ldots, N_T\}. \tag{6}$$

Clearly, the set $\mathcal{F}$ of feasible controls can also be written as:

$$\mathcal{F} = \{u \in \Theta \ : \ G_i(u) = 0, \ \ i = 1, \ldots, N\}. \tag{7}$$

Using the smoothing technique of Jennings and Teo (1989), we replace the non-smooth functions $\max\{0, g_i(t, x(t|u))\}$ by the smooth ones $g_{i,\epsilon}(t, x(t|u))$, where $\epsilon > 0$ and

$$g_{i,\epsilon}(t, x(t|u)) = \begin{cases} g_i(t, x(t|u)), & \text{if } g_i(t, x(t|u)) > \epsilon \\ (g_i(t, x(t|u)) + \epsilon)^2/4\epsilon, & \text{if } -\epsilon \leq g_i(t, x(t|u)) \leq \epsilon \\ 0, & \text{if } g_i(t, x(t|u)) < -\epsilon \end{cases} \tag{8}$$

For each $i = 1, \ldots, N$, define

$$G_{i,\epsilon}(u) = \int_0^T g_{i,\epsilon}(t, x(t|u))\, dt. \tag{9}$$

Then we consider the following approximate problem.

Problem $P_{\epsilon,\tau}$. The problem P with (3a) replaced by ($\tau > 0$)

$$G_{i,\epsilon}(u) - \tau \leq 0, \ \ i = 1, \ldots, N. \tag{10}$$

Let $\mathcal{F}_{\epsilon,\tau}$ be the feasible region of $P_{\epsilon,\tau}$ defined by

$$\mathcal{F}_{\epsilon,\tau} = \{u \in \Theta : G_{i,\epsilon}(u) - \tau \leq 0, \ \ i = 1, \ldots, N\}. \tag{11}$$

4. Control Parameterization.

As in Goh and Teo (1988), and Teo and Goh (1989), the control parameterization method will be used to solve the problem P. To be more precise, let $\{\mathcal{I}_p\}_{p=1}^{\infty}$ be a sequence of partitions of the interval $[0,T]$ such that $\mathcal{I}_p$ has n_p+1 elements, $\mathcal{I}_{p+1}$ is a refinement of $\mathcal{I}_p$ and $\|\mathcal{I}_p\| \to 0$ as $p \to \infty$, where $\|\mathcal{I}_p\|$ is the length of the largest sub-interval in the partition $\mathcal{I}_p$. In this paper, we assume that

$$\mathcal{I}_p = \{I_j^p\}_{j=1}^{N_p},$$

where

$$I_j^p = [t_{j-1}^p, t_j^p), \quad 0 = t_0^p < t_1^p < \cdots < t_{N_p}^p = T.$$

Then

$$\|\mathcal{I}_p\| = \max_{1 \le j \le N_p} \ell(I_j^p), \quad \ell(i_j^p) = t_j^p - t_{j-1}^p.$$

Let $\mathcal{U}^p$ be the subset of admissible controls which are piecewise constant and consistent with the partition $\mathcal{I}_p$. It is clear that each $u^p \in \mathcal{U}^p$ can be written as

$$u^p(t) = \sum_{k=1}^{N_p} \sigma^{p,k} \chi_k^p(t), \quad t \in [0,T]$$

where $\sigma^{p,k} \in R^r$ and χ_k^p denotes the characteristic function of I_k^p. This means that each control $u^p \in \mathcal{U}^p$ can be identified uniquely with a control parameter vector σ^p and vice versa, where

$$(\sigma^p)^\top = [(\sigma^{p,1})^\top, \cdots, (\sigma^{p,N_p})^\top].$$

Thus when no confusion can arise, we interchangeably refer to $u^p \in \mathcal{U}^p$ and $\sigma^p \in \mathcal{U}^p$. This convention will also be used in all similar situations.

Given a particular control parameter $\sigma^p \in \mathcal{U}^p$, we can re-define the functions and sets used in the previous section. Let $x(t|\sigma^p)$ be the solution to the system $\dot{x} = f(t, x(t), \sigma^p)$ with initial conditions $x(0) = x_0$. Then, we define

$$G_i(\sigma^p) = \int_0^T \max\{0, g_i(t, x(t|\sigma^p))\}\, dt \tag{12}$$

and

$$G_{i,\epsilon}(\sigma^p) = \int_0^T g_{i,\epsilon}(t, x(t|\sigma^p))\, dt. \tag{13}$$

The feasible sets used previously become:

$$\Theta^p = \{\sigma^p \in \mathcal{U}^p : \phi_i(x(T|\sigma^p)) \le 0 \quad i = 1, \ldots, N_T\}, \tag{14}$$

$$\mathcal{F}^p = \{\sigma^p \in \Theta^p : G_i(\sigma^p) = 0, \quad i = 1, \ldots, N\}, \tag{15}$$

$$\mathcal{F}^p_{\epsilon,\tau} = \{\sigma^p \in \Theta^p : G_{i,\epsilon}(\sigma^p) - \tau \le 0, \quad i = 1, \ldots, N\}. \tag{16}$$

We re-define the two control problems used previously but now parameterized by p.

Problem $P(p)$. Find a control vector $\sigma^p \in \mathcal{F}^p$ such that the cost functional

$$J(\sigma^p) = \phi_0(x(t|\sigma^p)) + \int_0^T \mathcal{L}_0(t, x(t|\sigma^p), \sigma^p)\, dt \tag{17}$$

is minimised.

Problem $P_{\epsilon,\tau}(p)$. Find a control vector $\sigma^p \in \mathcal{F}^p_{\epsilon,\tau}$ such that the cost functional $J(\sigma^p)$ is minimised.

To solve the problem P by the control parameterization method, we need to solve a sequence of problems $\{P(p)\}_{p=1}^{\infty}$. However, each of these approximate problems contains constraints which are non-smooth (c.f. Eqs (12) and (15)). Thus, following the idea of Jennings and Teo (1989), we have constructed a further sequence of problems: $\{P_{\epsilon,\tau}(p) : \epsilon > 0, \ \tau > 0\}$, corresponding to each $p = 1, 2, \ldots$.

For each $p = 1, 2, \ldots$, $\epsilon > 0$ and $\tau > 0$, the problem $P_{\epsilon,\tau}(p)$ is a standard optimal parameter selection problem in which gradients of the cost functional and the constraints can be calculated. The details of these gradient calculations can be found in the Appendix of Teo *et al* (1986). Thus, for each $p = 1, 2, \ldots$, and $\tau > 0$, the problem $P_{\epsilon,\tau}(p)$ can be viewed as a mathematical programming problem, and hence can be solved by existing standard optimization software packages, such as NLPQL (Schittkowski (1985)). Then, by appropriate adjusting of the parameters ϵ and τ we can compute a sequence of suboptimal control parameters to the problem $P(p)$ with each of them in $\mathcal{F}^p$. The details are given in the following algorithm.

Algorithm

Step 0. Choose an initial ϵ $(= 10^{-2})$ and τ $(= \epsilon T/4$, say).

Step 1. Solve $P_{\epsilon,\tau}(p)$ to give $\sigma^p_{\epsilon,\tau}$.

Step 2. If $\sigma^p_{\epsilon,\tau}$ is not in $\mathcal{F}^p$ then set $\tau \leftarrow \tau/2$ and goto step 1.
else if $\epsilon > 10^{-6}$ then $\epsilon \leftarrow \epsilon/10$ and $\tau \leftarrow \tau/10$ and goto step 1.
else stop.

Remark 4.1. With trivial modification to the proof of a theorem in Jennings and Teo (1989), we can show that the halving process of τ in step 2 need only be done a finite number of times. Thus, for each $p = 1, 2, \ldots$, the algorithm produces a sequence (in ϵ), of suboptimal parameter vectors to the problem $P(p)$. Let $\{\sigma^p_{\epsilon,\tau}\}$ be the sequence in ϵ of the suboptimal control parameters produced by the above algorithm. Furthermore, let $\{u^p_{\epsilon,\tau}\}$ be the corresponding sequence of controls in $\mathcal{F}^p_{\epsilon,\tau}$.

Remark 4.2. In step 1 of the algorithm, we are required to solve the problem $P_{\epsilon,\tau}(p)$. This is essentially a nonlinear programming problem in the control parameters, and hence can be solved by any standard nonlinearly constrained optimization software.

Let $\{u^p_{\epsilon,\tau}\}$ be as defined in remark 4.1. Then under certain reasonable assumptions, we can show that:

(i) $J(u^p_{\epsilon,\tau}) \to J(u^{p,*})$, where $u^{p,*}$ is an optimal control to the problem $P(p)$;

(ii) $J(u^{p,*}) \to J(u^*)$, where u^* is an optimal control to the problem P; and

(iii) If $u^{p,*}$ converges to $\overline{u}$ almost everywhere in $[0, T]$ then $\overline{u}$ is an optimal control to the problem P.

The proofs of these results are more of a mathematical interest. They appear rather involved and hence are to be the subject of another paper (Teo and Jennings (1989)).

5. Fishery Optimal Control Problem.

This problem is taken from Ryan (1987) where dynamic programming techniques were used to compute solutions. In Goh and Teo (1989), this example was computed for a number of different cases to show the effectiveness of the control parameterization technique. Let $N(s)$ be the population of the fishery, $U(s)$ the effort expended function and T the planning horizon. The optimal control problem is:

$$\min_U J(U) = \int_0^T -e^{-\delta s}\big\{[p_0 q + q\theta(1 - e^{-\lambda s})]U(s)N(s) - c\big(U(s)\big)\big\}\, ds,$$

where the cost of effort is modelled by the quadratic or linear function of effort,

$$c(U) = c_1 U + c_2 U^2,$$

and subject to

$$\frac{dN}{ds} = rN(s)\left\{1 - \frac{N(s)}{K}\right\} - qU(s)N(s), \quad N(0) = N_0,$$

with constraints on the effort,

$$0 \leq U(s) \leq U_{max},$$

and a lower bound on the population,

$$-N(s) + N_{min} \leq 0.$$

The parameters are:

δ : depreciation (inflation) of dollar value per year.

λ : reciprocal of the average time in years to the price increase,

p_0 : current price in dollars.

q : catchability coefficient.

θ : new price.

K : carrying capacity of the population.

For computational purposes we scale the three variables, time, population and effort.

1. Scale the time interval to $[0, 1]$,

$$t = s/T \text{ so that } ds = T\,dt.$$

2. Scale the population by the carrying capacity,

$$x(t) = N(s)/K, \text{ so that } \frac{dx}{dt} = \frac{dN}{ds}\frac{T}{K}.$$

3. Scale the effort by the ratio r/q, the effort value which makes the coefficient of the linear term of the differential equation zero,

$$u(t) = U(s)q/r.$$

For convenience, let the price increase be measured by a parameter β_1 where $\theta = \beta_1 p_0$, so that a 25% increase in price gives $\beta_1 = 0.25$.

The scaled problem is

$$\min_u J(u) = \int_0^1 -e^{-\alpha_1 t}\left\{A_1 h(t)u(t)x(t) - b_1 u(t) - b_2\big(u(t)\big)^2\right\} dt,$$

where $h(t) = 1 + \beta_1(1 - e^{-\alpha_2 t})$, $\alpha_1 = \delta T$, $\alpha_2 = \lambda T$, $A_1 = p_0 rKT$, $b_1 = Tc_1 r/q$ and $b_2 = Tc_2(r/q)^2$. The differential equation and constraints simplify to:

$$\frac{dx}{dt} = A_2\left\{(1-u(t))x(t) - (x(t))^2\right\}, \quad x(0) = N_0/K, \quad \text{where } A_2 = Tr,$$

$$0 \le u(t) \le \beta_2, \quad \text{where } \beta_2 = U_{max}q/r,$$

$$-x(t) + \beta_3 \le 0, \quad \text{where } \beta_3 = N_{min}/K.$$

The parameter values used in Ryan (1987) are: $T = 10$ years, $r = 5 \times 10^{-2}$ per year, $K = 4 \times 10^5$ blue whale units (BWU), $N_0 = 1.6 \times 10^5$ BWU, $p_0 = 7 \times 10^3$ \$ per BWU, $q = 1.3 \times 10^{-5}$ per catcher day, $\lambda = 0.5$ per year, $\delta = 0.1$ per year, $\beta_1 = 0.25$, $c_1 = 5 \times 10^3$ \$ per catcher day per year, $c_2 = 0.975$ or 0 \$ per (catcher day)^2 per year. Hence the parameters of the scaled problem are (with remaining dimensions): $\alpha_1 = 1.0$, $\alpha_2 = 5.0$, $\beta_1 = 0.25$, $\beta_2 = 1.0$ or 2.6, $\beta_3 = 0.4$, $A_1 = 1.4$ (billion dollars), $A_2 = 0.5$, $b_1 = 0.1923$ (billion dollars), $b_2 = 0.1442$ or 0 (billion dollars) and $x(0) = 0.4$.

We report the optimal solutions for three cases where the constraint is that the state variable (population) has to remain above $0.4K$ at all times. In one case the population starts at $0.4K$ and the effort is bounded by r/q. The other two cases have $N(0) = 0.45K$ and two different bounds on the effort; one is r/q the other $5r/q$. These three cases were run for three different depreciation (inflation) rates 0%, 5%, 10%, but with the same price increase distribution and size as in Ryan (1987). We chose the linear cost functional as we do not require a quadratic term for stability of the algorithm as in Ryan (1987). It is important to note that our optimal solution is only valid for the 10 year period chosen. Table 1 shows the objective function values for the optimal solutions for all nine cases reported.

%	Case 1.	Case 2.	Case 3.
0%	0.3030	0.4416	0.4660
5%	0.2241	0.3343	0.3358
10%	0.1786	0.2735	0.2768

Table1. Depreciated Profit of Fishery (\$ billion)

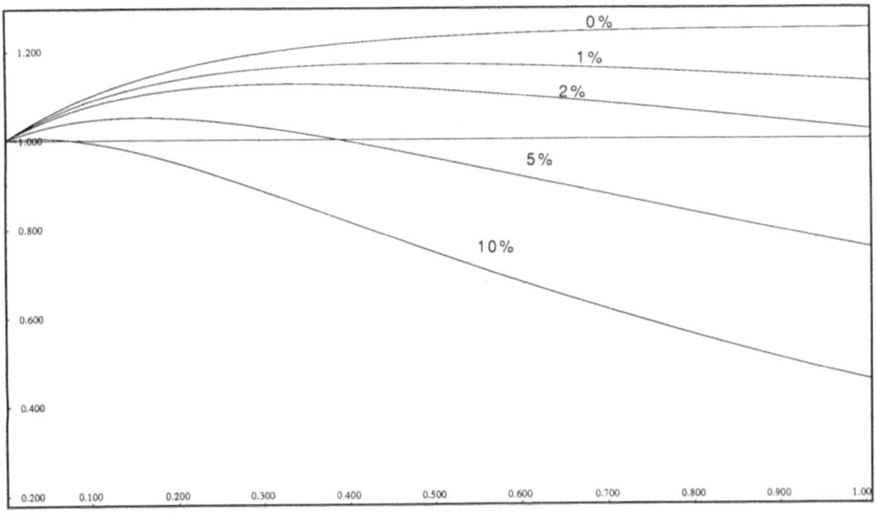

Figure 1. Depreciated price per BWU.

In Figure 1 the term appearing in the objective function,

$$e^{-\alpha_1 t}(1 + 0.25(1 - e^{-5t}))$$

for $\alpha_1 = 0,\ 0.1,\ 0.2,\ 0.5$, and 1.0, corresponding to a depreciation of 0%, 1%, 2%, 5% and 10% is plotted. Note that at the 10% level we need to make money while we can and the optimal strategies reflect this.

In figure 2 the case $N(0) = 0.4K$, $U_{max} = r/q$ shows that for small depreciation values we can allow the population to grow and then increase effort towards the end of the period and thus force the population back to $0.4K$ at the final time. For 10% depreciation we harvest at a rate where the population constraint is active for most of the time.

In figure 3 the case $N(0) = 0.45K$, $U_{max} = r/q$ shows the effect of taking the lower limit slightly lower than the starting population. With high depreciation we cannot allow the population to grow, but with lower depreciation and the price increase down the track, we can allow the population to grow. Note that the final population is always as low as it can be for this fixed 10 year period. The small aberration at the end of the 10 years shows the effects of the finite dimensionality of the computed problem. A starting population of $0.5K$ is enough to allow an effort of r/q for all 10 years, irrespective of the depreciation rate.

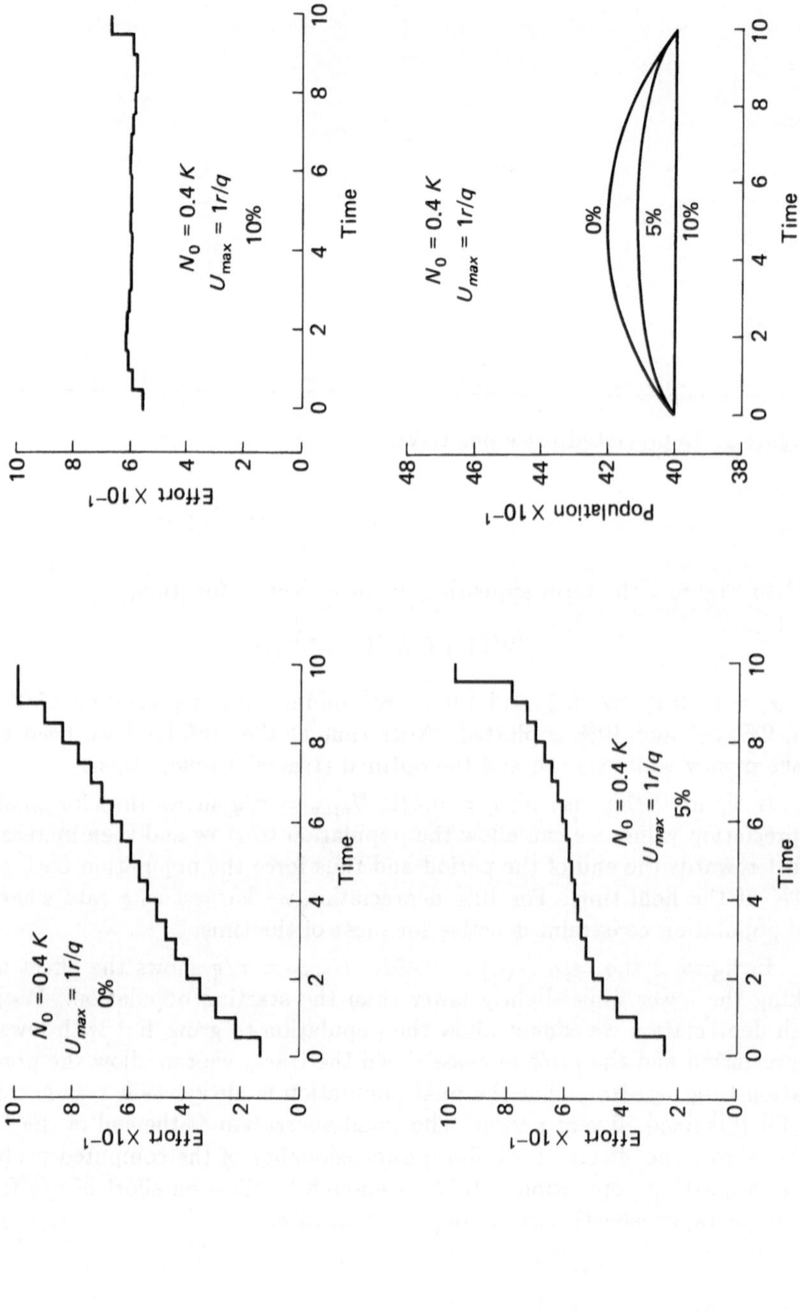

Figure 2. Case 1 – Population to remain greater than initial population.

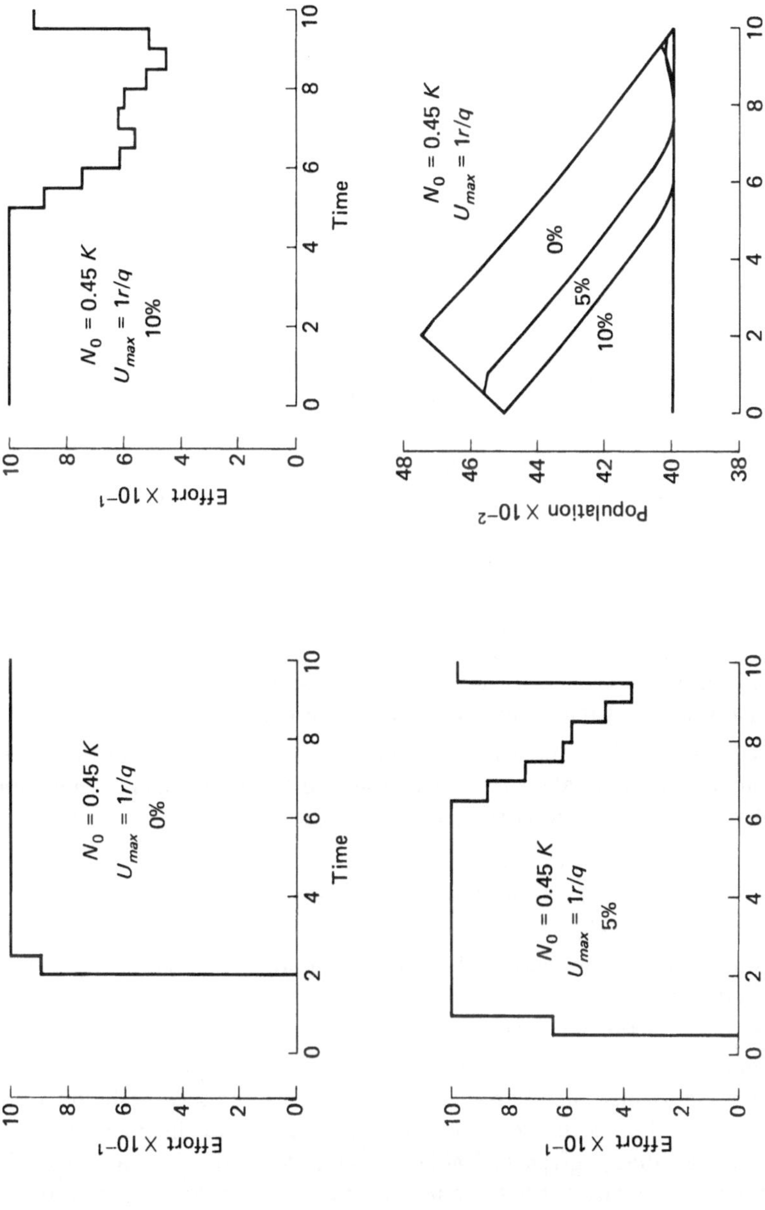

Figure 3. Case 2 – Initial population greater than final.

Figure 4 shows the effect of allowing a large effort (unlimited) with a starting population greater than the minimum. Obviously at zero depreciation one allows the population to flourish and then harvest at the end of the period. At 5% depreciation the optimal effort does not reach the maximum allowed and harvesting occurs around the expected time of the price increase. The software does not compute a 'smooth' effort towards the end of the time period. At 10% depreciation the maximum harvest occurs before the expected time to the price increase but nowhere near the maximum effort allowed. There is still the oscillation of effort in the latter half of the time period.

From our experience with optimal control problems we have noted that, close to the solution, relatively large changes in control parameters can sometimes produce only small changes in objective function. In other words, the exact value of the control does not matter greatly in some problems. But this phenomenon, while of value to the system designer, produces ill-conditioning in the computation of solutions to optimal control problems. It does mean that to get close to the optimal control we sometimes have to push the software to its limits. That is, we set the stopping criteria as small as we possibly can. We have noticed a tendency of some users of software to set stopping criteria too large, leading to premature exiting froom the software with subsequent non-optimal control values, or worse, constraint violation by more than one would wish. In making camparisons of algorithms and software it is important to report this push to the limit and also to report the different measures of success or failure — objective function values and constraint violations.

To illustrate these remarks we report case 1 at 10% depreciation where we compare the new algorithm with the one proposed in Teo and Goh (1987), which uses the constraint transcription of $g_i(t, x(t|u)) \leq 0$ to

$$\int_0^T \{\max(0, g_i(t, x(t|u))\}^2 \, dt = 0.$$

This constraint does not satisfy the normal constraint qualification, because the deviations from zero are squared. In Table 2, we present the 20 controls and 21 values of the state variable (equally spaced in time).

We used an accuracy requirement of 10^{-6} as the convergence criteria. The older algorithm failed to converge as it found an uphill search direction. The values of the two integral measures of violation of the constraint are (in scaled variables), given in Table 3. While the variations are small in this case note the large variations in the optimal controls given in Table 2.

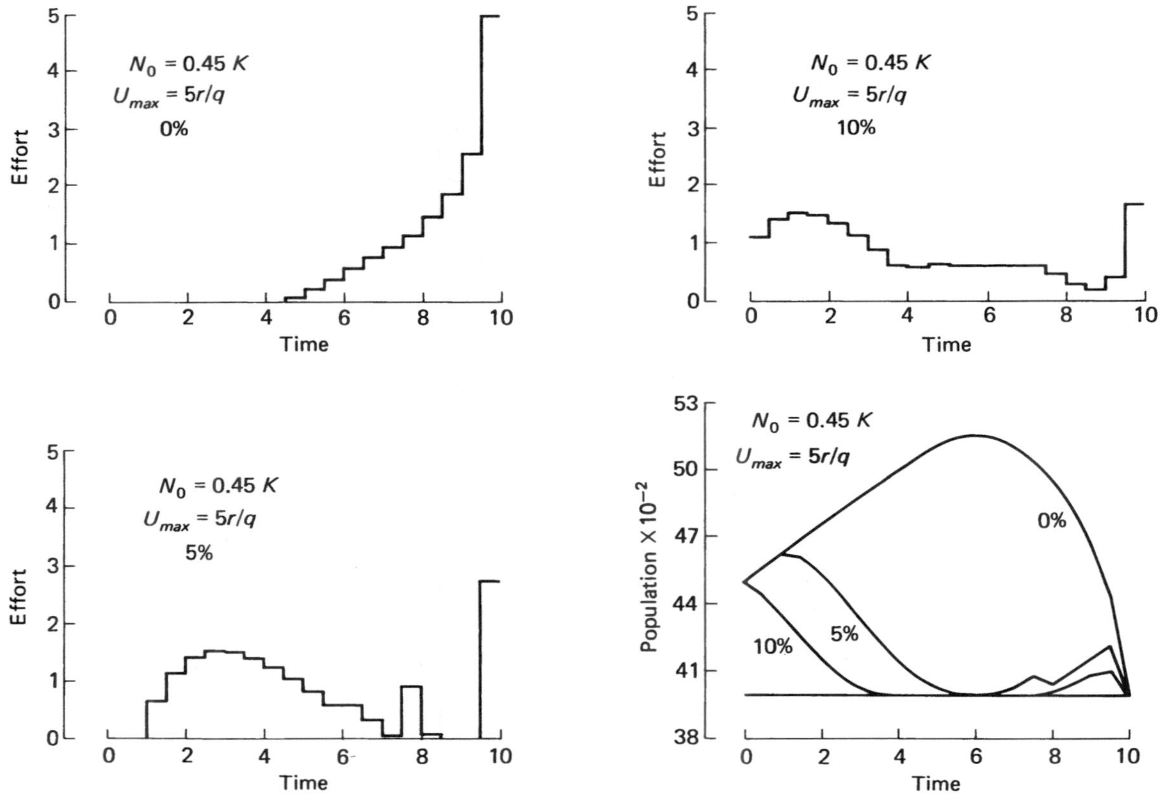

Figure 4. Case 3 – Allow greater effort in case 2.

	New Algorithm		Old Algorithm	
Time	Control	State	Control	State
0.00	0.5567	0.4000	0.5791	0.4000
0.05	0.5930	0.4004	0.5996	0.4002
0.10	0.6110	0.4005	0.6097	0.4002
0.15	0.6167	0.4004	0.6129	0.4001
0.20	0.6139	0.4002	0.6119	0.4000
0.25	0.6058	0.4001	0.6083	0.3999
0.30	0.5996	0.4000	0.6036	0.3998
0.35	0.5996	0.4000	0.5984	0.3997
0.40	0.5990	0.4000	0.5935	0.3998
0.45	0.6019	0.4000	0.5893	0.3998
0.50	0.5990	0.4000	0.5862	0.3999
0.55	0.5995	0.4000	0.5844	0.4001
0.60	0.6011	0.4000	0.5843	0.4002
0.65	0.6004	0.4000	0.5867	0.4004
0.70	0.5931	0.4000	0.5919	0.4005
0.75	0.5856	0.4001	0.6006	0.4006
0.80	0.5805	0.4002	0.6135	0.4006
0.85	0.5802	0.4004	0.6338	0.4004
0.90	0.5944	0.4006	0.6679	0.4001
0.95	0.6700	0.4007	0.7411	0.3994
1.00		0.3999		0.3980

Table 2. Comparison of Solutions

	New Algorithm	Old Algorithm
L_2	$< 2 \times 10^{-11}$	0.52×10^{-6}
L_1	0.63×10^{-5}	0.13×10^{-3}

Table 3. Measures of Constraint Violation

BIBLIOGRAPHY

[1] 1. L. Cesari,1983. *Optimization — Theory and Applications*, Springer Verlag, New York, New York.

[2] 2. C.J. Goh, K.L. Teo, 1987. MISER: An Optimal Control Software, *Applied Research Corporation*, National University of Singapore, Kent Ridge, Singapore.

[3] 3. C.J. Goh, K.L. Teo, 1988. Control parameterization: a unified approach to optimal control problems with general constraints, *Automatica* **24(1)**, pp. 3–18.

[4] 4. C.J. Goh, K.L.Teo, 1989. Species preservation in an optimal harvest model with random prices, To appear in *Mathematical Biosciences.*

[5] 5. L.S. Jennings, K.L. Teo, 1989. A computational algorithm for functional inequality constrained optimization problems, to appear in *Automatica.*

[6] 6. A.Miele, 1975. Recent advances in gradient algorithms for optimal control problems, *J. Optim. Theory Applic.* **17**, pp. 361–430.

[7] 7. A.Miele, R.E.Pritchard, J.N. Damoulakis, 1970. Sequential gradient-restoration for optimal control problems, *J. Optim. Theory Applic.* **5**, pp. 235–282.

[8] 8. A. Miele, T. Wang, V.K. Basapur, 1986. Primal and dual formulations of sequential gradient restoration algorithms for trajectory optimization problems, *Acta Astronautica* **13**, pp. 491–505.

[9] 9. D. Ryan, 1987. Effort fluctuations in a harvest model with random prices, *Math. Biosciences* **86**, pp. 171–181.

[10] 10. K. Schittkowski, 1985. NLPQL: a FORTRAN subroutine for solving constrained nonlinear programming problems, *Operations Research Annuls* **5**, pp. 485–500.

[11] 11. K.L. Teo, B.W. Ang, C.M. Wang, 1986. Least weight cables: Optimal parameter selection approach,*Engineering Optimization,* **9**, pp. 249–264.

[12] 12. K.L.Teo, C.J. Goh, 1987. A simple computational procedure for optimization problems with functional inequality constraints, *IEEE Trans. Automatic Control* **AC–32**, pp. 940–941.

[13] 13. K.L. Teo, C.J. Goh, 1989. A computational method for combined optimal parameter selection and optimal control problems with general constraints, *J. Australian Math. Soc., Ser. B* **30(3)**, pp. 350–364.

[14] 14. K.L. Teo, L.S. Jennings, 1989. Nonlinear optimal control problems with continuous state inequality constraints, to appear in *J. Optim. Theory Applic.*.

[15] 15. K.H. Wong, D.J. Clements, K.L. Teo,1986. Optimal control computation for nonlinear time-lag systems, *J. Optim. Theory Applic.* **47**, pp. 91–107.

PARTICIPANT'S COMMENTS

A considerable literature on numerical techniques for problems in optimal control theory has built up in recent years. This paper extends the current literature in this area by presenting a procedure for handling continuous state constraints thus greatly extending the range of problems which can be successfully solved by numerical techniques.

A wide range of biologically related problems, in particular those involving harvesting of resources, can be formulated as optimal control problems. One such problem is solved by the authors to show the effectiveness of their method. Of course, the main criticism with problems formulated in this way is the lack of precision in the models themselves. Nevertheless, important insight into the real world can often be obtained from solutions to these mathematical problems.

Mike Fisher

The paper presents a numerical algorithm for state constrained optimal control problems and applies the algorithm to an optimal harvesting problem for a fishery management system. The algorithm incorporates a new smoothing approximation to handle the fact that the state constraint inequalities are nonsmooth.

As a numerical optimal control procedure applied to a blue whale fishery, the paper fits in very well with the theme of the proceeding. In addition, there is a possible connection with the paper on discretization chaos by Grantham and Athalye. In particular, Jennings and Teo report that some numerical instability occurs for one of their examples. This instability may be related to chaos, either caused by the numerical method itself or by the fact that their fishery model is analogous to a logistic equation and the discrete version of the logistic equation is a classical chaotic system for certain parameter values.

W. Grantham

Part III

Games

Strategy Dynamics and the ESS

THOMAS L. VINCENT

Abstract

In a constant environment, evolutionary process should result in evolutionarily stable strategies for the creatures which ultimately survive to inhabit that environment. It is shown here that by adding a strategy dynamic to a previously developed theory, one not only obtains a convenient way of determining evolutionarily stable strategies, but interesting features about the process itself can be observed. Of particular interest, as demonstrated here, progression to evolutionarily stable strategies can take place even when the population densities are experiencing chaotic motion.

Introduction

In Maynard Smith's words (1982), "An ESS [evolutionarily stable strategy] is a strategy such that, if all members of a population adopt it, then no mutant strategy could invade the population under the influence of natural selection." There is an extensive literature on translating this verbal concept into a mathematical setting [See for example the reviews by Hines (1987) and Riechert and Hammerstein (1983)]. This includes our own work on the development of an ESS theory for determining ESS strategies (Vincent and Brown, 1984; Brown and Vincent, 1987a). In our previous theoretical developments we did not explicitly include strategy dynamics. Thus while our theory can be used to predict ESS strategies it does not explicitly take into account any mechanism for producing the ESS. Indeed it is not necessary, since the theory deals not with how an ESS evolves but rather the conditions under which it may exist. Using the theory to find an ESS, the resultant strategies can be tested against arbitrary "mutant" strategies by simply examining the dynamics of individual population densities as time goes on.

Since the fitness of each individual organism in a biological community is affected by the survival strategies of all other organisms, a struggle for survival exists. This struggle is an evolutionary game where the individual organisms (players) inherit their survival strategies (phenotypic characteristics) from a continuous play of the game from generation to generation. The evolutionary game includes both ecological and evolutionary processes. The ecological process involves the interaction between individuals and the environment for the determination of fitness. The evolutionary process is the dynamical process which takes the fitness of an individual and translates it into changes in the number and frequency of individuals using a particular strategy. Through appropriate models, the evolutionary game may be given a mathematical setting. Most commonly, the strategies are assumed to be constants associated with certain parameters in the model. The ESS is a particular constant (or vector of constants) which provide the stability property described by Maynard Smith.

In our previous work with evolutionary games, we assumed that individuals of a given strategy would breed true. Under this assumption, the strategies remain constant from generation to generation. The strategies themselves do not evolve. In such a setting, the evolutionary process simply tests the strategies against each other. This allows for the experimental determination of an ESS in a given evolutionary game by continually "seeding" the modeled community with random strategies (Brown and Vincent, 1987a). However this procedure is not computationally efficient nor is it likely to occur in nature, except perhaps in some island community developments. The breed true assumption does not hinder in the development of necessary conditions for determining the ESS, it simply sweeps the issue of how an ESS might come about under the evolutionary rug. Indeed, the analytical necessary conditions provided by the theory can be used to directly determine ESS strategies no matter how the ESS strategies are actually achieved.

In what follows, by contrast, it will be assumed that individual survival strategies are constant within a generation time period but not necessarily constant from generation to generation.

Population Dynamics via the G Function

In keeping with the notion that like begets like, we need some way of identifying who gets to beget with whom. We will assume here that individuals may not only be identified by the strategies they have employed for survival but also by other factors which keep an identifiable population of individuals breeding with each other. We will call such a population a "species" and identify various species with a numerical label. For example we use x_1 to identify the density of species 1 and u_1 to refer to the survival strategy used by all individuals of species 1. While this represents a departure in concept from our previous work where individuals were identified strictly by the strategies they were using, it does not require a change in notation. This departure is necessary, for we are now particularly interested in tracing strategy changes within a given species.

We will assume here that the density dynamics of a given species within a community of r species can be described by means of difference equations of the form

$$x_i(t+1) = x_i(t)\{1 + H_i[u, x(t)]\} \tag{1}$$

where t is a generation counter, x_i is the population density of species i, $(1 + H_i)$ is the fitness function of species i,

$$x = [x_1, \ldots, x_r] \tag{2}$$

is the vector of all species densities in the community under consideration and

$$u = [u_1, \ldots, u_r] \tag{3}$$

is the vector of strategies used by all of the species in the community. The strategies u_i may be either scalars or vectors. Henceforth, we will drop the t from variables [e.g., $x(t)$] which are evaluated at generation time t. Alternate difference equation models or differential equation models are also possible and will yield similar results (Vincent and Fisher, 1988).

For analysis it is more convenient to express population dynamics in terms of frequency dynamics. The total number of individuals in the entire community at any generation time t is given by

$$N = \sum_{i=1}^{r} x_i. \tag{4}$$

A measure of how well individuals of a given species are doing at any generation time t is related to their population size x_i. However, if we wish to measure how well one species is doing relative to all others, then the frequency of those individuals of a given species as defined by

$$p_i = \frac{x_i}{N} \tag{5}$$

is the proper measure to use. Using these definitions we have the following equivalent representation of (1).

$$p_i(t+1) = \frac{p_i(1 + H_i(u,p,N)}{\overline{H}} \tag{6}$$

and

$$N(t+1) = N\overline{H}, \tag{7}$$

where

$$\overline{H} = 1 + \sum_{i=1}^{r} p_i H_i(u,p,N), \tag{8}$$

and

$$p = [p_1, \ldots, p_r]. \tag{9}$$

Note that the functions $H_i(u,x)$ have been reformulated to be expressed as $H_i(u,p,N)$ and that (8) defines the average fitness for community as a whole.

We have previously introduced the notation of a G-function called a *fitness generating function* (Vincent and Brown, 1984; Brown and Vincent, 1987a; Vincent and Brown, 1988) to aid in the development of necessary conditions for an ESS. We need this concept here also. In general there will be s $(0 < s \leq r)$ species in the community which will share the same set of evolutionarily feasible strategies and which share the same ecological consequences from using those strategies. A group of such species will be said to share the same G-function. A function $G(v,u,p,N)$ is said to be a G-function for s species in the community if

$$G(u_i, u, p, N) = H_i(u,p,N) \tag{10}$$

for all the indices i corresponding to the s species. In a given community there could be several G-functions or only one. In this presentation we will assume that the community can be described in terms of a single G-function $(s = r)$. The determination of ESS strategies under several different G-functions has been characterized elsewhere (Brown and Vincent, 1987b; Vincent and Brown, 1989).

The G-function has the property that the fitness of an individual (i.e. $1 + H_i$) using one of the strategies of u is determined when v is replaced by that individual's strategy. Thinking of the variables u, p, and N as defining the current environment of the individual, it follows that the fitness of an individual then depends on its "choice" for v.

Necessary Conditions for an ESS

We need to distinguish between strategies which form an ESS and any other strategies outside the ESS. The vector of the first σ strategies of u is defined to be a coalition vector $u^c = [u_1, \ldots, u_\sigma]$ where $\sigma \geq 1$. The composite of the remaining $r - \sigma$ strategies is called a mutant vector and is designated by $u^m = [u_{\sigma+1}, \ldots, u_r]$. The total number of all individuals using coalition strategies is given by

$$N_c = \sum_{i=1}^{\sigma} x_i \tag{11}$$

and the total number of individuals using mutant strategies is given by,

$$N_m = \sum_{i=\sigma+1}^{r} x_i \tag{12}$$

The corresponding frequencies are given by

$$p_c = \frac{N_c}{N} \tag{13}$$

and

$$p_m = \frac{N_m}{N} \tag{14}$$

An ESS may now be defined in terms of a coalition vector:

Definition. A coalition vector u^c is said to be an ESS if there exists an ϵ in the interval $0 < \epsilon < 1$ such that for all mutant strategies u^m, initial frequencies p_i where $p_i > 0$ for $i = 1, \ldots, \sigma$ and $p_i \geq 0$ for $i = \sigma + 1, \ldots, r$ with $1 - \epsilon < p_c(0) < 1$, and initial densities $N(0) > 0$,

$$\lim_{t \to \infty} p_c(t) = 1 \tag{15}$$

with

$$\lim_{t \to \infty} p_i(t) > 0 \quad \text{for } i = 1, \ldots, \sigma \tag{16}$$

Under this definition any number of mutant species greater than one are allowed. Also the ESS may be a coalition of any number of species. The definition is local in the sense that $p_c(0)$ is allowed to be arbitrarily close to 1. If the definition holds for ϵ close to one then the ESS would be a global attractor in frequency space. Since $p_c + p_m = 1$, the mutant dynamics must obey

$$\lim_{t \to \infty} p_m(t) = 0. \tag{17}$$

This definition is consistent with the word definition given earlier. Species identified by the strategies in the coalition vector u^c will persist through time no matter how many mutant strategies are introduced. Moreover, individuals using the mutant strategies will die out with time.

We have previously shown (Vincent and Fisher, 1988) that this definition, along with the definition of the G-function, will lead to the following:

ESS necessary condition: Let $G(v, u, p, N)$ be a generating function for the community. If u^c is an ESS such that $\{p_c(t)\}$ is a monotone increasing sequence for all $t \geq t_m \geq 0$ and if p^* and N^* are asymptotically stable, then in the limit as $t \to \infty$, $G(v, u, p^*, N^*)$ must take on a maximum with respect to v at $u_1, \ldots, u_\sigma$.

It should be noted that a more general condition can be given which does not require asymptotic stability in either p^* or N^* (Vincent and Brown, 1987). Also if the ESS is independent of N, then stability in N^* is not required. This will be demonstrated below.

Strategy Dynamics

Roughgarden (1983) avoids genetic mechanisms in his models for coevolution by proposing a process in which trait change is proportional to the gradient of a species fitness function. In a fashion somewhat similar to this, but keeping in mind the above necessary conditions, strategy dynamics will be modeled here by assuming that trait change is proportional to the gradient of the G-function evaluated at equilibrium. With several different species in the population, all modeled by the same G-function, each species trait is assumed to evolve according to

$$u_i(t+1) = u_i + \alpha \frac{\partial G(u_i, u, p^*, N^*)}{\partial v}. \tag{18}$$

If each strategy is to evolve to an ESS then the ESS necessary condition must be satisfied by this strategy. In particular when u_i is an ESS strategy the gradient of the G-function will be zero (assuming an unbounded strategy set). The parameter α may be thought of as a mutation rate as it will control how rapidly the strategies change. For biological systems, α should be quite small. Clearly if $\alpha = 0$ then no change in strategy is possible.

The dynamical system is now defined by (6) - (8) and (18). By solving this system of equations together we not only obtain the population dynamics but as time goes on all surviving species should be at the ESS (with $\alpha > 0$). In other words we should be able to obtain the ESS numerically. Unfortunately, both p^* and N^* must be known in order to evaluate the partial derivative given in (18). Generally neither of these will be known. However if the ESS is a coalition of one and if the ESS is independent of N then $p^* = [1, 0, ..., 0]$ and the current value of N may be used in (18). For this case, provided that an ESS exists we should be able to find it using this numerical scheme. I will leave for future study the properties of such a system and what its implications are for finding ESS's when these two assumption are not satisfied.

In the following examples, the model is borrowed from a previous study (Brown and Vincent, 1987) where it is known that the ESS is composed of a coalition of one, and that the ESS is independent of N.

Examples

In this section we will examine strategy dynamics in the difference-equation form of the Lotka-Volterra competition model. We will use a form of the model which has been extensively studied in an evolutionary setting by Case (1982), Rummel and Roughgarden (1983, 1985), and Vincent and Brown (1987). Using the notation of equation (1), the H_i function for a given species i is given by

$$H_i[u, x] = R - \frac{R}{K(u_i)} \sum_{j=1}^{r} \alpha(u_i, u_j) x_j \tag{19}$$

where r is the total number of species currently in the community, R is the intrinsic rate of growth common to all species, $K(u_i)$ is the carrying capacity of the species i and $\alpha(u_i, u_j)$ is the competitive effect of species j using strategy u_j on the fitness of individuals of species i. In terms of frequency (19) becomes

$$H_i[u, p, N] = R - \frac{R}{K(u_i)} N \sum_{j=1}^{r} \alpha(u_i, u_j) p_j. \tag{20}$$

with the corresponding dynamics given by (6) - (8). Fitness as determined by (20) may be expressed in terms of the following G-function

$$G[v, u, p, N] = R - \frac{R}{K(v)} N \sum_{j=1}^{r} \alpha(v, u_j) p_j. \tag{21}$$

According to the ESS theorem, G is to take on a maximum with respect to v at equilibrium ($p_1 = 1, N^* = N_1$) and when $v = u_1$ (the ESS strategy). The gradient of G with respect to v under these conditions yields

$$\left.\frac{dK}{dv}\right|_{v=u_1} = N^* \left.\frac{\partial \alpha}{\partial v}\right|_{v=u_j=u_1} \tag{22}$$

and the equilibrium condition yields

$$K(u_1) = N^* \alpha(u_1, u_1). \tag{23}$$

Consider now the specific relation

$$K(v) = 100 \exp\left[-\frac{1}{2}\left[\frac{v}{\sigma_k}\right]^2\right] \tag{24}$$

with an asymmetric competition coefficient given by

$$\alpha(v, u_j) = 1 + \exp\left[-\frac{1}{2}\left[\frac{v - u_j + \beta}{\sigma_\alpha}\right]^2\right] - \exp\left[-\frac{1}{2}\left[\frac{\beta}{\sigma_\alpha}\right]^2\right] \tag{25}$$

Applying (22) and (23) yields

$$\frac{N^* u_1}{\sigma_k} = \frac{N^* \beta}{\sigma_\alpha} \exp\left[-\frac{1}{2}\left[\frac{\beta}{\sigma_\alpha}\right]^2\right] \tag{26}$$

and

$$N^* = 100 \exp\left[-\frac{1}{2}\left[\frac{u_1}{\sigma_k}\right]^2\right] \tag{27}$$

Notice that N^* cancels from (26) making u_1 independent of N^* and a function only of the parameters β, σ_α and σ_k. In conformity with our previous work we now set $\beta = \sigma_k = \sigma_\alpha = 2$. From (26) and (27) we obtain $u_1 = 1.213$ and $N^* = 83.199$.

Figure 1 illustrates typical operation of an ESS strategy when it is "played" together with two mutants. The mutant strategies were taken to be $u_2 = 0$ and $u_3 = 2$. All three species were started at a population size of 10 and all strategies were assumed to be fixed ($\alpha = 0$) for all time. The intrinsic growth rate for all species is $R = .25$. After approximately 300 generations the ESS is approaching its equilibrium value of $N^* = 83.2$ with both of the mutant strategies dying out. Even if the mutant strategies were very close to the ESS strategy they would eventually die out, it would simply take longer.

Figure 2a illustrates the effect of introducing strategy dynamics as given by (18) on an isolated population. The initial strategy is assumed to be $u_1 = 0$ with an initial population density of $N_1 = 10$ and $R = .25$. This population, without strategy dynamics, would die out in the presence of an ESS strategy as already noted in Figure 1. However by itself, even with a zero mutation rate, ($\alpha = 0$) it will grow asymptotically approaching the carrying capacity of $N^* = 100$ as indicated by the upper curve. This same population with strategy dynamics ($\alpha = 1$) will grow according to the lower curve until it reaches its ESS carrying capacity of $N^* = 83.2$. Figure 2b illustrates the corresponding changes in strategy. With $\alpha = 0$ there is no change in the initial strategy of $u_1 = 0$. However with $\alpha = 1$, the population evolves to the ESS of $u_1 = 1.213$ in less than 200 generations. Clearly the rate at which the population approaches the ESS is a function of α. For any non-zero α a similar result is obtained, only the time scale differs.

If there are no ESS strategies in a population of many strategies and if there is no strategy dynamics ($\alpha = 0$), then coexistence is uncertain. For example, Figure 3a illustrates two populations which can not coexist. Both are started at the same initial density of $N_1 = N_2 = 10$ with $R = .25$. The population which survives has a strategy of $u_1 = .75$ and the population which dies out has a strategy of $u_2 = 0$. Figure 3b illustrates two populations which can coexist. Both populations are started with $N_1 = N_2 = 10$ and $R = .25$. The N_1 population is using a strategy of $u_1 = 0$ and the N_2 population is using a strategy of $u_2 = 2$. Of course, in either of these cases, if the community were invaded by an ESS strategy, these species would not survive. In this particular model, the ESS is a single strategy and no coexistence is possible if an ESS is present.

The effect of introducing strategy dynamics on two populations, neither of which are initially using the ESS strategy, is illustrated in Figures 4a-d. All of these Figures were obtained using $R = .25$. The first situation depicted in Figure 4a corresponds to the same initial conditions of Figure 3a except now each species has the same mutation rate $\alpha = .1$. Because of the slow mutation rate, the outcome in terms of populations density is essentially the same as before (Figure 3a). Except now, as the population N_1 strategy approaches the ESS strategy (Figure 4b) the ESS equilibrium population is obtained. The strategy of N_2 also changes, but not fast enough, as the population is essentially zero after 200 generations. The situation changes significantly with an increase in mutation rate. Figure

4c illustrates the dynamics of the two populations with $\alpha = 1$. With this increase in mutation rate, both species may now coexist as both achieve the ESS strategy before one of them goes extinct. The sum of the equilibrium populations ($N_1^* + N_2^* = 83.2$) is at the ESS equilibrium density. Figure 4d displays the corresponding changes in strategies. After about 400 generations both populations are using the same strategy and they become indistinguishable from this point of view. This is what allows for a stable coexistence. The two populations need not represent the same species however, since other characteristics may prevent breeding across the two populations (as was assumed).

In Figures 5a-d, the intrinsic growth rate in the G-function (20) has been sufficiently increased ($R = 2.6$) to produce deterministic chaos in the population densities. Otherwise the initial population densities ($N_1 = N_2 = 10$) and the initial strategies ($u_1 = .75$, $u_2 = 0$) of Figure 5 are the same as used in Figure 4. Figures 5a-b re-examines the effect of a "low" mutation rate ($\alpha = .1$). Comparing Figures 4a and 5a, it follows that species 2 dies out within a few generations and comparing Figures 4b and 5b, we see that species 1 achieves the ESS much more rapidly under chaotic population dynamics. Evolution appears to have been "sped up". The effect appears to be greater than one would attribute to the increase in R alone. This represents an area of intriguing further study. By increasing the mutation rate to $\alpha = 1$, both species can evolve fast enough so that they end up coexisting. This is illustrated in Figures 5c-d. Note that the average density of species 2 in Figure 5c is less than that obtained in Figure 4c. Species 2 is able to evolve fast enough (Figure 5d) to avoid extinction, but its relative initial disadvantage with respect to strategy becomes magnified under chaos.

Conclusion

The introduction of strategy dynamics into our evolutionary game model has yielded three new features:

(i) One may obtain evolutionarily stable strategies as a byproduct from a sufficiently long simulation run.

(ii) There exist threshold values for the mutation rate below which coexistence of non ESS species is not possible.

(iii) One may study the effect of system parameters on the rate of evolution even when the system is chaotic.

While the ESS necessary condition can be used to find ESS strategies analytically, it is not always easy to do so. For the general case of multiple G- functions and non-stable equilibrium in p and N, the analytical determination of ESS strategies can be tedious at best. The introduction of strategy dynamics hold promise as a tool for numerical determination of the ESS. Much work remains to be done however, for example, it has not been shown that convergence to the ESS will be guaranteed when the ESS is N dependent.

It would appear that there are interesting connections between mutation rates and survival when none of the species in a community are at the ESS. In a "new" environment with many species attempting to establish themselves those with high mutation rates would be at a distinct advantage. However in well established environments, with most of the species at their ESS's, a low mutation rate (just large enough to track environmental changes) would suffice. In fact a high mutation rate would be disadvantageous as this would put resources in to unsuccessful offspring.

Since the dynamics of many biological system seem to exhibit chaotic motion it would be of interest to further study this case. In particular, the parameter which produces chaos here (i.e. R) was not an evolutionary variable. By making R dependent on strategy, a community with strategy dynamics, could go into or out of chaotic motion as it evolves.

REFERENCES

[1] Brown, J.S. and T.L. Vincent. 1987a. A theory for the evolutionary game. *Theoretical Population Biology* **31**, pp. 140-166.

[2] Brown J.S. and T.L. Vincent. 1987b. Predator-prey coevolution as an evolutionary game. *Lecture Notes in Biomathematics Vol 73*, pp. 83-101.

[3] Case, T.J. 1982. Coevolution in resource-limited competition communities, *Theoretical Population Biology* **21**, pp. 69-91.

[4] Hines, W.G.S. 1987. Evolutionary stable strategies: A review of basic theory, *Theoretical Population Biology* **31**, pp. 195-272.

[5] Maynard Smith, J. 1982. *Evolution and the theory of games.* Cambridge, Cambridge University Press.

[6] Riechert, S.E. and P. Hammerstein. 1983. Game theory in the ecological context. *Annual Review Ecological System* **14**, pp. 377-409.

[7] Roughgarden, J. 1983. The theory of coevolution, in D.J. Futuyma and M. Slatkin (eds.), *Coevolution*, Sinauer, Sunderland, MA. pp. 383-403.

[8] Rummel, J.D. and J. Roughgarden. 1983. Some differences between invasion- structured and coevolution-structured competitive communities: A preliminary theoretical analysis, *Oikos* **41**, pp. 477-486.

[9] Rummel, J.D. and J. Roughgarden. 1985. A theory of faunal buildup for competition communities. *Evolution* **39**, pp. 1009-1033.

[10] Vincent, T.L. and J.S. Brown. 1984. Stability in an evolutionary game. *Theoretical Population Biology* **26**, pp. 408-42.

[11] Vincent, T.L. and J.S. Brown. 1987. Evolution under nonequilibrium dynamics, *Mathematical Modelling* **8**, pp. 766-771.

[12] Vincent, T.L. and J.S. Brown. 1988. The evolution of ESS theory. *Annual Review of Ecology and Systematics* **19**, pp. 423-443.

[13] Vincent, T.L. and M.E. Fisher. 1988. Evolutionarily stable strategies in differential and difference equation models. *Evolutionary Ecology* **2**, pp. 321-337.

[14] Vincent, T.L. and J.S. Brown. 1989. The Evolutionary response to a changing environment, *Applied Mathematics and Computation*, In Press.

PARTICIPANT'S COMMENTS

The idea of evolutionary stable systems is a fascinating one and has an intuitive niceness to it that explains why Maynard Smith's work has been so well received. Nevertheless, it is only by trying to model the idea mathematically that one can reveal the power as well as the limitations of the idea. In this respect, Tom's paper is particularly welcome, as it shows that one can get even closer to representing real problems than he and his colleagues have done in earlier studies.

The principal point of the paper is that if the strategy of a population varies in time, it still makes sense to talk about evolutionary stability. In control theoretic terms, we are now talking about a control function instead of a fixed control parameter. The fact that it is still possible to study ESS, and the fact that the results do not differ greatly from previous ones, encourage me to believe that the whole approach is very robust, as it must be to survive the modelling errors that are inevitable in ecology. It is also encouraging that the conclusions about the effect of mutation rate are in accord with one's expectations.

There are many questions left open for future work. One of the most interesting is whether the discovery that chaos can sometimes speed up the approach to evolutionary dominance can be understood in more detail. It is not entirely clear to me from the paper whether the chaos is a result of changed time scales in the model, which themselves cause the speedup, or whether the chaos is involved in a more direct causal way.

Other things that might be tried are to look at problems such as ecological succession, say in forest regrowth after a fire, in which the species change the environment as experienced by others, and these others then outcompete the original species. This is a case where one is looking at transients rather than stable solutions, but it still ought to be possible to obtain the right sort of behaviour.

I look forward to more study of this problem in the future; for now, this paper is exciting as well as interesting.

Alistair Mees

In a series of recent papers, Vincent and Brown, Brown and Vincent, and most recently Vincent and Fisher, have developed a simple yet powerful approach to ecological coevolution, providing a very natural dynamic formulation of Maynard Smith's pregnant notion of an "evolutionary stable strategy" (ESS). While many dynamic–game formulations of evolution have been proposed (see the 1988 textbook of Hofbauer and Sigmund), theirs' seems to me to be particularly natural and satisfying. The notion of Vincent *et al* does not require that the winning evolutionary "coalition" tend to demographic equilibrium, only that it persist and be stable against invasions.

The mathematical constructs of the theory have direct biological interpretation. In particular this is true of the fitness-generating function $G(v, u, p, N)$, which may be interpreted as the individual fitness of a single organism of phenotype v, which is functioning in a biological environment consisting of a mixed–species population of phenotypes u in numbers $x_i = p_i N$, for $i = 1, 2, \ldots, r$.

In line with this interpretation, one might argue for modelling evolutionary changes in a species' phenotype, not by equation (18) but instead by

$$u_i(t+1) - u_i(t) = \alpha \frac{\partial G(u_i, u, p, N)}{\partial v}.$$

That is, the phenotype will change, not in reference to an eventual demographic equilibrium p^*, N^*, but rather in response to the system's contemporary state. After all, natural selection acts through the current biotic and abiotic environment, and even a wily "selfish gene" will not necessarily gain by prematurely embracing the ultimately inevitable ESS.

J.Hofbauer and K. Sigmund. 1988. *The Theory of Evolution and Dynamical Systems*, London Mathematical Society.

Robert McKelvey

REPLY

Author's Reply (to Bob McKelvy)

One of the features of our approach is that the mechanism (genetic or otherwise) for producing an ESS need not be known, only that it exists. It was on this basis that equation (18) was chosen. Any equilibrium point obtained using (18) will satisfy the ESS necessary conditions. Equation (18) was not intended to imply a specific mechanism for strategy dynamics. Otherwise, as Bob points out, how could individuals respond to an unknown equilibrium state? This is an useful observation and it would be of interest to seek a simple mechanistic strategy dynamic which would guarantee that the system arrives at the ESS.

Tom Vincent

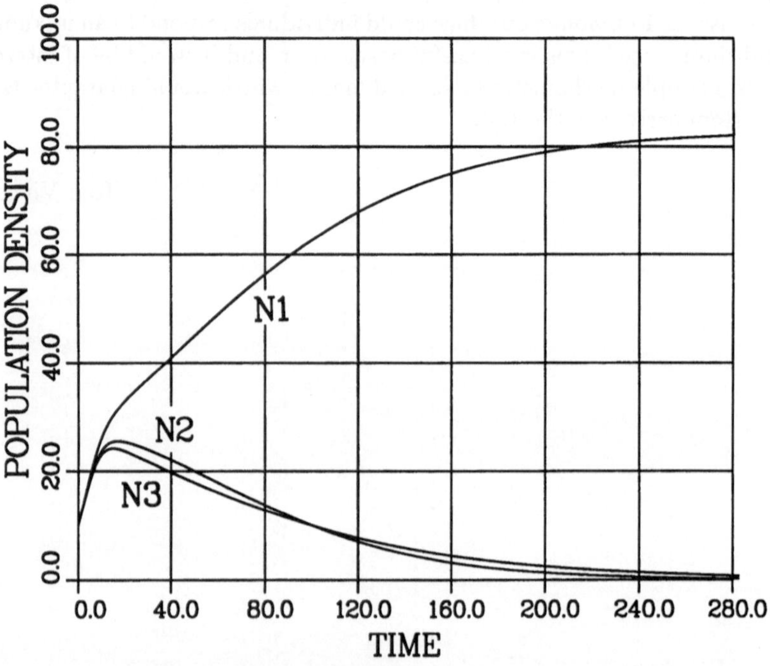

Figure 1. An ESS with two mutants.

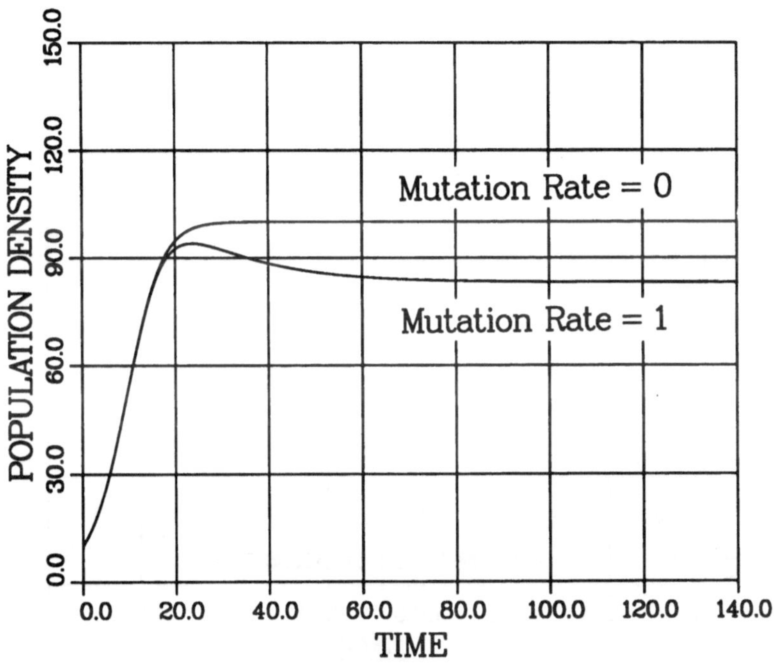

Figure 2a. An initial non ESS strategy with and without strategy dynamics. Population change with time.

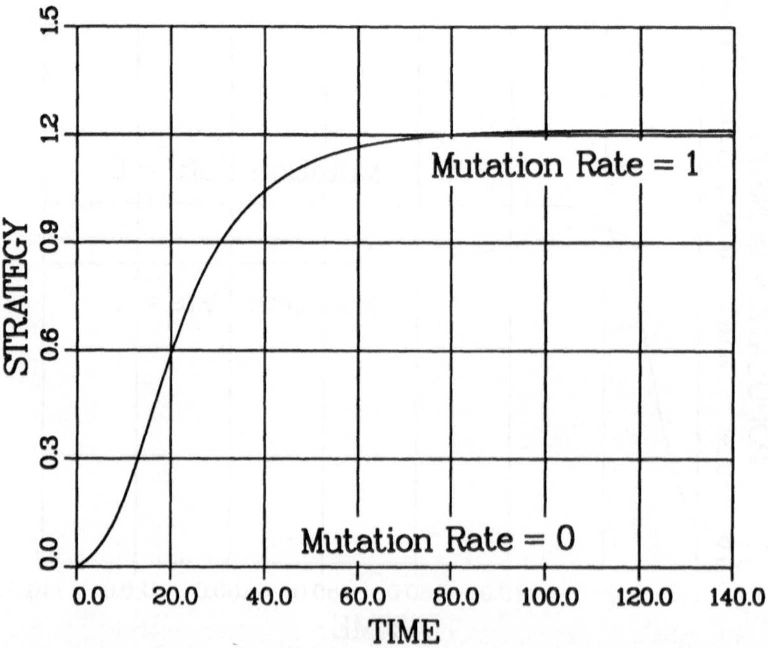

Figure 2b. An initial non ESS strategy with and without strategy dynamics. Strategy change with time.

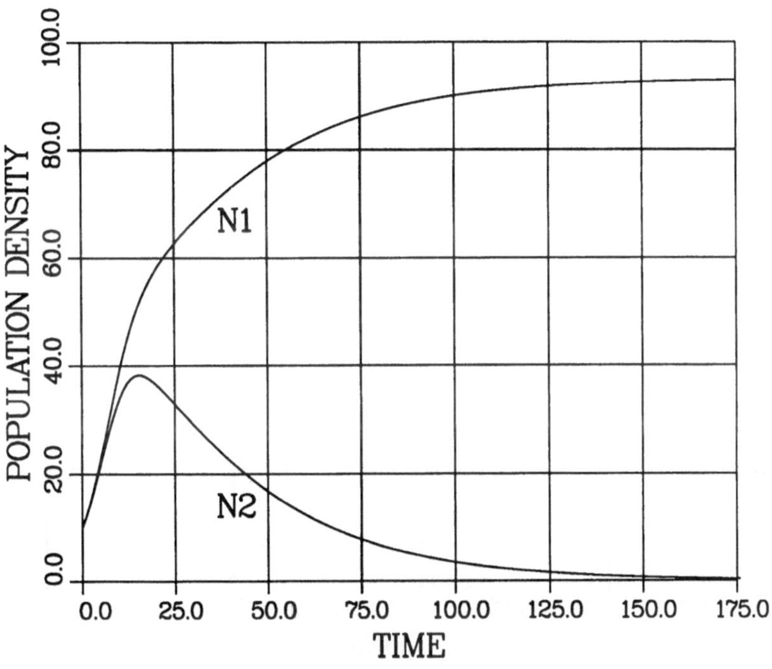

Figure 3a. Two non ESS strategies which do coexist.

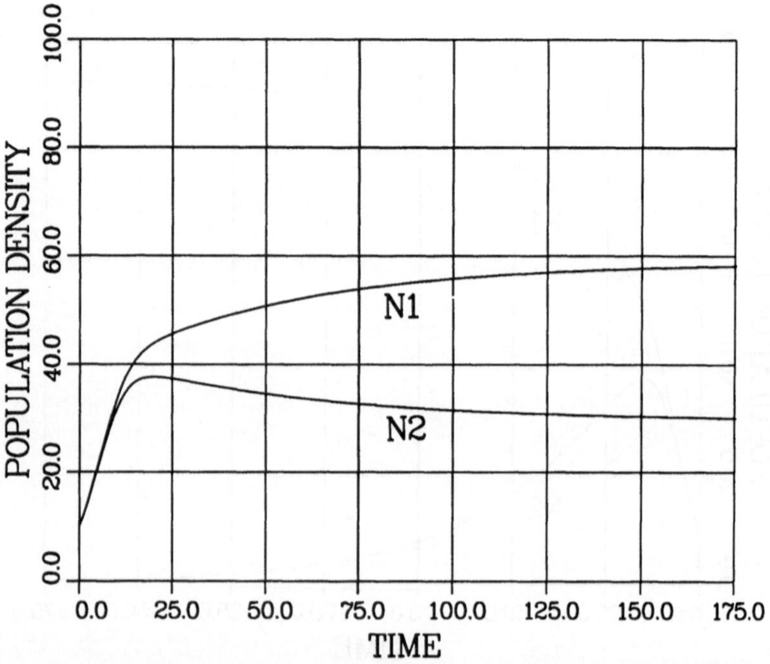

Figure 3b. Two non ESS strategies which do not coexist.

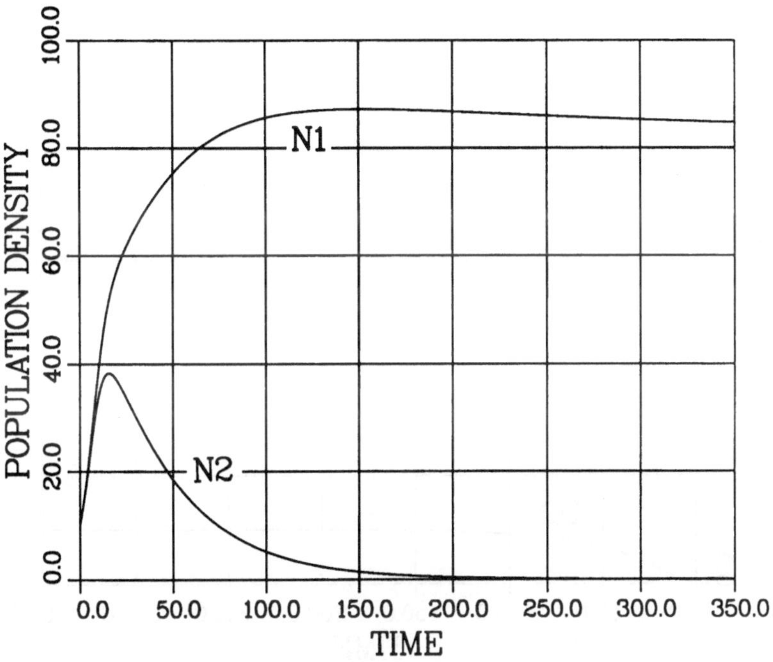

Figure 4a. Two initially non ESS strategies with strategy dynamics. Population change with time (mutation rate $\alpha = .1$).

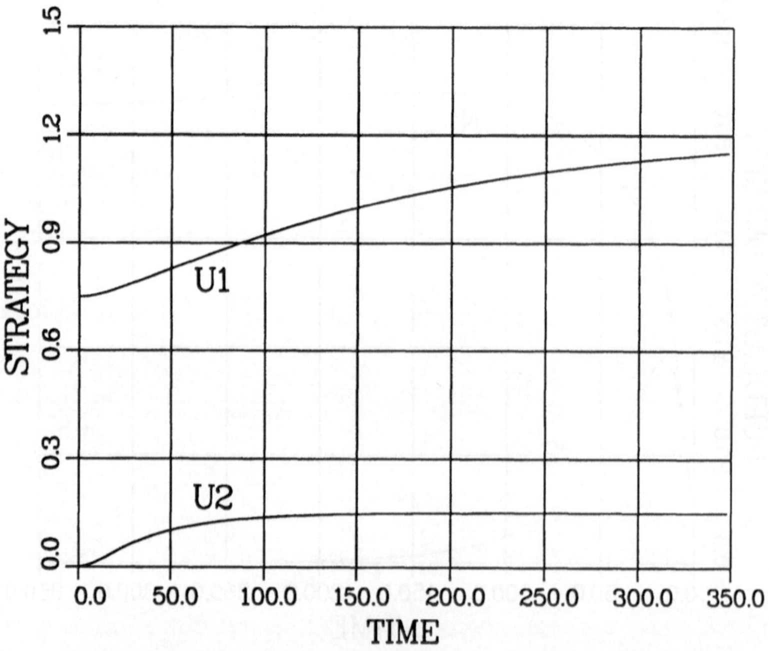

Figure 4b. Two initially non ESS strategies with strategy dynamics. Strategy change with time (mutation rate $\alpha = .1$).

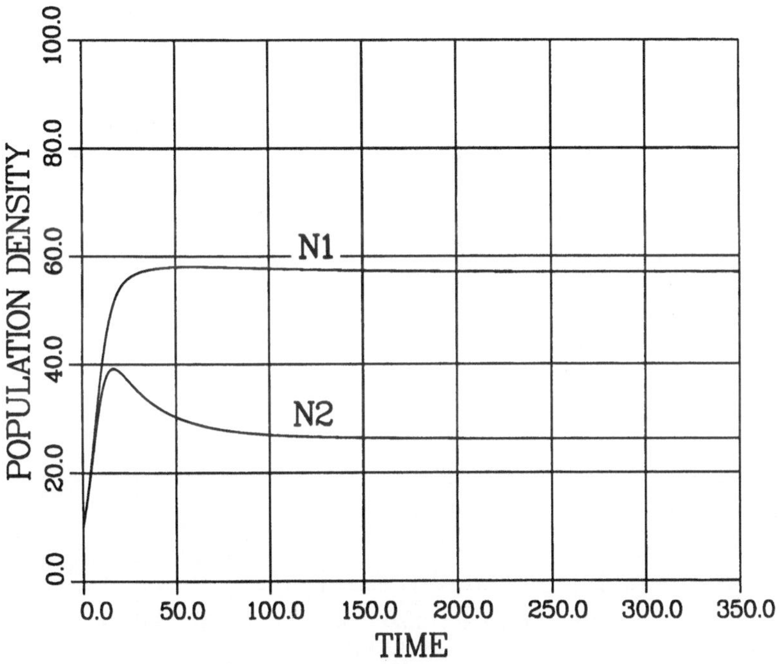

Figure 4c. Two initially non ESS strategies with strategy dynamics. Population change with time (mutation rate $\alpha = 1$).

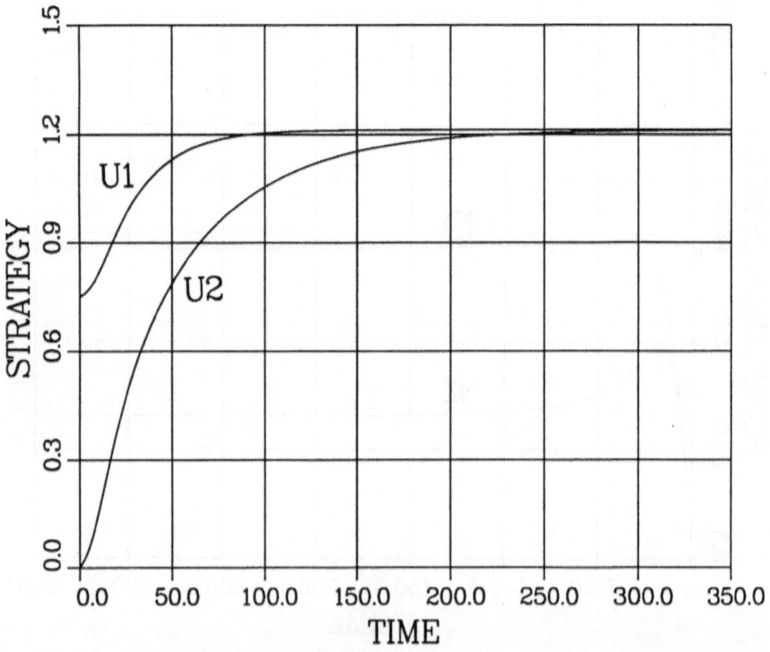

Figure 4d. Two initially non ESS strategies with strategy dynamics. Strategy change with time (mutation rate $\alpha = 1$).

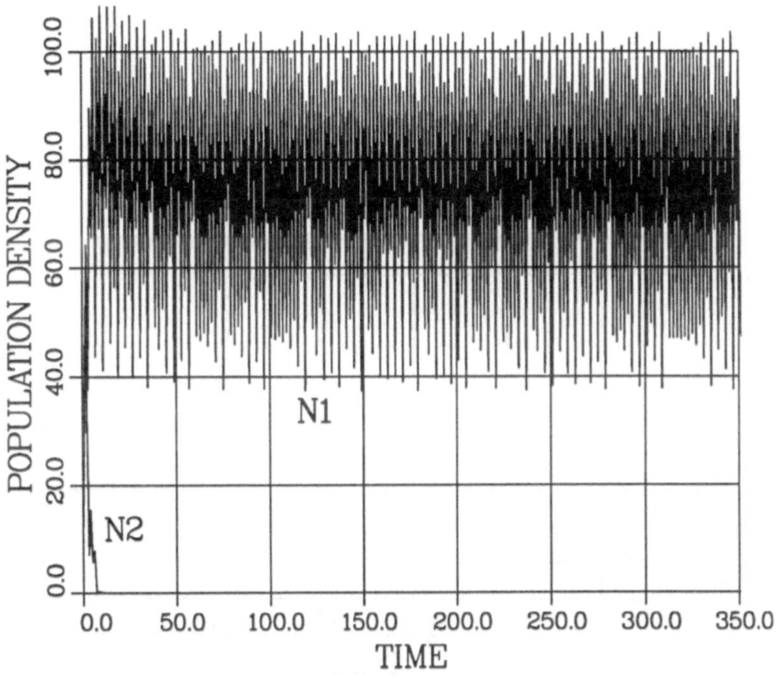

Figure 5a. Two initially non ESS strategies under chaotic motion. Population change with time (mutation rate $\alpha = .1$).

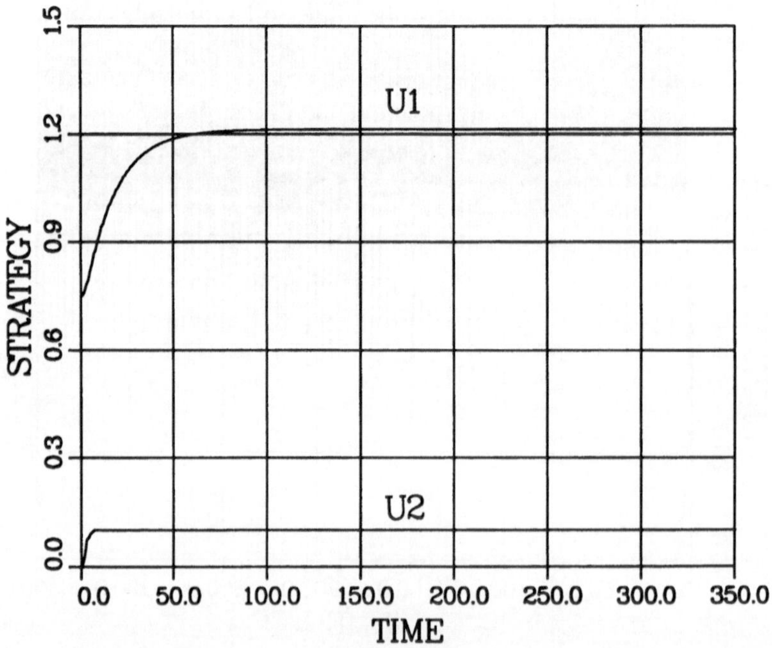

Figure 5b. Two initially non ESS strategies under chaotic motion. Strategy change with time (mutation rate $\alpha = .1$).

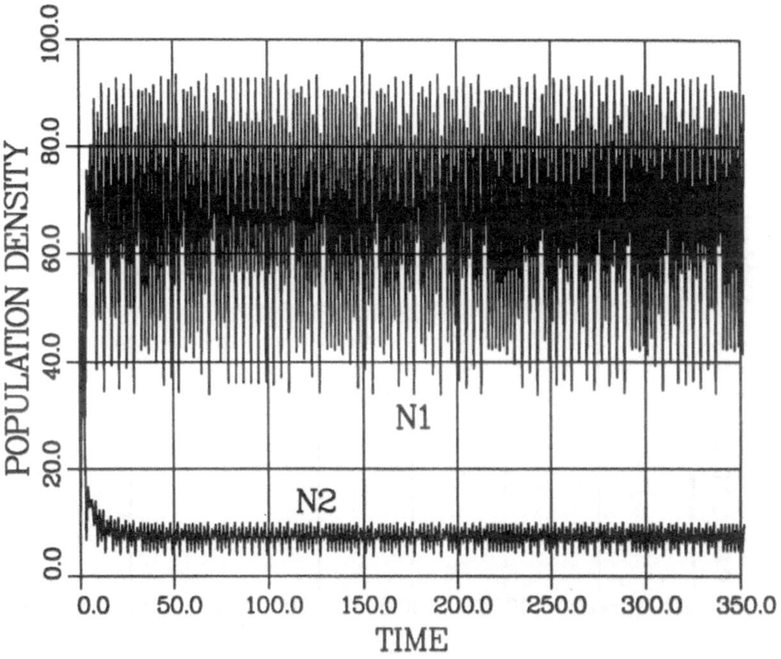

Figure 5c. Two initially non ESS strategies under chaotic motion. Population change with time (mutation rate $\alpha = 1$).

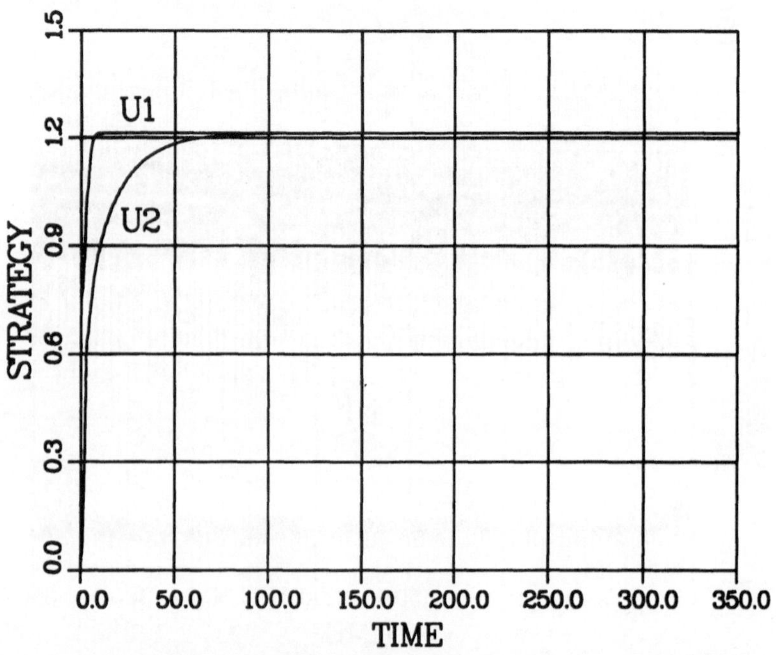

Figure 5d. Two initially non ESS strategies under chaotic motion. Strategy change with time (mutation rate $\alpha = 1$).

Community Organization Under Predator-Prey Coevolution

JOEL S. BROWN

Abstract

Here, I consider a simple predator-prey model of coevolution. The number of species comprising the ESS is influenced by a parameter that determines the predator's niche breadth. Depending upon the parameter's value the evolutionarily stable strategies (ESS) may contain any number of prey and predator species. Evolutionarily, these different ESS's all emerge from the same model. Ecologically, however, these ESS's result in very different patterns of community organization. In all cases, the removal of a species from the community results in evolutionary instability. However, from the perspective of the human lifetime this may never be perceived. The removal of a species also has ecological implications in that, depending upon the ESS, some species are keystone. Hence, to understand and effectively manage different communities, we may need to know as much about the evolutionary as the ecological contexts of the constituent species.

Introduction

Evolutionary ecology, that harmonious blend of what is evolutionarily feasible with what is ecologically acceptable, has often been difficult to achieve. The study of evolutionary processes has generally been the domain of population genetics where the focus is on the forces that result in changes in gene frequency. In contrast, students of ecological processes generally focus on factors that influence population sizes. Blending evolutionary processes via genes with ecological processes via population sizes has often proven awkward.

Evolutionary game theory provides an alternative approach to evolutionary ecology (Vincent and Brown, 1988). The focus is on the heritable phenotypes (henceforth referred to as strategies) of individuals. Evolution is defined as change in strategy frequency and the focus of ecological processes is on the per capita growth rates of strategies. The evolutionary and ecological processes are brought together in the concept of evolutionarily stable strategies (ESS; Maynard Smith and Price, 1973). An ESS is a strategy or set of strategies which when common in the population cannot be invaded by rare alternative strategies. At an ESS, the strategies are ecologically stable in that they can persist together through time, and they are evolutionarily stable in that no alternative strategies can increase in frequency when rare.

Here, I consider a simple predator-prey model of coevolution. Elsewhere, we have studied this model in the context of harvesting a natural resource (Vincent and Brown, 1987) and as a model of the coevolution of two prey species in the face of a predator (Brown and Vincent, 1987a). I extend our analysis of this model by varying a parameter that determines the predator's niche breadth. In such an evolutionary analysis the number of strategies or species that comprise the ESS are not fixed, but rather, emerge from the game itself. As such, varying the predator's niche breadth results in a variety of ESS's and predator-prey communities. Evolutionarily, these different ESS's all emerge from the same model. Ecologically, however, these ESS's result in very different patterns of community organization.

Predator-Prey Coevolution

The Model

I envision a community of prey and predators (the model follows that of Brown and Vincent, 1987a). I assume that the prey possess a strategy that influences their competitive interaction with other prey individuals and that influences their capture susceptibility by predators. I assume that different strategies among the prey denote different species and that collectively the prey, regardless of species, are evolutionarily identical. The prey are evolutionarily identical if they share the same evolutionary strategy set and if the evolutionary consequences of those strategies are the same for all prey individuals (see Brown and Vincent, 1987a for a formal definition of this term). I assume that the predators possess a strategy that influences their capture success on prey individuals. As in the case of the prey, I assume that different strategies among the predators denote different species and that all predator individuals are evolutionarily identical ; although the prey and predators are not.

Let the per capita growth rate, in discrete time, of prey individuals using strategy u be given by:

$$G_1(u, \underline{u}, \underline{v}, \underline{N}, \underline{P}) = 1 + r_1 \left(\frac{K(u) - \sum_{i=1}^{m} a(u, u_i) N_i}{K(u)} \right) - \sum_{j=1}^{n} b(u, v_j) P_j \quad (1)$$

and let the per capita growth rate of predator individuals using strategy v be given by:

$$G_2(v, \underline{u}, \underline{v}, \underline{N}, \underline{P}) = 1 + r_2 \left(1 - \frac{\sum_{j=1}^{n} P_j}{c \sum_{i=1}^{m} b(u_i, v) N_i} \right) \quad (2)$$

The above functions contains the following elements: 1) $u \in E^1$ is the strategy of the focal prey species, 2) $v \in E^1$ is the strategy of the focal predator species, 3) $\underline{u} = (u_1, ..., u_m)$ is the vector of different strategies found among the prey where m denotes the number of prey species and u_i is the strategy of the ith species, 4) $\underline{v} = (v_1, ..., v_n)$ is the vector of different strategies found among the predators where n denotes the number of predator species and v_j is the strategy of the jth predator species, 5) $\underline{N} = (N_1, ..., N_m)$ is the vector of prey species' population sizes where N_i gives the population size of prey species u_i, and 6) $\underline{P} = (P_1, ..., P_m)$ is the vector of predator species' population sizes where P_j gives the population size of predator species v_j.

Equations (1) and (2) are the fitness generating functions (Vincent and Brown, 1984, 1988; Brown and Vincent, 1987c) of the prey and predators, respectively. They are so called because the fitness of prey or predator individuals using strategy u_i or v_j are determined by setting $u = u_i$ or $v = v_j$, respectively. The fitness generating function of the prey, for instance, shows that the expected fitness of an individual is influenced by its own strategy, u, the strategies of other prey and predators, $\underline{u}$ and $\underline{v}$, and the frequency and abundances of those strategies among the prey and predators, $\underline{N}$ and $\underline{P}$ (Brown and Vincent, 1987b). Because the prey are assumed to be evolutionarily identical they all share the same fitness generating function given by (1). By the same assumption, all predators share the same fitness generating function given by (2). However, because prey and predator are not assumed to be evolutionarily identical; the expected fitness of a predator individual cannot be found by substituting its strategy for u in (1), or vice-versa for prey individuals.

The fitness generating functions have been written as difference equations and as such the prey and predator dynamics are assumed to occur in discrete time. They could have been written as differential equation analogues without altering the techniques and results of the analyses to follow (see Vincent and Fisher, 1988). The population dynamics of prey species i and predator species j, respectively, are given by

$$N_i(t+1) = N_i(t)G_1(u_i, \underline{u}, \underline{v}, \underline{N}, \underline{P})$$
$$P_j(t+1) = P_j(t)G_2(v_j, \underline{u}, \underline{v}, \underline{N}, \underline{P})$$

where t and $t+1$ represent successive time units. Throughout, I assume that the predator and prey population dynamics lead to stable equilibria.

The prey's fitness generating function assumes that they compete among themselves according to Lotka-Volterra competition equations, and that the probability of a given prey individual being captured by predators is independent of prey density, i.e. the predators exhibit a Type I functional response curve (Holling, 1965). The strategy of the individual prey influences: its carrying capacity, $K(u)$, the competitive effect of any given prey species, $a(u, u_i)$, and the probability of being captured by any given predator species, $b(u, v_j)$. In what follows, I ascribe the following functional forms to the carrying capacity, competition coefficients, capture probabilities, and parameters:

$$K(u) = 100\exp(-(u^2)/2)) \tag{3a}$$
$$a(u, u_i) = \exp(-(u-u_i)^2/4) \tag{3b}$$
$$b(u, v_j) = 0.15\exp(-(u-v_j)^2/S) \tag{3c}$$
$$c = r_1 = r_2 = 0.25 \tag{3d}$$

The predator's fitness generating function assumes that the predators interfere with each other in such a way that their fitness is a decreasing function of the total number of predators. This interference effect is assumed to be independent of the predators' strategies. The strategy of an individual predator only influences its probability of capturing any given prey species, $b(u_i, v)$. From the predator's viewpoint the capture probability function (3c) has the following form:

$$b(u_i, v) = 0.15\exp(-(u_i - v)^2/S$$

The functions (3a) - (3c) have the following forms. Expression (3a) assumes that the carrying capacity of an individual prey takes on a maximum at $u = 0$ and declines normally as u deviates from 0. Expression (3b) assumes that individuals compete most with individuals using similar strategies. As such, the competitive effect of species u_i is greatest when $u = u_i$ and declines normally as u deviates from u_i or as u_i deviates from u. Expression (3c) assumes that the predators possess a best matching strategy to that of the prey. I assume that this matching strategy has been scaled so that the probability of a predator capturing a prey of species u_i is greatest when $v = u_i$; otherwise the probability of capture declines normally as v deviates from u_i or as u deviates from v_j.

The parameter S in (3c) gives the niche breadth of the predators. A small value indicates a narrow niche breadth, the capture probability of a predator for a given prey species declines rapidly as the predator's strategy deviates from the prey's strategy. A large value for S indicates a broad niche breadth. In what follows, I consider the nature of the ESS's that emerge from this model as the niche breadth varies. In turn, I consider a very broad niche breadth ($S = 10$), a broad niche breadth ($S = 4$), a narrow niche breadth ($S = 1$), and a very narrow niche breadth ($S = 0.75$). (The niche breadth distinctions are arbitrary!)

Very Broad Niche Breadth

Let the predator's niche breadth be broad, i.e. $S = 10$. I begin by seeking an ESS that has a single prey species, u_1, and a single predator species, v_1. To satisfy the ESS criterion for evolutionary stability the prey and predator strategies must maximize individual fitness when these strategies are common; i.e. $N_i = 0$, and $P_j = 0$ for $i, j > 1$ (see Vincent and Brown, 1988). Since, the scalar strategy sets of the prey and predator are unbounded, necessary conditions to maximize G_1 at $u = u_1$, and G_2 at $v = v_1$ are given by

$$\frac{\partial G_1(u_1, \underline{u}, \underline{v}, \underline{N}^*, \underline{P}^*)}{\partial u} = 0 \tag{4a}$$

$$\frac{\partial G_2(v_1, \underline{u}, \underline{v}, \underline{N}^*, \underline{P}^*)}{\partial v} = 0 \tag{4b}$$

where $\underline{N}^* = (N_1^*, 0, ..., 0)$ and $\underline{P}^* = (P_1^*, 0, ..., 0)$ satisfy the conditions for ecological stability. By the assumption of equilibrial population dynamics, $\underline{N}^*$ and $\underline{P}^*$ satisfy the equilibrium conditions

$$G_1(u_1, \underline{u}, \underline{v}, \underline{N}^*, \underline{P}^*) = 1 \tag{5a}$$

$$G_2(u_1, \underline{v}, \underline{v}, \underline{N}^*, \underline{P}^*) = 1 \tag{5b}$$

where ($5a$) and ($5b$) are evaluated at the same points as ($4a$) and ($4b$).

The four expressions for evolutionary and ecological stability can be used to solve for four unknowns: the prey and predator strategies (u_1 and v_1), and their equilibrium population sizes (N_1^* and P_1^*). Substituting equations (1) and (2) for the fitness generating functions, and equations ($3a - d$) for the specific functional forms into ($4a, b$) and ($5a, b$) yields the following ESS candidate solution: $u_1 = v_1 = 0$, $N_1^* = 30.8$, and $P_1^* = 1.15$.

This candidate solution can be shown graphically as prey and predator frequency dependent adaptive landscapes (Fig. 1). Such a landscape plots G_1 and G_2 as functions of u and v, respectively, where the other terms have been set to their ESS values (Brown and Vincent, 1987b; see Wright, 1931, for the origin and application of adaptive landscapes in population genetics). Because the prey and predator adaptive landscapes take on global maxima at $u = u_1$ and $v = v_1$, respectively, the above candidate solution is an ESS. In fact, it can be shown that this ESS is global with respect to strategy frequencies. Hence, the ESS is unique; it is the sole ESS for the model when $S = 10$.

When the predator has a very broad niche breadth the ESS contains a single prey and a single predator species. The single predator species is not surprising. Because there is only a single prey species, selection on the predators is stabilizing; all predator individuals are selected to possess the best matching strategy for that prey. The reason for the single prey species is not so obvious. The prey are subjected to two sources of disruptive selection. First, intra-specific competition selects for individuals to be non-conformist in strategy. Second, the single predator species with the perfect matching strategy also selects for non-conformity. In other words, an individual prey can reduce the competitive effect of others and reduce its mortality rate to predators by adopting a strategy other than $u = u_1$. What selects for conformity? The prey's ESS strategy maximizes carrying capacity. Hence, the advantage of having a higher carrying capacity acts as a source of stabilizing selection. When the predator possesses a very broad niche breadth the disruptive selection exerted by the predators is small in relation to the stabilizing selection of a higher carrying capacity. The resultant ESS has a single prey species.

Broad Niche Breadth

Let the predator's niche breadth be broad, i.e. $S = 4$. Again, I begin by seeking an ESS that has a single prey species, u_1, and a single predator species, v_1. Conditions ($4a, b$) and ($5a, b$) determine the following ESS

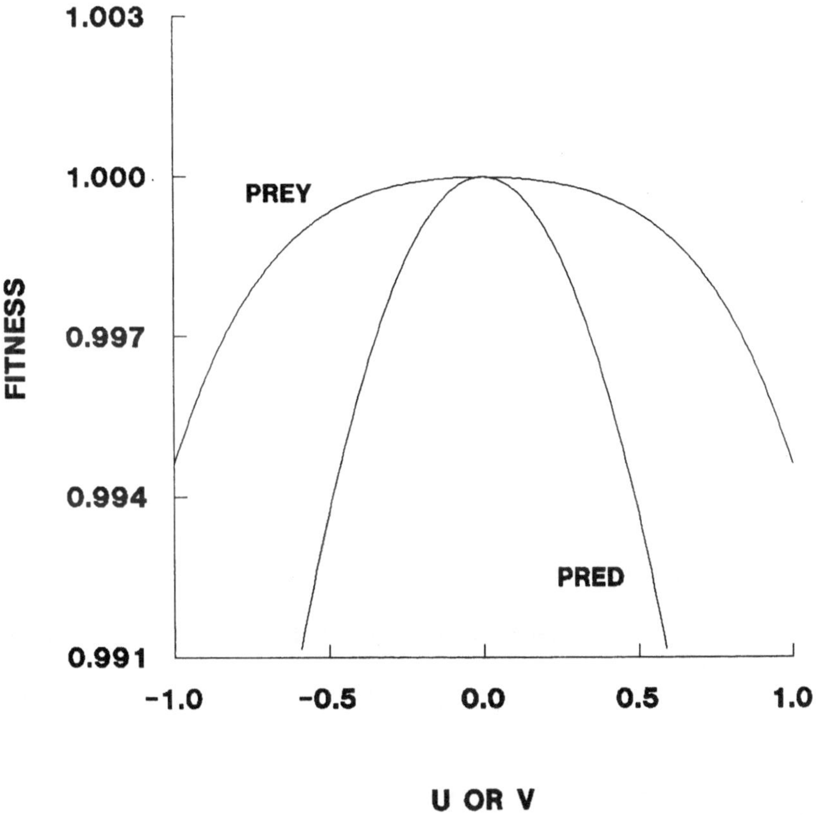

Figure 1. The prey (PREY) and predator (PRED) adaptive landscapes for the model when the predator's niche is very broad (S = 10). Each landscape gives the per capita growth rate of an individual (FITNESS) as a function of its strategy (u or v). The strategies of others, $\underline{u}$ and $\underline{v}$, and their densities in the population, $\underline{N}$ and $\underline{P}$, have been set to those of the ESS candidate solution with one prey and one predator species. This candidate solution ($u_1 = v_1 = 0$) is an ESS.

candidate solution: $u_1 = v_1 = 0$, $N_1^* = 30.77$, and $P_1^* = 1.15$. Adaptive landscapes can be used to evaluate this candidate solution (Fig. 2). In this case, the predator's adaptive landscape takes on a global maximum at $v = v_1$; however, the prey's adaptive landscape no longer takes on a global maximum at $u = u_1$. In fact, the prey's candidate strategy is now a local minimum. The two peaks in the prey's adaptive landscape suggest an ESS with two prey strategies.

Consider now an ESS with two prey species, u_1 and u_2, and one predator species, v_1. To satisfy the ESS criterion for evolutionary stability the prey and predator strategies must maximize individual fitness when these strategies are common; i.e. $N_i = 0$ for $i > 2$, and $N_j = 0$ for $j > 1$. Again, because the strategy sets are unbounded, the necessary conditions to maximize G_1 at u_1 and u_2, and G_2 at v_1 are given by

$$\frac{\partial G_1(u_1, \underline{u}, \underline{v}, \underline{N}^*, \underline{P}^*)}{\partial u} = 0 \tag{6a}$$

$$\frac{\partial G_1(u_2, \underline{u}, \underline{v}, \underline{N}^*, \underline{P}^*)}{\partial u} = 0 \tag{6b}$$

$$\frac{\partial G_2(v_1, \underline{u}, \underline{v}, \underline{N}^*, \underline{P}^*)}{\partial v} = 0 \tag{6c}$$

where $\underline{N}^* = (N_1^*, N_2^*, 0, ..., 0)$ and $\underline{P}^* = (P_1^*, 0, ..., 0)$ satisfy the conditions for ecological stability. As such, they satisfy the equilibrium conditions

$$G_1(u_1, \underline{u}, \underline{v}, \underline{N}^*, \underline{P}^*) = 1 \tag{7a}$$

$$G_1(u_2, \underline{u}, \underline{v}, \underline{N}^*, \underline{P}^*) = 1 \tag{7b}$$

$$G_2(u_1, \underline{v}, \underline{v}, \underline{N}^*, \underline{P}^*) = 1 \tag{7c}$$

where $(7a - c)$ are evaluated at the same points as $(6a - c)$.

The six expressions for evolutionary and ecological stability can be used to solve for six unknowns: the prey and predator strategies (u_1, u_2, and v_1), and the prey's and predator's equilibrium population sizes (N_1^*, N_2^*, and P_1^*). Substituting equations (1) and (2) for the fitness generating functions, and equations $(3a - d)$ for the specific functional forms into $(6a - c)$ and $(7a - c)$ yields the following ESS candidate solution: $u_1 = 0.90$, $u_2 = -0.90$, $v_1 = 0$, $N_1^* = N_2^* = 19.35$, and $P_1^* = 1.19$.

This candidate solution can be shown graphically as prey and predator adaptive landscapes (Fig. 3). Because the prey and predator adaptive landscapes take on global maxima at $u = u_1$ and $u = u_2$, and $v = v_1$,

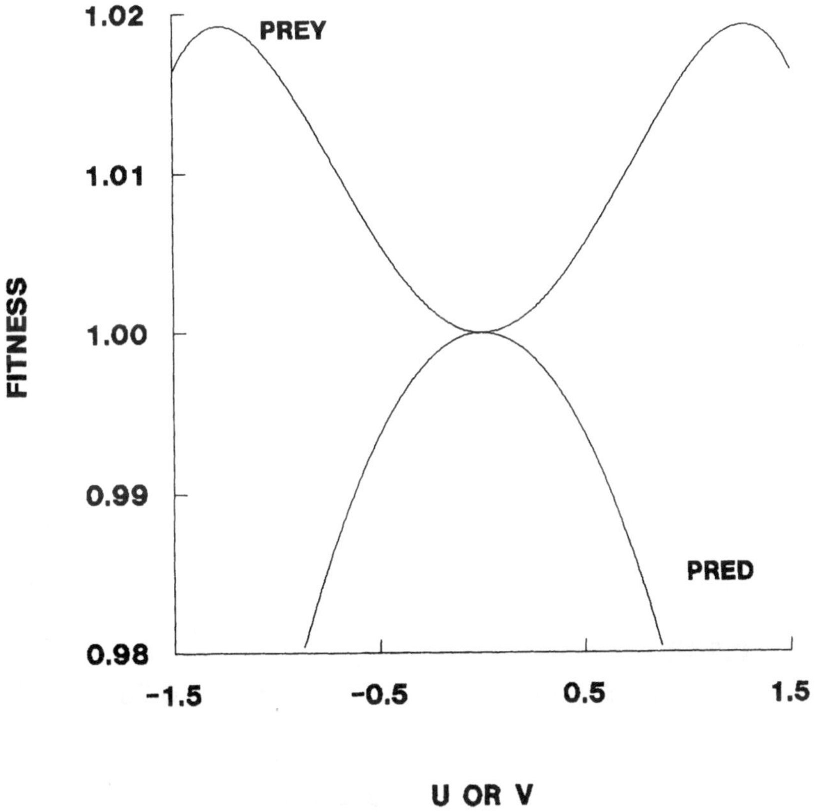

Figure 2. The prey and predator adaptive landscapes for the model when the predator's niche is broad ($S = 4$). The strategies of others, $\underline{u}$ and $\underline{v}$, and their densities in the population, $\underline{N}$ and $\underline{P}$, have been set to those of the ESS candidate solution with one prey and one predator species. This candidate solution ($u_1 = v_1 = 0$) is not an ESS.

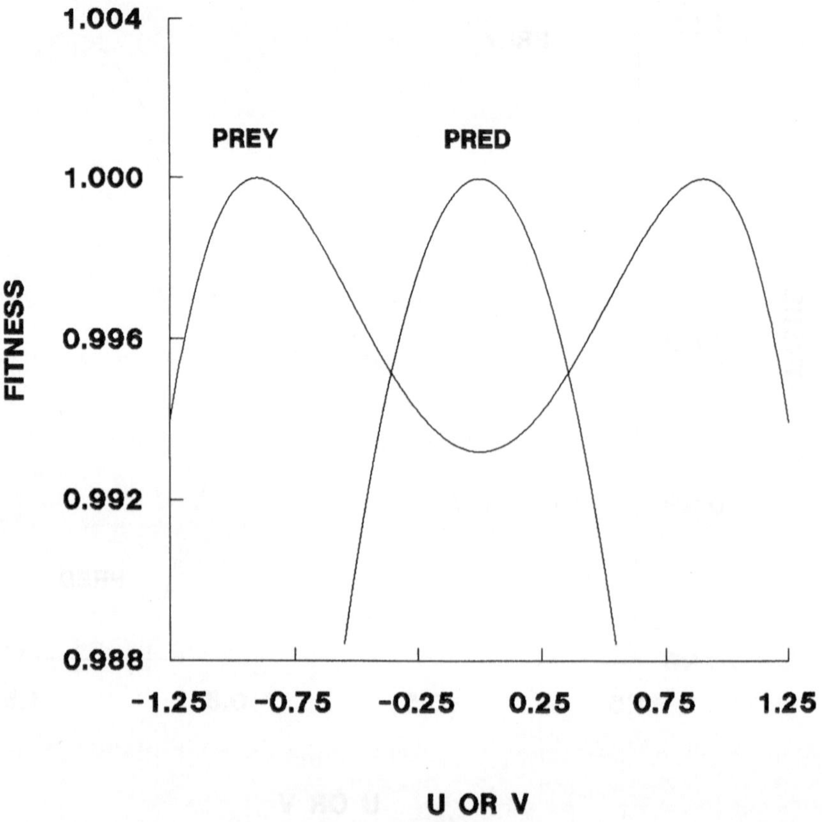

Figure 3. The prey and predator adaptive landscapes for the model when the predator's niche is broad ($S = 4$). The strategies of others, $\underline{u}$ and $\underline{v}$, and their densities in the population, $\underline{N}$ and $\underline{P}$, have been set to those of the ESS candidate solution with two prey and one predator species. This candidate solution ($u_1 = 0.90, u_2 = -0.90, v_1 = 0$) is an ESS.

respectively, the above candidate solution is an ESS. It can be shown that this ESS is global with respect to strategy frequencies.

When the predator has a broad niche breadth the ESS contains two prey and a single predator species. In this case, the disruptive selection exerted by the the single predator species on the prey is sufficiently strong to overcome the stabilizing effect of maximizing carrying capacity. The predator species is the selective force behind the diversification of prey species. Now, however, the presence of the single predator species is no longer obvious. The two prey species exert disruptive selection on the predator to diverge and "specialize" on the two prey species. Because of the broad niche breadth this disruptive selection is quite weak; the benefits to a predator of altering its strategy towards one of the prey species does not compensate the predator for its loss of ability on the other prey species.

The ESS when $S = 4$ is a more diverse and interesting ecological community than when $S = 10$. By removing one or the other species in this community, it is possible to examine its organization from the perspective of an ecologist. Consider the effects on the population sizes of the remaining two species when one of the prey or predator species is removed while holding the strategies of the remaining species fixed. This assumes that the ecological process of changes in population sizes occurs much more rapidly than evolutionary changes in the species' strategy. Such species removal experiments are performed by ecologists to examine the significance of species interactions in organizing communities and permitting species coexistence (Schoener, 1983; Connell, 1983).

Under the community of the present ESS, the removal of prey species u_1 from the community results in a large density increase of prey species u_2 from 19.35 to 33.33, and a small density decrease in the predator species v_1 from 1.19 to 1.02. Without further knowledge of the system, an ecologist might conclude that the two prey species are intense competitors and that prey species 1 is relatively unimportant to the predator species. The removal of predator species v_1 results in an increase in prey population sizes to $N_1 = N_2 = 46.15$. Interestingly, the ecological result of removing the predator species suggests that the predator is unimportant in the organization of this community or the coexistence of the two prey species. This insignificance of the predator, however, is only in "ecological" time. In the absence of the predator, the two prey species would no longer experience as much disruptive selection, and both would evolve towards the common strategy of $u_1 = u_2 = 0$. The predator is evolutionarily necessary for the maintenance of the two prey species but is not ecologically necessary for their coexistence (Brown and Vincent, 1987a).

Narrow Niche Breadth

Let the predator's niche breadth be narrow, i.e. $S = 1$. In consideration of the previous analysis, I begin by seeking an ESS that has two prey species, u_1 and u_2, and a single predator species, v_1. Conditions $(6a-c)$ and $(7a-c)$ determine the following ESS candidate solution: $u_1 = 0.94$, $u_2 = -0.94$, $v_1 = 0, N_1^* = N_2^* = 33.65$, and $P_1^* = 1.05$. In this case, the prey's adaptive landscape takes on global maxima at $u = u_1$ and $u = u_2$; however, the predator's adaptive landscape no longer takes on a global maximum at $v = v_1$ (Fig. 4). The predator's candidate strategy is now a local minimum. The two peaks in the predator's adaptive landscape suggest the possibility of an ESS with two prey strategies and two predator strategies.

Consider now an ESS with two prey species, u_1 and u_2, and two predator species, v_1 and v_2. To satisfy the ESS criterion for evolutionary stability the prey and predator strategies must maximize individual fitness when these strategies are common; i.e. $N_i = 0$ and $N_j = 0$ for $i, j > 2$. The necessary conditions to maximize G_1 at u_1 and u_2, and G_2 at v_1 and v_2 are given by

$$\frac{\partial G_1(u_i, \underline{u}, \underline{v}, \underline{N}^*, \underline{P}^*)}{\partial u} = 0 \tag{8a}$$

$$\frac{\partial G_2(v_j, \underline{u}, \underline{v}, \underline{N}^*, \underline{P}^*)}{\partial v} = 0 \tag{8b}$$

for $i, j = 1, 2$ and where $\underline{N}^* = (N_1^*, N_2^*, 0, ..., 0)$ and $\underline{P}^* = (P_1^*, P_2^*, 0, ..., 0)$ satisfy the conditions for ecological stability. As such, they satisfy the equilibrium conditions

$$G_1(u_i, \underline{u}, \underline{v}, \underline{N}^*, \underline{P}^*) = 1 \tag{9a}$$

$$G_2(v_j, \underline{v}, \underline{v}, \underline{N}^*, \underline{P}^*) = 1 \tag{9b}$$

for $i, j = 1, 2$.

The eight expressions for evolutionary and ecological stability can be used to solve for the prey and predator strategies (u_1, u_2, v_1, and v_2), and the prey's and predator's equilibrium population sizes (N_1^*, N_2^*, P_1^*, and P_2^*). Substituting equations (1) and (2) for the fitness generating functions, and equations $(3a-d)$ for the specific functional forms into $(8a, b)$ and $(9a, b)$ yields the following ESS candidate solution: $u_1 = 0.79$, $u_2 = -0.79$, $v_1 = 0.56$, $v_2 = -0.56$, $N_1^* = N_2^* = 28.71$, and $P_1^* = P_2^* = 0.60$.

The prey and predator adaptive landscapes at this candidate solution take on global maxima at $u = u_1$ and $u = u_2$, and $v = v_1$ and $v = v_2$,

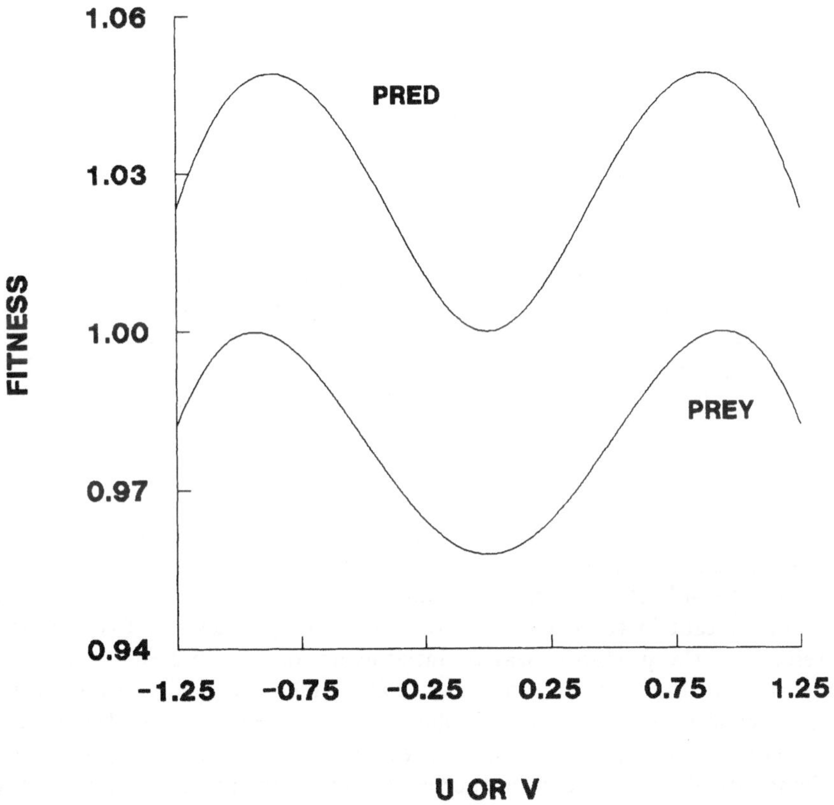

Figure 4. The prey and predator adaptive landscapes for the model when the predator's niche is narrow ($S = 1$). The strategies of others, $\underline{u}$ and $\underline{v}$, and their densities in the population, $\underline{N}$ and $\underline{P}$, have been set to those of the ESS candidate solution with two prey and one predator species. This candidate solution ($u_1 = 0.94,\ u_2 = -0.94,\ v_1 = 0$) is not an ESS.

respectively; the above candidate solution is an ESS (Fig 5). It can be shown that this ESS is global with respect to strategy frequencies.

When the predator has a narrow niche breadth the ESS contains two prey and two predator species. In this case, the disruptive selection imposed by the two prey species on the single predator species is sufficiently strong to promote diversification of the predators into two species. Each predator species "specializes" on a different prey species. However, each predator strategy is not as extreme as its corresponding prey's strategy. Each predator strategy reaches a balance where the benefits of altering its strategy towards one of the prey species does not compensate the predator for its loss of ability on the other prey species.

The ESS when $S = 1$ forms a more diverse ecological community than when $S = 4$. As before, the properties of this community in ecological time can be investigated by examining the consequences of removing a prey or a predator species while holding the strategies of the remaining species fixed. Under the present community, the removal of prey species u_1 from the community results in a small density increase of prey species u_2 from 28.71 to 29.46, a large density increase of predator species v_2 from 0.60 to 1.05, and the extinction of predator species v_1. The removal of a prey species results in the extinction of its corresponding specialist predator species.

The removal of predator species v_1 results in a density increase of prey species u_1 from 28.71 to 63.00, a decrease of prey species u_2 from 28.71 to 9.54, and an increase of predator species v_2 from 0.60 to 0.72. The ecological result of removing a predator species suggests that the predator is unimportant in the coexistence of the two prey species. Of course, the presence of the predators was essential evolutionarily and the removal of any species results in a community that, while ecologically stable, is no longer evolutionarily stable. Ecologically, the prey are essential to the coexistence of the two predator species; as such they are keystone species (Paine, 1966). The predator species, however, are not necessary for the coexistence of the two prey species.

Very Narrow Niche Breadth

As a final scenario, let the predator's niche breadth be very narrow, i.e. $S = 0.75$. In consideration of the previous analysis, consider an ESS candidate that possesses two prey and two predator species. Applying conditions $(8a-b)$ and $(9a-b)$ yields the following ESS candidate solution: $u_1 = 0.72$, $u_2 = -0.72$, $v_1 = 0.57$, $v_2 = -0.57$, $N_1^* = N_2^* = 29.54$, and $P_1^* = P_2^* = 0.60$. Inspection of the adaptive landscapes of this candidate

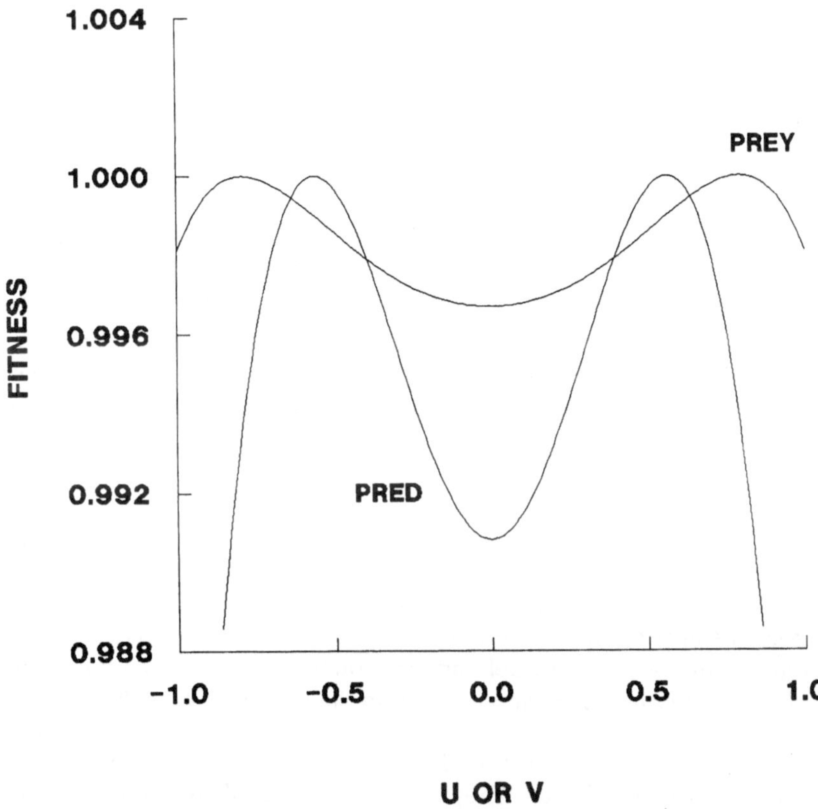

Figure 5. The prey and predator adaptive landscapes for the model when the predator's niche is narrow ($S = 1$). The strategies of others, $\underline{u}$ and $\underline{v}$, and their densities in the population, $\underline{N}$ and $\underline{P}$, have been set to those of the ESS candidate solution with two prey and two predator species. This candidate solution ($u_1 = 0.79$, $u_2 = -0.79$, $v_1 = .56$, $v_2 = -.56$) is an ESS.

solution reveals that it is not an ESS (Fig. 6). The prey's landscape does not take on global maxima at $u = u_1$ and $u = u_2$. The three peaks in the prey's adaptive landscape, however, suggest the possibility of an ESS with three prey strategies and two predator strategies.

The ESS candidate solution for three prey and two predator species can be found by applying necessary conditions ($8a-b$) and ($9a-b$) for $i = 1, 2, 3$ and for $j = 1, 2$. These conditions yield ten equations that can be used to solve for the prey and predator strategies and their respective equilibrium densities. Substituting equations (1) and (2) for the fitness generating functions, and equations ($3a-d$) for the specific functional forms into ($8a, b$) and ($9a, b$) yields the following ESS candidate solution: $u_1 = 0.82$, $u_2 = -0.82$, $u3 = 0$, $v_1 = 0.55$, $v_2 = -0.55$, $N_1^* = N_2^* = 25.61$, $N_3^* = 9.18$, and $P_1^* = P_2^* = 0.59$. Inspection of the adaptive landscapes shows this candidate solution to be an ESS: the prey's takes on global maxima at $u = u_1$, u_2, and u_3, and the predator's at $v = v_1$ and v_2 (Fig. 7). It can be shown that this ESS is global with respect to strategy frequencies.

As the predator's niche breadth narrows the amount of disruptive selection exerted by each specialist predator on its respective prey species intensifies. This eventually selects for a third prey species that re-attains the strategy that maximizes carrying capacity. Each predator species adopts a strategy that balances its ability to capture this third prey species and one of the other prey species, i.e. u_3 and u_1, or u_3 and u_2, respectively.

The ESS when $S = 0.75$ forms a more diverse ecological community than when $S = 1$. As before, the properties of this community in ecological time can be investigated by examining the consequences of removing a prey or a predator species while holding the strategies of the remaining species fixed. The removal of prey species u_1 from the community results in a large density increase of prey species u_3 from 9.18 to 49.67, a large increase of predator species v_2 from 0.59 to 1.25, and the extinction of prey species u_2 and predator species v_1. Prey species 1 is critical to the organization of the community and the coexistence of several of the species; its removal causes the community to collapse to a single prey and predator species. The removal of prey species u_3 results in density increases of species u_1 and u_2 from 25.61 to 31.13, and slight decreases in predator species v_1 and v_2 from 0.59 to 0.58. Ecologically, prey species 3 contributes little to the organization of this community.

The removal of predator species v_1 results in a density decrease of prey species u_1 from 25.61 to 20.78, a large increase of prey species u_3 from 9.18 to 53.24, a large increase of predator species v_2 from 0.59 to 1.60, and the extinction of prey species u_2. Unlike the previous communities,

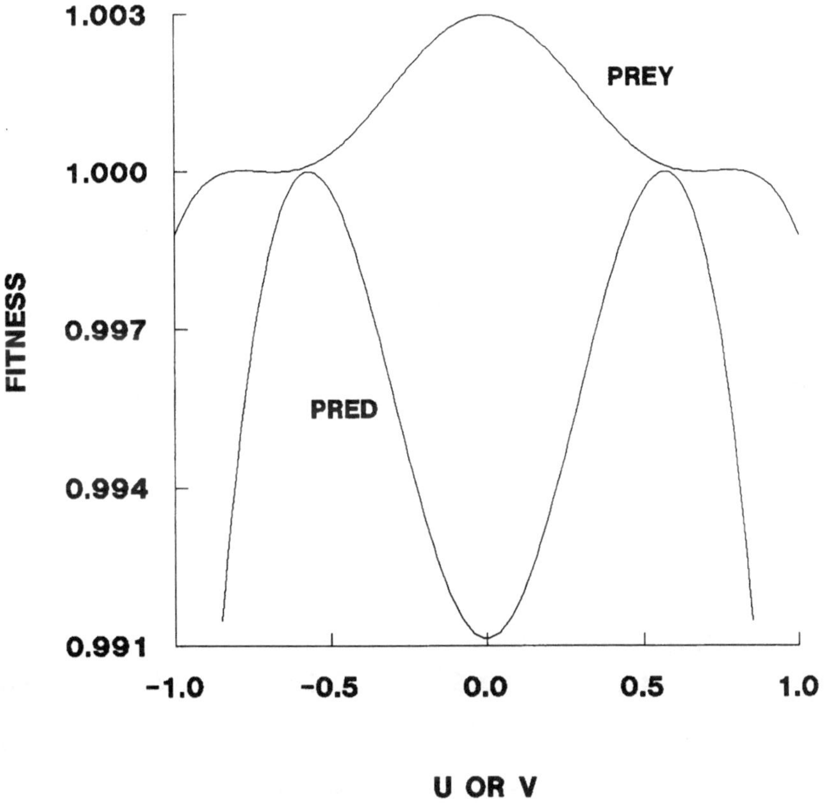

Figure 6. The prey and predator adaptive landscapes for the model when the predator's niche is very narrow ($S = 0.75$). The strategies of others, $\underline{u}$ and $\underline{v}$, and their densities in the population, $\underline{N}$ and $\underline{P}$, have been set to those of the ESS candidate solution with two prey and two predator species. This candidate solution ($u_1 = 0.72,\ u_2 = -0.72,\ v_1 = .57,\ v_2 = -.57$) is not an ESS.

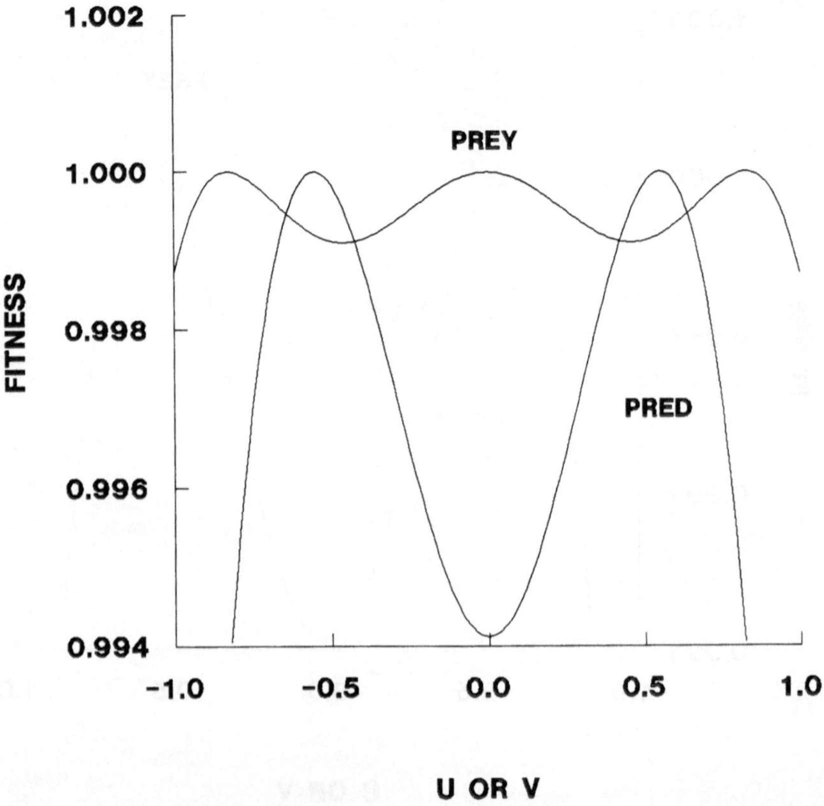

Figure 7. The prey and predator adaptive landscapes for the model when the predator's niche is very narrow ($S = 0.75$). The strategies of others, $\underline{u}$ and $\underline{v}$, and their densities in the population, $\underline{N}$ and $\underline{P}$, have been set to those of the ESS candidate solution with three prey and two predator species. This candidate solution ($u_1 = 0.82,\ u_2 = -0.82,\ u_3 = 0,\ v_1 = .55,\ v_2 = -.55$) is an ESS.

the predator species appears to be keystone (Paine 1966). The predator species is important for prey species coexistence. Prey species u_3 is the superior competitor because it possesses a strategy that maximizes carrying capacity. However, it suffers the highest predator mortality rate because both predator species are effective on u_3. The removal of predator species v_1 results in a large increase in the population size of prey species u_3 which consequently causes the decline in prey species u_1 and the competitive exclusion of prey species u_2. In summary: ecologically, the different species contribute variously to the community's organization; evolutionarily, all are essential to the ESS.

If the predator's niche breadth is narrowed still further from 0.75, the process of adding species to the ESS would continue; the next community would contain 3 prey and 3 predator strategies, followed by 4 prey and 3 predator strategies, etc. Presumably, these different communities would possess somewhat different community organizations that are emergent properties of the actual ESS's. In other words, simply knowing the number of species comprising the ESS, or knowing the underlying fitness generating functions does not immediately provide insights into the ecological roles played by the constituent species.

Discussion

A number of conceptual and mathematical frameworks have been applied to the organization of communities under natural selection. Inspired by the empirical observations that coexisting competitors often show regular differences in some ecologically relevant trait such as body size (Hutchinson, 1959), MacArthur and Levins (1967) asked how similar could a number of species be to each other without competitive exclusion occurring. From this initial analysis emerged the theory of limiting similarity (May and MacArthur, 1972; see May, 1974). Models of limiting similarity are primarily ecological and possess little evolutionary motivation for the niche positions of the coexisting species. In particular, empirical evidence for character divergence – competitor species when they occur together are often more different than when they occur apart – suggest a strong role for natural selection in shaping the interactions between competitors.

A body of theory has developed around the coevolution of competitors. Models have used quantitative genetic analyses (Slatkin, 1980; Taper and Case, 1985; Case and Taper, 1986), and the ESS concept (Lawlor and Maynard Smith, 1976; Abrams, 1986, 1987) to model the reciprocal evolutionary effects that pairs of competitors have on each other. In most of these models the number of species is held fixed and it is their ecological relationships to one another that evolve. Roughgarden (1983) has developed a framework for modeling coevolutionary processes that permits the number of species to emerge from the evolutionary model itself. Because evolutionary stability is determined by maximizing a group criterion of fitness, the domain of application of this approach is limited to systems that lack most forms of frequency dependence (Brown and Vincent, 1987b; Taper, 1988; but see Roughgarden, 1987 for a reply).

The coevolutionary framework advocated here applies the ESS concept to models built around fitness generating functions . The number of strategies or species comprising the ESS and their ecological relationships to one another emerge from the evolutionary dynamics. Any model of population dynamics can be given an evolutionary dimension by the inclusion of a scalar or vector-valued strategy that influences relevant parameters of the model. When applied to models of predator-prey dynamics the transformation yields a model of predator-prey coevolution.

Predator-prey coevolution has inspired a number of theories of evolution that differ somewhat from those usually applied to the coevolution of competitors. Specifically, the prey and predator are sometimes viewed as caught in an evolutionary rat race (Red Queen Hypothesis of Van Valen, 1973). The prey are under constant selection pressure to better evade predators, and the predators under constant selection to better capture prey. An evolutionary Red Queen may result; the prey and predators are constantly evolving while their ecological relationship becomes stationary (Rosenzweig and Schaffer, 1978; Schaffer and Rosenzweig, 1978). The conditions under which a Red Queen emerges is quite restrictive: first, the strategy sets of the prey and predator must be unbounded, and second, the strategy must not be involved in ecological tradeoffs or compromises - a more extreme value of the phenotype is always better. If these conditions are not satisfied then the evolutionary process can be expected to result in combinations of strategies or species that are ESS (Rosenzweig *et al* , 1987).

The application of evolutionary game theory to models of predator and prey not only permits investigation of the evolutionary impact of a predator on its prey (Vincent and Brown, 1987a) but, more significantly, it allows for models of coevolution within and between trophic levels. In the model of this paper, the diversity of species comprising the ESS are generated both within and between fitness generating functions. A diversity of species within a fitness generating function represents competitors at the same trophic level. Species from different fitness generating functions represent predator and prey on different trophic levels. As such, the application of evolutionary game theory in this paper breaks the traditional dichotomy between the coevolution of competitors and the coevolution of predator-prey systems. The analysis advocates an approach in which natural selection influences the organization of entire food webs.

The diversity of ESS's that emerge from the model of this paper are best understood in their evolutionary context. The number of species comprising the ESS result from the sources and intensities of disruptive and stabilizing selection. For the prey, the form of the competition coefficient, $a(u, u_i)$, is always a source of disruptive selection, and the functional form of the carrying capacity, $K(u)$, is always a source of stabilizing selection towards the strategy that maximizes carrying capacity. Depending upon the number and strategies of the predator species, the capture probability function, $b(u, v_j)$, is either a source of disruptive or directional selection. The intensity of this selective pressure depends upon the predator's niche breadth. As the predator's niche breadth contracts, the species composition of the ESS increases systematically. Starting from one prey and one predator species, each successive ESS alternatively adds one prey or one predator species.

From an ecological perspective, the organizations of the different ESS communities are quite different. The intensities of competitive and predator-prey interactions between species emerge as a by-product of the ESS. The intensities of interactions, in turn, determine to what extent a particular species of prey and predator plays an ecological role that is minor or keystone (see Table 1). In some communities, a given prey or predator species contributes little to the abundance and persistence of the other species, while in other communities the same species plays an ecological role that is essential for the coexistence of other species. Without some knowledge or hypotheses of the evolutionary context, it would be impossible for an ecologist to relate these communities to one another.

Evolutionary game theory provides a tool for integrating evolutionary and ecological processes into a more complete theory of evolutionary ecology. The use of different fitness generating functions permits considerations of species that may currently share very different strategy sets and ecologies. The consideration of ESS's that may contain several strategies from the same fitness generating function recognizes the possibility that many extant species may share the same set of evolutionarily feasible strategies. Finally, the ESS concept itself combines these ecological and evolutionary processes in predicting the community level consequences of evolution by natural selection.

Table 1. The ecological roles played by specific predator and prey species within the different ESS communities that emerge from varying the predator's niche breadth, S. The strategies u_1 and u_3 represent prey species and the strategy v_1 represents a predator species. A species is "keystone" if its removal from the community results in the extinction of one or more of the remaining species. A species' role is "minor" if its removal does not result in the extinction of any of the remaining species. If the species is not present in the particular ESS, its role is indicated by "N/A".

S	u_1	u_3	v_1
10	keystone	N/A	minor
4	minor	N/A	minor
1	keystone	N/A	minor
0.75	keystone	minor	keystone

Acknowledgements: I am grateful to the organizers, Drs. Alistair Mees and Thomas Vincent, for hosting a most enjoyable and stimulating workshop. Their comments and the comments of other participants improved the presentation of these ideas. I particularly want to thank A. Mees and Michael Deakin for their helpful reviews of the manuscript.

REFERENCES

[1] Abrams, P. 1986. Character displacement and niche shift analyzed using consumer-resource models of competition. *Theor. Pop. Biol.*, **29**, pp. 107-160.

[2] Abrams, P. 1987. Alternate models of character displacement. I. Displacement when there is competition for nutritionally essential resources. *Evolution* **41**, pp. 651-661.

[3] Brown, J.S. and T.L. Vincent. 1987a. Predator-prey coevolution as an evolutionary game. In Y. Cohen (ed.), *Lecture Notes in Biomathematics Vol. 73*, pp. 83-101. Springer-Verlag, Berlin.

[4] Brown, J.S. and T.L. Vincent. 1987b. Coevolution as an evolutionary game. *Evolution* **41**, pp. 66-79.

[5] Brown, J.S. and T.L. Vincent. 1987c. A theory for the evolutionary game. *Theor. Pop. Biol.* **31**, pp. 140-160.

[6] Case, T.J. and M.L. Taper. 1986. On the coexistence and coevolution of asexual and sexual competitors. *Evolution* **40**, pp. 366-387.

[7] Connell, J.H. 1983. On the prevalence and relative importance of interspecific competition: evidence from field experiments. *Amer. Natur.* **122**, pp. 661-696.

[8] Holling, C.S. 1965. The functional response of predators to prey density and its role in mimicry and population regulation. *Mem. Ent. Soc. Can.* **45**, pp. 1-60.

[9] Hutchinson, E.G. 1959. Homage to Santa Rosalia, or why are there so many kinds of animals? *Amer. Natur.* **93**, pp. 145-159.

[10] Lawlor, L.R. and J. Maynard Smith. 1976. The coevolution and stability of competing species. *Amer. Natur.* **110**, pp. 79-99.

[11] MacArthur, R.H. and R. Levins. 1967. The limiting similarity, convergence, and divergence of coexisting species. *Amer. Natur.*, **101**, pp. 377-385.

[12] May, R.M. and R.H. MacArthur. 1972. Niche overlap as a function of environmental variability. *Proc. Nat. Acad. Sci USA* **69**, pp. 1109-1113.

[13] May, R.M. 1974. *Stability and Complexity in Model Ecosystems, 2nd Edition*, Princeton University Press, Princeton, New Jersey, 265 p.

[14] Maynard Smith, J. and G.R. Price. 1973. The logic of animal conflicts. *Nature* **246**, pp. 15-18.

[15] Paine, R.T. 1966. Food web complexity and species diversity. *Amer. Natur.* **100**, pp. 65-67.

[16] Rosenzweig, M.L., J.S. Brown and T.L. Vincent. 1987. Red Queens and ESS: The coevolution of evolutionary rates. *Evol. Ecol.*, **1**, pp. 59-94.

[17] Rosenzweig, M.L. and W.M. Schaffer. 1978. Homage to the Red Queen. II. Coevolutionary responses to enrichment of exploitation ecosystems. *Theor. Pop. Biol.* **14**, pp. 158-163.

[18] Roughgarden, J. 1983. The theory of coevolution. In D.J. Futuyma and M. Slatkin (eds.), *Coevolution*, Sinauer, Sunderland, Massachusetts, pp. 33-64.

[19] Roughgarden, J. 1987. Community coevolution: A comment. *Evolution* **41**, pp. 1130-1134.

[20] Schaffer, W.M. and M.L. Rosenzweig. 1978. Homage to the Red Queen. I. Coevolution of predators and their victims. *Theor. Pop. Biol.*, **14**, pp. 135-157.

[21] Schoener, T.W. 1983. Field experiments on interspecific competition. *Amer. Natur.* **122**, pp. 240-285.

[22] Slatkin, M. 1980. Ecological character displacement. *Ecology,* **61**, pp. 163-177.

[23] Taper, M.L. 1988. The coevolution of resource competition: appropriate and inappropriate models of character displacement. *Bull. Soc. Popul. Ecol.* **44**, pp. 45-53.

[24] Taper, M.L. and T.J. Case. 1985. Quantitative genetic models for the coevolution of character displacement. *Ecology,* **66**, pp. 355-371.

[25] Van Valen, L. 1973. A new evolutionary law. *Evolutionary Theory* **1**, pp. 1-30.

[26] Vincent, T.L. and J.S. Brown. 1984. Stability in an evolutionary game. *Theor. Pop. Biol.* **26**, pp. 408-427.

[27] Vincent, T.L. and J.S. Brown. 1987. An evolutionary response to harvesting. In T.L. Vincent, Y. Cohen, W.J. Grantham, G.P. Kirkwood, and J.M. Skowronski (eds.), *Lecture Notes in Biomathematics*, Springer- Verlag, Berlin, Vol. 72, pp. 80-95.

[28] Vincent, T.L. and J.S. Brown. 1988. The evolution of ESS theory. *Ann. Rev. Ecol. Syst.* **19**, pp. 423-443.

[29] Vincent, T.L. and M.E. Fisher. 1988. Evolutionarily stable strategies in differential and difference equation models. *Evol. Ecol.* **2**, pp. 321-337.

[30] Wright, S. 1931. Evolution in Mendelian populations. *Genetics* **16**, pp. 97-159.

PARTICIPANT'S COMMENTS

What is most impressive about this study is that a really quite simple model generates a variety of complex and subtle behaviour. The different situations summarised in Table 1 and the diagrams come about as a result of the variation of only a very small number of parameters. The case $S = 4$ is particularly intriguing. It would be interesting to see how the results depend on the form assumed by Equations (3). Are there perhaps qualitative robustness theorems to be discovered?

I do, however have one serious misgiving about this type of endeavour. This is the almost total neglect of *genetics* — the assumption that it is the *phenotypes* which are heritable. It was the wedding of Darwinian concepts to Mendelian genetics that made our modern evolutionary theories viable. But the detailed analysis of simple genetic models has meant that a number of rather counter–intuitive results have appeared. The simplistic views of literal Darwinism can lead us astray. Thus Moran (*Ann. Human Genetics*, **27**,383) has shown that selection does not always increase mean fitness, and Deakin (*Aust. J. Biol. Sc.*, **26**,1443) demonstrated that in the standard two–locus model the fittest gametotype cannot be the most frequent.

As Moran's work occurs in the context of debunking Sewell Wright's "adaptive topography" concept, the term "adaptive landscape" in the present paper may be unfortunate. The analysis is, of course, correct. What could still be in doubt is the *biological relevance* of the concept.

Michael Deakin

In my comments on the paper "Strategy Dynamics and the ESS" by Tom Vincent elsewhere in this volume, I make some general remarks on the concept of evolutionary stability and why it is attractive. Joel's paper displays some of the power that comes from having a mathematical model that is in accord with ecological concepts. Its success in modelling diversity, or its lack, in predator-prey communities offers strong evidence that the ESS approach is going to be of major importance in the future.

The variable which he calls niche breadth turns out to be a bifurcation parameter, resulting in different numbers of predator and prey species as it takes on different values. There is certainly room here for a detailed bifurcation study, but already we can see interesting features. As in all bifurcation problems, there are at least two timescales, called here evolutionary (slow) and ecological (faster). It is particularly interesting that under certain conditions the system can possess ecological stability yet not evolutionary stability.

As an applied mathematician, I see an immediate important question as being whether the specific functional forms are critical to the results. The answer is almost certainly no, but some research into this seems called for.

Alistair Mees

REPLY

Michael Deakin is concerned about the lack of explicit genetic mechanisms in most strategic models of evolution. In fact, the strength of strategic (or phenotypic) models of evolution derive from their ability to side-step the morass of genetics that underlie heritability. I feel the rediscovery of Mendel's Laws was a two edged sword. On the one hand it greatly enhanced our understanding of inheritance and permitted precise models of expected outcomes of breeding experiments. On the other, it created a preoccupation with "elementary particles" of inheritance. This preoccupation has yielded much fruitful research (witness the results of genetic engineering), but it has stymied our understanding of the consequences of natural selection. Evolution by natural selection is a more robust notion than specific genetic mechanisms, for natural selection neither presupposes a particular genetic process nor requires one. That traits are heritable is essential, that genes are the vehicle of inheritance is only of secondary importance. Changing our definition of evolution from "change in gene frequency" to "change in strategy frequency" may breath new vitality into investigations of the consequences of natural selection.

Joel Brown

The Exploiters Conservationists Game: How to be an Effective Conservationist

YOSEF COHEN

Abstract

An ecosystem model, with exploiters and conservationists is examined in a game-theoretic set up. The Nash-equilibrium solution is derived for two conservation strategies: conservation of the resource and interference with the exploiters' effort. I show that direct interference leads to a more effective conservation, whereas conservation of the resource, particularly when it is modeled as a rate, leads to de facto subsidy of the exploiters.

§1. Introduction: the E-C Game

The following general game is analyzed: we have an ecosystem model, which describes the trajectories of its various compartments (e.g., species populations) as a function of time, the rate of exploitation effort, and the rate of conservation effort. Each of the exploiters (E) and conservationists (C) has a cost function which is to be minimized. The exploiters and conservationists do not cooperate. The question is then what would be the fate of the ecosystem under such a game. This question is posed more specifically in §2. In §3 I will discuss a discrete-time version of the game, with a slight modification of the way in which C interferes with E's exploitation effort. I will then show - by numerical example - that the success of C very much depends on the way they choose to interfere with E's effort. The game, which I dub the E-C game is then modeled in continuous-time as follows:

$$\frac{dx_i}{dt} = x_i(t) \left(f_i \left(x_1(t), \ldots, x_S(t) \right) - \sum_{j=1}^{M} \pi_i^j p_i^j + \sum_{j=1}^{N} \theta_i^j q_i^j \right) \qquad (1.1)$$

where:

- $x_i f_i$ models the interaction of each component of the ecosystem with the rest of the ecosystem;
- $x_i(t)$ is the quantity of compartment i of the ecosystem (e.g., population of species i) at time t;
- S is the number of ecosystem components;
- π_i^j is the efficiency of harvest of component i, by exploiter j;
- M is the number of exploiters;
- p_i^j is the rate of exploitation effort of the i-th compartment by the j-th exploiter;
- θ_i^j is the efficiency of conservation of component i, by conservationist j;
- N is the number of conservationists;
- q_i^j is the rate of conservation effort of the i-th compartment by the j-th conservationist.

The exploiters act in a competitive market, and I use a modification of the well known Cournot's quantity model (Cournot 1838; see also Levitan and Shubik 1971), where it is assumed that the exploiters are maximizing their profit. The demand from a specific exploiter depends on the price that the exploiter charges; i.e.,

$$d_i^j = \frac{\beta_i}{M}\left[V_i - a_i^j - \gamma_i\left(a_i^j - \bar{a}_i\right)\right] \tag{1.2}$$

where:

- d_i^j is the demand for product i, which is derived from compartment i, from exploiter j;
- β_i is the price sensitivity of the overall demand for product i;
- V_i is the price at which the demand for product i is zero;
- a_i^j is the price charged by exploiter j for product i;
- γ_i is the sensitivity of the demand to deviation of price from the mean market price for product i;
- $\bar{a}_i$ is the average market price for product i.

The parameters above could incorporate interactions of demands and prices for the various products, as well of the various exploiters and conservationists. I do not include such interactions for the sake of notational simplicity. Next, assume that the cost of harvest rate increases with decrease in the level of compartment i, and that each exploiter is maximizing his/her net income. Thus, the objective, P_i^j , of the j-th exploiter with regard to the i-th ecosystem component is to

$$\text{minimize } P_i^j = -\left[a_i^j - g_i(x_i)\right] d_i^j ; i = 1, \dots, S; j = 1, \dots, M. \tag{1.3}$$

where $g_i(.)$ is a decreasing function of its argument.

The conservationists seek to maximize the quantity of each of the i-th ecosystem component, and the cost (in units of utility) increases with the conservation effort rate. The objective, Q_i^j, of the j-th conservationist with regard to the i-th component is then to

$$\text{minimize } Q_i^j = -h_i\left(x_i, q_i^j\right); i = 1, \dots, S; j = 1, \dots, N \tag{1.4}$$

where $h_i(.,.)$ is an increasing function of its first argument and a decreasing function of its second. Equations (1.3) and (1.4) define a so-called n-person, non-zero sum game. To find the solution for the game (with the assumption that it exists), define the price charged for a product by a particular exploiter as his/her "control" variable. The Nash-equilibrium solution is then found by solving the set of equations

$$\frac{\partial P_i^j}{\partial a_i^j} = \frac{\partial Q_i^j}{\partial q_i^j} = 0. \tag{1.5}$$

Again, for the sake of simplicity no constraints on p_i^j and q_i^j are placed (any model which claims to be realistic will become very complex even for an ecosystem model with just a few states and thus numerical solutions will be necessary; it is therefore useless to discuss the abstract model above in more details than necessary). Since the solution at equilibrium is to be examined, x_i can be expressed as a function of p and q by setting the rhs of equation (1.1) to zero. If one assumes that the exploiters meet the demand, then $d_i^j = \xi_i^j x_i$ where ξ_i^j is some scaling parameter. The rhs of equations (1.3) and (1.4) become functions of p and q only and equation (1.5) can be solved with respect to p and q. When the various functions are defined such that equations (1.3) and (1.4) are quadratic in p and q, and with the coefficient of $(p_i^j)^2$ in equation (1.3) and $(q_i^j)^2$ in equation (1.4) both positive, then the game has a unique solution, which gives the optimal exploitation rate by E, the optimal conservation rate by C, and the level of each ecosystem component at equilibrium. This is the so-called Nash-equilibrium solution of a quadratic, $N+M$ person non-zero sum game (Intriligator 1971, Basar and Ho 1974, Friedman 1977, Case 1979).

§2. Example: Single Species with M Exploiters and N Conservationists

To demonstrate a specific potential application, consider a single species ecosystem with M exploiters and N conservationists. In the sequel, parameters (all positive constants) shall be denoted by lower-case Greek and upper case letters. Variables shall be denoted by lower-case letters. To further distinguish between variables and parameters, the former shall be super-scripted and the latter sub-scripted (where necessary). Special attention to interpretation of the parameters is paid (see Table 1). Suppose that there are M exploiters and N conservationists, all concerned with a particular species. Furthermore, suppose that the species population behaves according to the logistic. Then

$$\frac{dx}{dt} = \rho x\left(1 - \frac{x}{K}\right) - x\sum_{j=1}^{M} \pi_j p^j + x\sum_{j=1}^{N} \theta_j q^j \tag{2.1}$$

where the various parameters and variables, along with their meaning and units of measurement are defined in Table 1.

The variables p_i and q_i are the so-called control variables - the exploiters and the conservationists decide on the level of application of their respective effort so that some criterion is minimized. The equations which tie the behavior of the exploiters become:

$$d^j = \frac{\beta}{M}\left[V - a^j - \gamma\left(a^j - \bar{a}\right)\right] . \tag{2.2}$$

Next, we assume that the cost of harvest increases linearly with decrease in population size at equilibrium, and that each exploiter is maximizing his/her net income. The objective, P^j, of the j-th exploiter is then to

$$\text{minimize } P^j = -\left[a^j - (\alpha - \eta_j x)\right] d^j, j = 1, \ldots, N. \tag{2.3}$$

The conservationists seek to maximize the number of animals per unit of time, and the cost of conservation effort rate increases quadratically. Thus, the objective, Q^j, of the j-th conservationist is to

$$\text{minimize } Q^j = -x^2 + \zeta_j (q^j)^2, j = 1, \ldots, M. \tag{2.4}$$

Equations (2.3) and (2.4) define the $M+N$-person, non-zero sum quadratic game.

Several questions arise with respect to the above game: (1) Suppose that the exploiters, E, and the conservationists, C, do not cooperate, each seeking to minimize his/her goal; then what would be the fate of species x at equilibrium? (2) If there is a positive equilibrium to the outcome of this game, how does it depend on the parameter values of the model? (3) What happens to the equilibrium outcome of the game when one of the players deviates from his/her optimal policy? (4) What happens when the number of participants increases?

When E and C do not cooperate, we seek the so-called Nash-equilibrium solution. Since the game defined above is quadratic, a unique Nash-equilibrium solution - for particular parameter values - exists. To gain some insight into the dependence of the solution on the various parameters, assume that both the exploiters and the conservationists enter the market symmetrically and that the parameter values result in strictly convex functions (this assumption will be addressed in some detail later). Thus, we set $\pi_i \equiv \theta_i \equiv 1, a \equiv a_i, \eta \equiv \eta_i, d \equiv d^i, \zeta \equiv \zeta_i, p \equiv p^i, q \equiv q^i, P \equiv P^i$, and $Q \equiv Q^i$. Then equations (2.1) to (2.4) become

$$\frac{dx}{dt} = \rho x \left(1 - \frac{x}{K}\right) - Mxp + Nxq, \tag{2.5}$$

$$d = \frac{\beta}{M}\left[V - a\right], \tag{2.6}$$

$$\text{minimize} \;\; P = -\left[a - (\alpha - \eta x)\right] d, \tag{2.7}$$

and

$$\text{minimize} \;\; Q = -x^2 + \zeta q^2. \tag{2.8}$$

This simplification renders the model less realistic. Yet, its salient features remain unchanged - in particular, the dependence of the various solutions on the parameter values. It should be emphasized that P in equation (2.7) is minimized at equilibrium: the game considered herein is static, as opposed to a dynamic game, where P is required to be minimized at each time.

When E does not know what action C takes (i.e., when q is not accessible to E), then E's strategy should be to minimize P in equation (2.7) with respect to any possible q. This leads to the so-called rational reaction set of E, which is denoted by D_E. Similar considerations lead to the rational reaction set of C, denoted by D_C. Since the game is quadratic, and since p and q are unconstrained (we are interested in the situation when the Nash-equilibrium exists and is positive–see below), D_E and D_C are derived from the relations

$$D_E = \left\{(a \mid \frac{\partial P}{\partial a} = 0\right\} \text{ and } D_C = \left\{(q \mid \frac{\partial Q}{\partial q} = 0\right\} \qquad (2.9)$$

respectively. The solution of equations (2.9) defines the Nash-equilibrium solution. Next, substitute the value of x at equilibrium in the equation for a, as derived from the relation (2.9), and solve for d in equation (2.6) by substituting the value of a from equation (2.9) to get:

$$d = \frac{\beta K\eta}{2M\rho}(Nq - Mp) + \frac{\beta}{2M}(V - \alpha + K\eta). \qquad (2.10)$$

Assuming that the exploiters are regulating their effort p so as to meet the demand, with some scaling parameter, ξ, so that $\xi p = d$, the rational reaction set for E in terms of p is then:

$$D_E = \left\{p \mid p = \frac{1}{M}\frac{\beta\rho(V - \alpha + K\eta)}{(K\beta\eta + 2\rho\xi)} + \frac{N}{M}\frac{\beta K\eta}{(K\beta\eta + 2\rho\xi)}q\right\}. \qquad (2.11)$$

The rational reaction set for C is:

$$D_C = \left\{(q \mid q = -\frac{K^2 N\rho}{K^2N^2 - \rho\zeta} + \frac{K^2MN}{K^2N^2 - \rho^2\zeta}p\right\}. \qquad (2.12)$$

Since eqs. (2.11) and (2.12) are linear in p and q, their solution, which gives the Nash-equilibrium, is unique:

$$\left.\begin{aligned} p^* &= \frac{\beta}{M}\frac{K^2(V-\alpha) - \frac{\rho^2\zeta}{N^2}(V - \alpha + K\eta)}{2K^2\xi - \frac{\rho\zeta}{N^2}(K\beta\eta + 2\rho\xi)} \\ q^* &= \frac{K^2}{N}\frac{\beta(V-\alpha) - 2\rho\xi}{2K^2\xi - \frac{\rho\zeta}{N^2}(K\beta\eta + 2\rho\xi)} \\ x^* &= \frac{K\rho\zeta}{N^2}\frac{\beta(V-\alpha) - 2\rho\xi}{2K^2\xi - \frac{\rho\zeta}{N^2}(K\beta\eta + 2\rho\xi)} \end{aligned}\right\} \qquad (2.13)$$

where p^* and q^* define the optimal rates of harvest effort and conservation effort, and x^* denotes the population equilibrium when the conservationists and the exploiters do not cooperate.

The dependence of the solution on the parameter values can now be examined. First, note that as M, the number of exploiters increases, the exploitation rate p^* decreases. Furthermore , neither the value of q^* nor that of x^* depend on M. The reason is simply because the way the E-C game is modeled (through equations 2.5-2.8): the conservationists are not concerned with the harvest effort directly; as the harvest effort increases, the conservationists increase their conservation effort (see equation 2.12) as applied to the animal population directly. This in turn allows higher effort by the exploiters (there are more animals and they are cheaper to exploit). Thus, there may be an important lesson for conservationists in choosing their strategy. It may be more effective to interfere with the objective of the exploiters directly (by increasing their costs or decreasing the demand for the product - see section §3) than indirectly by subsidizing the exploiters through support of their resource. Next, note that as the number of exploiters increases, the effort of each, p^*, decrease. As the number of conservationists increases, their effort q^* decreases. Finally, because efforts are modeled as rates (see Table 1), the number of animals at the Nash-equilibrium, x^*, decreases as the conservationists effort increases. This again is the result of the fact that by conserving the resource directly the conservationists are in fact subsidizing the exploiters. The sensitivity of p^*, q^*, and x^* to changes in M and N is summarized in Fig. 1: p^* decreases as both M and N increase, but is more sensitive to changes in M than to changes in N; q^* and x^* are not affected by M, but decrease as N increases.

The next question pertains to the stability of the Nash equilibrium. To explain the concept of stability of the Nash-equilibrium solution in the E-C game, consider the following: Suppose that one of the Es deviates from the Nash-equilibrium strategy. The rest of the players now observe E's move, and change their strategy accordingly, so that their costs are minimized. Now the specific E changes his/her strategy to minimize P, and the rest of the players react to E's latest move, and so on. If the infinite sequence that is generated by any "strategy perturbation" converges back to the Nash-equilibrium, we say that the Nash-equilibrium solution is stable; if convergence occurs for small strategy perturbations only, we say that the Nash-equilibrium is locally stable; otherwise it is unstable. The corresponding alternating moves just described lead to the iterative relations (see Basar and Olsder 1982):

$$p_{k+1} = Ap_k - B \tag{2.14}$$

when one of the Es is the starting player, where

$$A = \frac{K^3\beta\eta}{2(K\beta\eta + \rho\xi)\left(K^2 - \frac{\rho^2\zeta}{N^2}\right)}, \tag{2.15}$$

and to

$$q_{k+1} = Aq_k - B' \tag{2.16}$$

when one of the Cs is the starting player (the explicit terms B and B' are messy functions of the parameters and are not given here). Since the Nash-equilibrium solutionunder the iterative procedure described above - is stable for $-1 < A < 1$, we conclude from equation (2.15) that the outcome of the game becomes more stable as N, the number of conservationists, increases.

§3. A More Successful Conservation Effort

In section §2 we saw that by conserving the resource directly, the conservationists are in fact subsidizing the exploiters. Next, let us see how the conservationists can become more effective, by interfering directly with the exploiters' effort. To do that, I will introduce a slightly modified model. Assume a discrete recruitment function and let

$$x(k+1) = \sigma x(k) + \rho x(k)\left[1 - \frac{x(k)}{K}\right] - Mp(k)x(k) \tag{3.1}$$

where now:

$x(k)$ - animal density at time k;
σ - the proportion of animals surviving from k to $k+1$;
ρ - intrinsic rate of growth;
K - carrying capacity;
M - number of exploiters;
$p(k)$ - proportion of animals harvested at time k by an exploiter;
$Mp(k)x(k)$ - number of animals harvested at time k.

Denote by $e(k)$ the effort by an exploiter. Then the total effort is $Me(k)$. Assume that the number of animals harvested by an exploiter, $p(k)x(k)$, increases linearly with the effort, and decreases with the number of exploiters:

$$p(k)x(k) = \alpha e(k)/M \tag{3.2}$$

where α - positive constant (number of animals harvested per unit effort). The cost of exploitation, by an exploiter, increases linearly with the effort by the exploiter, and with the effort by the conservationists:

$$c(k) = \Gamma e(k) + N\phi q(k) = \frac{\Gamma}{\alpha} M p(k)x(k) + N\phi q(k) = \delta M p(k)x(k) + N\phi q(k) \tag{3.3}$$

where:

δ	- a positive constant;
ϕ	- positive constant;
N	- number of conservationist;
$q(k)$	- conservation effort at time k.

The demand function (given that the exploiters enter the market symmetrically), from an exploiter is

$$d(k) = \frac{\beta}{M}[V - a(k)] \tag{3.4}$$

where:

β	- overall market sensitivity;
V	- price at which demand is zero;
$a(k)$	- price charged by an exploiter at time k.

The exploiter is interested in maximizing profits with respect to the price charged:

$$\text{minimize} \;\; P = -[a(k) - c(k)]d(k) \tag{3.5}$$

$$= [a(k) - \delta M p(k)x(k) - N\phi q(k)]\frac{\beta}{M}[V - a(k)].$$

With the assumption that the exploiter meets the demand we get that

$$d(k) = p(k)x(k). \tag{3.6}$$

We can therefore solve for P with respect to the price, a, substitute it in the equation for demand, and solve equation (3.6) to determine $p(k)$ and $x(k)$.

The conservationist tries to maximize P for the exploiter, with respect to $q(k)$. At the same time, he/she tries to minimize the conservation effort, which increase quadratically with the effort:

$$\text{minimize} \;\; Q = -P + \gamma[q(k)]^2. \tag{3.7}$$

where γ is a positive constant. Under the definitions (3.1) through (3.7) we no longer have a quadratic game, and the solution with respect to p and q can be achieved numerically. First, the Nash-equilibrium is obtained by solving

$$\frac{\partial P}{\partial a} = \frac{\partial Q}{\partial q} = 0$$

which results in

$$a^* = \frac{2\gamma MV + \beta\phi^2 N^2 V + 2\delta\gamma M^2 p^* x^*}{4\gamma M + \beta\phi^2 N^2} \tag{3.8}$$

$$q^* = \frac{\beta\phi N(V - \delta M p^* x^*)}{4\gamma M + \beta\phi^2 N^2} \tag{3.9}$$

with the effect of the parameters on the Nash-equilibrium solution readily interpretable. To see how p^* (and therefore in turn x^*) changes with respect to M and N, we solve (3.1) with respect to x^* and substitute in (3.8) and (3.9). We then use the assumption that $d^* = p^* x^*$, solve the demand equation (3.4) for a^*, and substitute in equation (3.8). Since the solution of the pair (p^*, q^*) can only be achieved numerically, its sensitivity (for a particular set of values of the parameters) with respect to changes in M and N is displayed in Figure 2. Note that as opposed to the model in section §2, the proportion of animals harvested by an exploiter decreases as both M and N increases. This is a result of the direct interference of the conservationists with the harvest effort–in this example through increasing the cost of exploitation.

§4. Conclusions

Some (e.g., Wilson 1988) argue that the exploiters and conservationists are engaged in a two-person-zero-sum game. The fact is that because the currencies for the two groups differ (the former measure goals in \$, the latter, usually, in satisfaction), because some resources are used by conservationists, as well as by exploiters, and because of other reasons, the E-C game is not zero sum. The game, as formulated in here, can lead to several extensions. Particularly interesting would be an investigation of the effect of various coalitions on the equilibrium solution, and an ESS analysis in the vein of Brown and Vincent (this volume).

The major conclusion is that to be effective, the conservationists must concentrate their efforts in interfering directly with the harvest effort of the exploiters. By concentrating their efforts on the resource itself, the conservationists are in effect subsidizing the exploiters. In section §3, for example, the conservationists interfere by increasing the cost of harvest. Recent events, particularly with regard to the conservation of whales and the tactics that Green Peace adopt lend some credibility to this conclusion. Direct interference, however, has its side effects: it may (in fact it did) lead to confrontation and possibly violence (even by governments; e.g., the French). Such violence incurs costs and may change the dynamics of the E-C game: a fact which I have not addressed.

Acknowledgements: Contribution No. 12,500 of the University of Minnesota Agricultural Experiment Stations. The manuscript benefited from discussions with participants during the workshop.

REFERENCES

[1] Basar, T., and G. J. Olsder. 1982. *Dynamic noncooperative game theory.* Academic Press, New York.

[2] Case, J. H. 1979. *Economics and the competitive process.* New York University Press, New York.

[3] Cournot, A. 1838. *Recherches sur les principes mathematiques de la theorie des richesses.* Hachette, Paris. (English edition published in 1960 by Kelley, New York, under the title "Researches into the mathematical principles of the theory of wealth, trans. by N. T. Bacon.)

[4] Friedman, J. W. 1977. *Oligopoly and the theory of games.* North-Holland, Amsterdam.

[5] Intriligrator, M. D. 1971. *Mathematical optimization and economic theory.* Prentice Hall, Englewood Cliffs, New Jersey.

[6] Levitan, R. E., and M. Shubik. 1971. *Noncooperative equilibria and strategy spaces in ologopolistic market.* Pages 429-448 in H. W. Kuhn, and G. P. Szego, eds. Differential games and related topcis. North-Holland, Amsterdam.

[7] Wilson, E. O. 1988. *Biodiversity.* National Academy Press, Washington, DC.

PARTICIPANT'S COMMENTS

This paper begins with an elaborate model which is reduced by acceptable assumptions. From this is drawn interesting insights into the interaction of conservationists and exploiters. Particularly important is how conservationists, usually at a disadvantage to exploiters, can most effectivly use their resources. The last thing a conservationist would want is to use a strategy which subsidises the exploiters. If conservationists are not already aware of the benefits of interference into the exploiters' effort this paper should be convincing.

This paper has implications, beyond the scope of the conference, on the nature of aid to 'developing countries'. Perhaps this is subsidising the exploitation of these countries? What should be the interference here?

Kevin Judd

This paper considers an ecosystem involving exploiters and conservationists. It is then examined in a game theoretic setting. Two conservation strategies are considered. The first one is conservation of the resource, while the second one is interference with the exploiter's effect. The Nash–equilibrium solution is derived for these two conservation strategies. The author's finding is interesting. The first strategy turns out to subsidise the exploiters through supporting their resource. The second strategy is, on the other hand, much more efficient. This finding provides important information to conservationists in choosing their strategy. The conclusion is logical and agrees well with commonsense.

K.L. Teo

Table 1 Parameters, variables, their notation, and their units of measurements as given in §2; a denotes number of animals. Greek and upper case letters denote parameters (all positive); lower case letters denote variables. Superscripts denote explicit (or implicit) control variables, and subscripts denote the subscripted player.

	Explanation	Units
x	number of animals at equilibrium	a
ρ	growth rate when the population is small	$\frac{\text{growth rate}}{\text{capita}}$
K	carrying capacity	a
π_i	exploitation efficiency of the i-th exploiter	$\frac{a \text{ harvested}}{\text{exploitation effort}}$
p^i	exploitation-effort rate by the i-th exploiter	$\frac{\text{effort}}{\text{time}}$
M	number of exploiters	
θ_i	conservation efficiency of the i-th conservationist	$\frac{a \text{ added}}{\text{conservation effort}}$
q^i	conservation-effort of the i-th conservationist	$\frac{\text{effort}}{\text{time}}$
N	number of conservationists	
d^i	demand from exploiter i	a demanded
β	price sensitivity of the overall demand	$\frac{\# \text{ exploiters} \bullet a \text{ supp}}{\$\text{per } a \text{ demanded}}$
a^i	price charged by exploiter i	$\frac{\$}{a \text{ supplied}}$
$\bar{a}$	average price charged by all exploiters	$\frac{\$}{a \text{ supplied}}$
V	price at chich the demand is zero	$\frac{\$}{a \text{ supplied}}$
γ	a positive constant which quantifies the rate at which the demand from exploiter i decreases as a function of the price charged (by exploiter i) over the average price.	
P^i	profit for exploiter i	$\frac{\$ \bullet a \text{ demanded}}{a \text{ supplied}}$
α	a large positive number	$\frac{\$}{a \text{ supplied}}$
η_i	a positive parameter which describes the rate at which the cost of obtaining animals drops as the number of animals increases, for exploiter i	$\frac{\$ \text{ per } a \text{ supplied}}{a \text{ at equilibrium}}$

Q^i	utility for conservationist i	utility
ζ_i	a positive parameter which describes the rate at which the utility of conservation decreases with increasing the conservation effort by conservationist i	

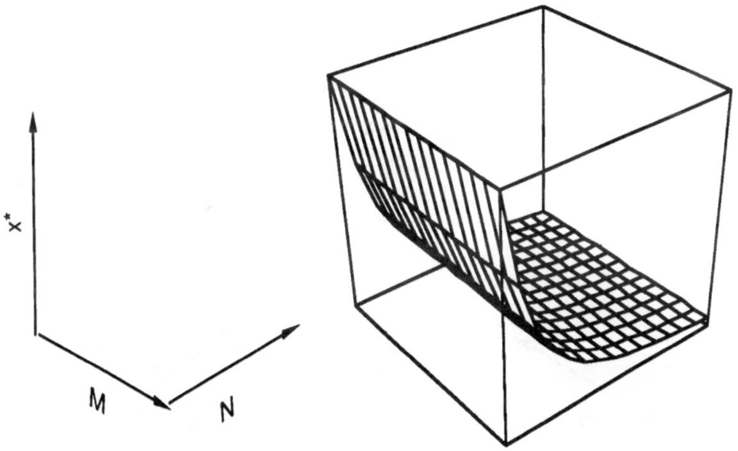

Figure 1.

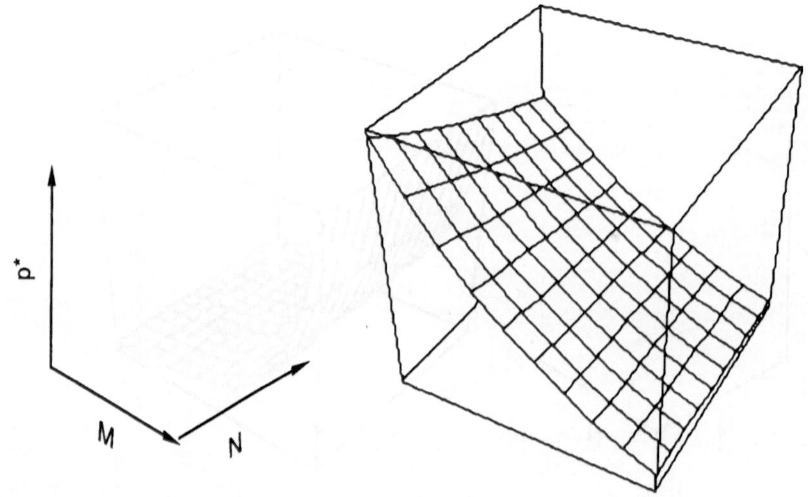

Figure 2.

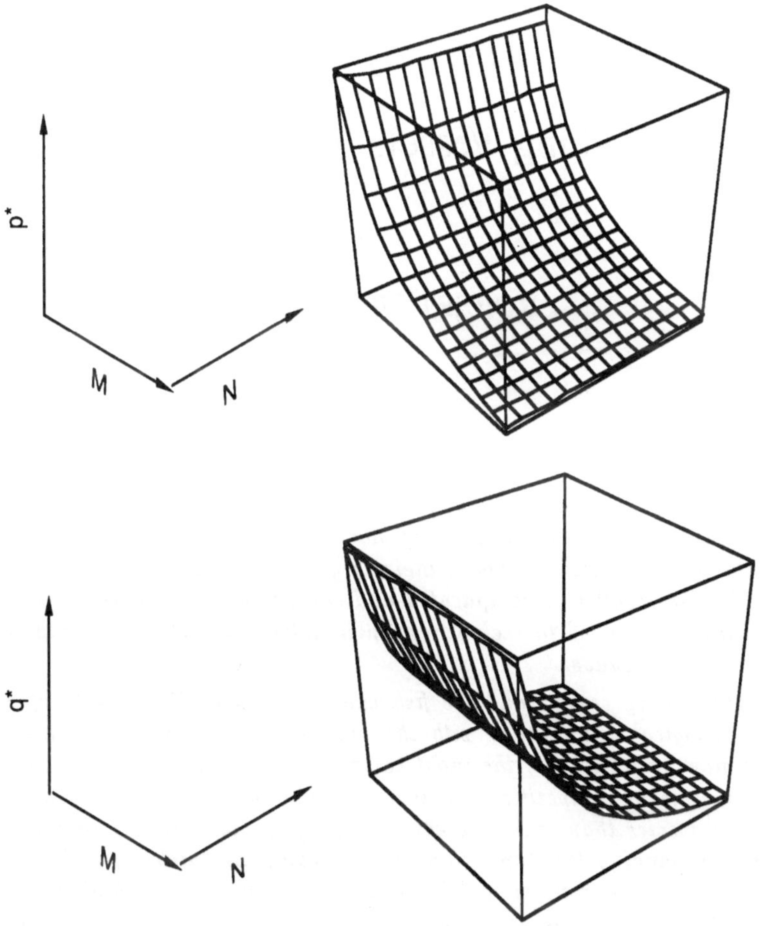

Figure 2. (contd)

Analyzing the Harvesting Game or Why Are There So Many Kinds of Fishing Vessels in the Fleet?

ROBERT MCKELVEY

Abstract

Harvesting of commercial marine fish stocks can be thought of as a game, with free entry of all kinds of players (the fishing vessels), and freedom for these vessels to choose their target species, and to switch targets at will. The result is a complicated and confusing mix of vessel types and activities. One is led to seek general principles, with which to bring some order into this chaos.

The fishing vessels and the fish can be thought of as a kind of artificial ecological community, with the vessels as predators, competing for a common prey. One of the most commonly cited doctrines in ecology is the principle of competitive exclusion: There can be no more competing predator species than there are common prey species-unless predators find ways of expanding the dimension of their common "niche space", for instance through asymmetric partitioning of space or time. I shall show that this ecological principle, made precise, has its direct analogue in resource economics. Thereby I hope to make sense of the "harvesting game."

I shall develop this theme by presenting a series of mathematical models. These will illustrate how bioeconomic forces can bring about modes of fishing vessel coexistence, involving respectively: time partitioning, resource partitioning, spatial partitioning, and risk partitioning. Technically the models, which have been developed in a series of articles over a period of several years, are of optimal control or differential game type - and they are quite different from the original models used to study the exclusion principle in ecology.

1. Introduction.

Harvesting of commercial marine fish stocks can be thought of as a game, with open entry to any number and all kinds of players (i.e., the fishing vessels), and freedom for these vessels to choose their target species and to switch targets at will. The result, in any major fishing ground of the world, is a complicated and confusing mosaic of vessel types and activities. This complexity of "industrial structure" not only poses an intriguing intellectual puzzle to sort out, but also introduces difficult and important issues for a fishery management agency, attempting to regulate the harvest.

From a biological perspective, the vessel's ability to move freely between fishing grounds, and to target different species at will, creates a kind of artificial predator-prey ecological community, with the fishermen as predators and various fish species, otherwise ecologically independent, as prey.

One of the most sacred of (classical) ecological theorems has been the "competitive exclusion" principle (Hardin, 1960; MacArthur and Levins, 1967; Levin, 1970; Armstrong and McGehee, 1976). This asserts that, in a resource-limited ecological community, there can be no more competing species than the dimension of the ecological niche that they share, and - unless there are special circumstances - this dimension is simply the number of species of prey. Special circumstances that enlarge the dimension of prey space include an asymmetrical partitioning of space or time or, with a randomly fluctuating environment, differing responses to risk.

We shall see that the idea of competitive exclusion carries over as a principle for understanding the structure of the artificial ecosystem which constitutes the fishery. I shall describe bioeconomic models, incorporating fishing vessels' technological characteristics and the fishermen's profit-maximizing objectives, along with the population biology of the fish. As in theoretical ecology, the models are made as simple as possible, while yet illustrating the general principle. They incorporate only one or two species of fish and only a few vessel types. However they do demonstrate how strict limits on the number of vessel types are set by the number of prey species, and how these limits are modified by incorporating spatial or temporal effects.

From an economic perspective, open entry and open switching create of the fish a "common property" resource, in that no one has exclusive rights to any part of the stocks. Consequently the harvest by each individual fishing vessel depletes the common resources of all. Not only does this diminish the effectiveness of all the fishing vessels, but it also leads to over-fishing and degradation of the biological system (Scott Gordon, 1954).

Public management agencies try to cure these problems by imposing regulations on the fishermen: that is, by changing the rules of the game (C.W. Clark, 1985). This may be done by direct restriction of output, or of input factors of production (e.g. by placing harvest quotas on fish or limiting the licensing of vessels to the fishery). Or it may be achieved by indirect means which provide economic incentives to change fishermen's behavior (e.g. levying landings taxes, or creating markets for vessel licenses, or for quotas on landings rights). There has been much discussion - and much controversy - in the management literature over these proposed regulatory devices.

For our models, the management issues can be approached in the standard way of bioeconomics (R. McKelvey, 1989). That is, one finds a cooperative solution to the fishing game, replacing the bioeconomic equilibrium of pure competition by an optimal solution. This maximizes the overall industry return, and may be achieved through decentralized regulation of the competitive fleet. Even in this cooperative mode, a form of exclusion principle still applies. We shall not, in the present article, take up these regulatory issues, preferring to concentrate instead on exposing the principles governing exclusion or coexistence in the purely competitive case.

There is one caveat. Competitive exclusion in ecology rests on an assumption that the ecological community is structured by resource limited competition, an assumption now being challenged on empirical grounds (Diamond and Case, 1986). In the bioeconomic setting this assumption may be more secure. Either way, competitive exclusion remains as a touchstone for testing any theory that seeks to explain the structural organization of a common property resource industry.

My own views, on the general organizing principles for a competitive fishing industry, have developed over a number of years. The model in section 2, of time partitioning, first appeared in McKelvey, (1983). (See also Smith and McKelvey, 1986). In the 1983 article I also investigated optimal risk strategies for a centrally managed fishing industry in a randomly fluctuating environment. The analysis in section 3, of risk-sharing in a competitive industry, is from McKelvey (1988). The two prey-species model of section 4, and its spatial interpretation in section 5, are new. A more extensive analysis of these models will be published elsewhere.

2. Coexistence through time partitioning.

In this first model, we shall examine the draw-down of a single stock of fish within a single season, by a mixed fleet of fishing vessels. We shall see that two vessel-types, relying on technologies that are respectively capital intensive and labor intensive, will be led to adopt differing temporal harvest profiles and will in this way succeed in coexisting on a single fish stock.

The stock $X(t)$ declines from the initial level, $X(0) = R$, according to

$$dX/dt = -h(t) \quad 0 \leq t \leq T,$$

where $h(t)$ is the rate of harvest and T is the length of the prescribed season. Thus, for simplicity, we are ignoring natural mortality during the season.

Harvest rate is

$$h(t) = E(t) \cdot X(t),$$

where $E(t)$ is fishing effort intensity. It follows immediately that the stock level declines according to

$$X(t) = R \cdot \exp - \int_0^t E(t) \cdot dt,$$

and total *seasonal* harvest is

$$H = \int_0^T h(t) \cdot dt = R - X(T).$$

Total effort intensity is the sum of the separate efforts of the two fleet components α and β:

$$E(t) = E_\alpha(t) + E_\beta(t).$$

Component effort levels are constrained by component fleet capacities (i.e. the number of vessels):

$$E_\alpha(t) \leq \Phi_\alpha, \quad E_\beta(t) \leq \Phi_\beta.$$

We shall assume that each α-vessel incurs two kinds of cost: an annual entry cost k_α for participating in the fishery, plus an operating cost rate c_α while actively harvesting. Likewise β-vessels incur costs k_β and c_β. We shall assume that

$$c_\alpha < c_\beta \text{ and } k_\alpha > k_\beta,$$

e.g. that α-vessels have high capital costs and low operating costs and that β-vessels have the opposite.

If p is the unit price received for the harvest, then the overall fleet profit for the season is

$$\Pi = \int_0^T [p \cdot h(t) - c_\alpha \cdot E_\alpha(t) - c_\beta \cdot E_\beta(t)] \cdot dt - k_\alpha \cdot \Phi_\alpha - k_\beta \cdot \Phi_\beta.$$

On reflection it should be clear that a social manager, coordinating fleet operations to maximize overall profit, would

a) choose a minimum fleet size to achieve any prescribed total seasonal effort, by operating all vessels at maximum capacity throughout the full season $0 \leq t \leq T$.

b) utilize only that vessel type which minimizes seasonal cost per unit effort, while operating at full capacity. Thus he would compare

$$c_\alpha T + k_\alpha \text{ with } c_\beta T + k_\beta.$$

The interesting circumstance (which will allow vessel coexistence under a competitive regime) occurs when the season length T is sufficiently long that

$$c_\alpha T + k_\alpha < c_\beta T + k_\beta.$$

In this case the social manager will maximize overall profit Π by employing a pure α-fleet.

By contrast let us now examine what would happen if, instead of being managed cooperatively, the vessels were behaving competitively. Let us first exclude β vessels by fiat and consider the operation of a pure α-fleet. Any α-vessel which enters the fishery will harvest at full capacity until the fish stock level drops to the point where the net return to effort is zero:

$$p\dot{h} - c_\alpha E = (pX - c_\alpha)E = 0.$$

This occurs at the time T_α when $X_\alpha = c_\alpha/p$. Thereafter it is no longer profitable to fish. Each vessel's seasonal profit, depending on the size Φ_α of fleet sharing the catch, is

$$\pi_\alpha(\Phi_\alpha) = p[R - x_\alpha]/\Phi_\alpha - [c_\alpha T_\alpha + k_\alpha].$$

By straightforward calculation (see McKelvey, 1983)

$$\pi_\alpha(\Phi_\alpha) = \gamma(\theta_\alpha)/\Phi_\alpha - k_\alpha/c_\alpha,$$

where

$$\theta_\alpha = pR/c_\alpha \triangleq \rho/c_\alpha \text{ and } \gamma(\theta_\alpha) \triangleq \theta_\alpha - \ln\theta - 1.$$

However, with no restrictions on entry, the size Φ_α of the fleet will expand to the *bionomic equilibrium* level Φ_α^o at which vessels only break even, i.e. $\pi_\alpha(\Phi_\alpha^o) = 0$. The bionomic fleet size Φ_α^o and the length T_α^o of active season both are functions of $\theta_\alpha = \rho/c_\alpha$, with $\rho = pR$ measuring the "richness" of the potential harvest. Φ_α^o is identically zero for $\rho \leq c_\alpha$ and increases monotonically thereafter. T_α^o on the other hand is monotone decreasing: i.e. the bionomic fleet size grows and the active season shortens as the fishery becomes "richer". Explicitly:

$$\Phi_\alpha^o = c_\alpha\gamma(\theta_\alpha)/k_\alpha, \qquad T_\alpha^o = k_\alpha \cdot \ln\theta/c_\alpha \cdot \gamma(\theta_\alpha).$$

A similar analysis holds for a pure β-fleet operating competitively. Note that for any ρ, since $c_\alpha < c_\beta$, the break-even stock levels satisfy $X_\alpha < X_\beta$.

In figure 1, I have plotted Φ_α^o and Φ_β^o against ρ, for fixed $c_\alpha < c_\beta$. Note that for small ρ the bionomic capacity of pure α-fleet is greater than that of a pure β-fleet, but that the curves cross at a point ρ_∞. Since

$$\Phi_\alpha = \frac{p[R - X_\alpha]}{c_\alpha T_\alpha + k_\alpha}, \quad \Phi_\beta = \frac{p[R - X_\beta]}{c_\beta T_\beta + k_\beta},$$

therefore for $\rho > \rho_\infty$, we have

$$c_\alpha T_\alpha + k_\alpha > c_\beta T_\beta + k_\beta.$$

That is, the shortening of the seasons has made the β-fleet more cost-effective than the α-fleet.

In mixed fleet operation one must distinguish two periods of the harvest: In the early period, so long as the stock level exceeds $X_\beta = c_\beta/p$, both type vessels can harvest profitably. However when X falls below X_β, only α-vessels can do so, until the stock falls to $X_\alpha(< X_\beta)$. Thus the incremental stock $X_\beta - X_\alpha$ is a reserved stock, accessible only to α-vessels.

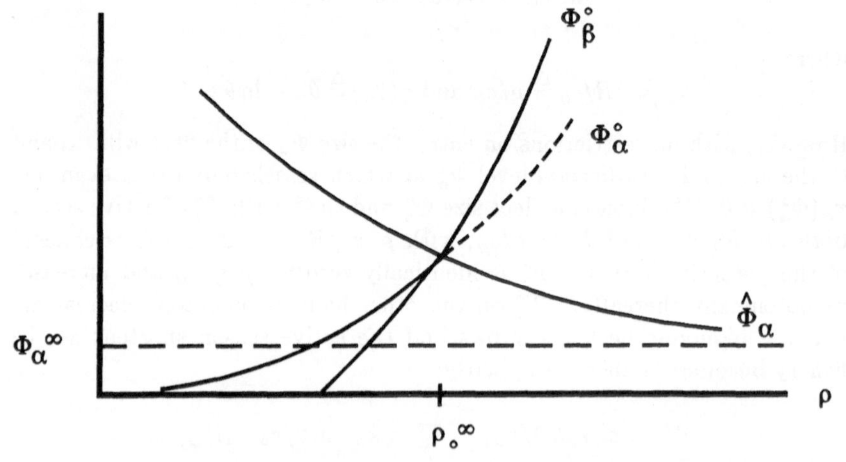

Figure 1. Coexistence through time partitioning. Equilibrium fleet capacity vs. "richness" $\rho = p \cdot R$. Here Φ^o_α and Φ^o_β are fleet capacities for a pure fleet. In mixed competition, α will exclude β to the left of ρ_o^∞. To the right of ρ_o^∞ coexistence will prevail, with *total* fleet capacity equal to Φ^o_β and α - capacity limited to $\hat{\Phi}_\alpha$. As ρ increases, α - capacity approaches asymptotically to a "refugium" level Φ^∞_α.

Consequently, the pure β-fleet operation described above is unstable, since it can always be invaded by α-vessels utilizing the reserve stock.

So long as $\rho < \rho_\infty$, the pure α-fleet *is* stable against invasion, since $\Phi^o_\alpha > \Phi^o_\beta$: Though $X > X_\beta$, so that a β-invader can operate, it is part of a fleet whose total size is too large to enable β - vessels to recover their fixed costs.

On the other hand, when $\rho > \rho_\infty$, a mixed fleet can coexist. Its total bionomic size must be Φ^o_β, with all vessels operating until time T_β, when $X = X_\beta$. Up to that time, α-vessels will have earned some income, though not enough to cover fixed costs. However a limited number of α-vessels, by operating until $X = X_\alpha$ and sharing the reserve stock, can make up the difference and break-even overall.

As the fishery's richness ρ grows, more and more β-vessels are able to enter, thereby lessening the share of the first-phase harvest available to the α-fleet. Hence, for very large ρ, α-vessels are confined almost completely to harvesting the reserve stock.

The contrast with optimal cooperative management is stark: while optimal management calls for utilizing a relatively small fleet of α-vessels, competitive open access harvesting is predicted to yield an entirely different regime: For small ρ, there is still a pure α-fleet, but excessive in size. For larger ρ, there will be a mixed fleet increasingly dominated by β-vessels, and operating during a drastically shortened season.

The details of the above analysis can be found in McKelvey [1983], section 7.

3. Coexistence through Risk Partitioning.

The preceeding model was entirely deterministic: both biological and economic factors were assumed to be constant and known throughout time. But in reality, marine fisheries are subject to enormous and unpredictable fluctuations in both biotic and economic environment. By choosing differing strategies for coping with these environmental risks, quite dissimilar vessels can manage to coexist.

In modeling the phenomenon, we shall abstract from the in-season details that were center-stage in §2. Instead we now specify the season fleet yield H as a function of total season effort E according to

$$H = \xi \cdot \mathbf{Y}(E).$$

Here $\mathbf{Y}(E)$ is monotone increasing, bounded, and concave on $E > 0$, with $\mathbf{Y}(0) = 0$. The factor ξ is a positive random variable, expressing uncertainty in the stock levels and hence in the yield. For an individual vessel, the harvest yield h is

$$h = \xi \cdot \mathbf{Y}(E)/E \triangleq \xi \cdot Y(E),$$

defining Y as a function of the *total* fleet effort E.

Fleet effort is due to the involvement of two subfleets, of vessel types α and β. Thus

$$E = E_\alpha + E_\beta, \quad \text{with} \quad 0 \leq E_\alpha \leq \Phi_\alpha \quad \text{and} \quad 0 \leq E_\beta \leq \Phi_\beta,$$

Here the capacity constraints Φ_α and Φ_β represent the total number of vessels present in the available subfleets. Effort levels E_α, E_β are chosen anew each season, while subfleet capacities Φ_α and Φ_β are determined over the long run.

Seasonal vessel return, for a vessel actively fishing during the season, is

$$\Pi_\alpha = P \cdot \xi Y(E) - C_\alpha \geq 0, \quad \text{or}$$
$$\Pi_\beta = P \cdot \xi Y(E) - C_\beta \geq 0, \quad \text{respectively.}$$

Here price P and costs C_α and C_β are all positive random variables.

It is assumed that all random variables have known independent probability distributions. Their actual values in a particular season are made known at the beginning of the season, and enter into the determination of effort levels E_α and E_β during the season. The in-season effort levels are constrained by the fixed capacity levels Φ_α or Φ_β, which are determined once and for all over the long run, and reflect a trade - off between the annualized cost K_α or K_β of acquiring a unit of capacity, and the expected increase in annual vessel profitability that would result from the added capacity.

Thus, for an open-access competitive fishery, the effort control variables adjust as follows:

a) In-season, subfleet effort levels expand so that the active components of the fleet either break even or are limited by capacity constraints. Knowing the current values of ξ, P, C_α, and C_β, one can determine the break-even levels. Break-even for α-vessels occurs when *total* active fleet size $E = E_\alpha + E_\beta$ equals $E_\alpha^\#$ which satisfies

$$P \cdot \xi Y(E_\alpha^\#) = C_\alpha.$$

Likewise break-even occurs for active β-vessels when total fleet size equals $E_\beta^\#$, where

$$P \cdot \xi Y(E_\beta^\#) = C_\beta.$$

Only when, because of capacity constraints, total fleet size E is $< E_\alpha^\#$ do α - vessels make a profit; and a parallel statement holds for the β-fleet.

Because of the capacity constraints there are a number of distinct fleet effort configurations that can occur, depending on the relative sizes of $\Phi_\alpha, \Phi_\beta, \Phi = \Phi_\alpha + \Phi_\beta, E_\alpha^\#$, and $E_\alpha^\#$. These are shown in figure 2, for a season when $C_\alpha < C_\beta$; switched configurations prevail when $C_\alpha > C_\beta$. We shall denote these subfleet effort levels as $E_\alpha^o(\Phi_\alpha, \Phi_\beta)$ and $E_\beta^0(\Phi_\alpha, \Phi_\beta)$,

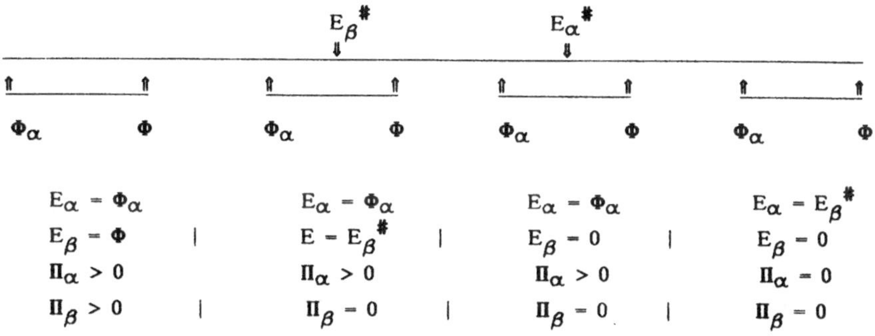

Figure 2. Coexistence through risk partitioning. The diagram shows in - season effort distribution (E_α, E_β) and profitability (Π_α, Π_β), as these depend on the current state configuration: This latter is characterized by the current break - even levels $E_\alpha^\#$ and $E_\beta^\#$ of *total* - fleet effort (i.e. break - even for α and β - vessels, respectively), relative to the equilibrium levels of fleet capacity: Φ_α, Φ_β, and $\Phi = \Phi_\alpha + \Phi_\beta$. This patterns changes from season to season, as recruitment and prices fluctuate.

resp. The corresponding profits are $\Pi_\alpha^0(\Phi_\alpha, \Phi_\beta)$ and $\Pi_\beta^0(\Phi_\alpha, \Phi_\beta)$, resp. (These all are *random* functions!)

b) Fleet capacities Φ_α and Φ_β are set in the long run at such levels that the *average* seasonal returns exactly balance the seasonal costs of maintaining capital; i.e.

$$\mathbf{E}\Pi_\alpha^0(\Phi_\alpha, \Phi_\beta) = K_\alpha, \quad \mathbf{E}\Pi_\beta^0(\Phi_\alpha, \Phi_\beta) = K_\beta,$$

where $\mathbf{E}$ denotes expected value.

Similarly, for the cooperatively managed fishery, the decision problem decomposes as follows:

a) In-season, given Φ_α and Φ_β, and knowing ξ, P, C_α, and C_β, choose $E_\alpha < \Phi_\alpha$ and $E_\beta < \Phi_\beta$ to maximise

$$\Pi = P \cdot \xi \mathbf{Y}(E) - C_\alpha E_\alpha - C_\beta E_\beta.$$

Note that the *un*constrained optimum for α-effort occurs when total fleet effort E equals $E_\alpha^{\#\#}$ for which

$$P \cdot \xi \mathbf{Y}'(E_\alpha^{\#\#}) = C_\alpha,$$

and the *un*constrained optimum for β-effort occurs when total fleet effort E is $E_\beta^{\#\#}$ for which

$$P \cdot \xi \mathbf{Y}'(E_\beta^{\#\#}) = C_\beta.$$

For the constrained problem, at most one subfleet component can operate at the unconstrained optimum (when $C_\alpha \neq C_\beta$); otherwise active fleets are constrained by capacity limits. In a season when $C_\alpha < C_\beta$, there are 4 configuration regimes, analogous to those for competitive open access, and depending on the size configuration of Φ_α and Φ versus $E_\alpha^{\#\#}$ and $E_\beta^{\#\#}$. We denote the optimal effort levels of the subfleets as E_α^* and E_β^*, and the corresponding in-season profit as Π^*. As before, these are random functions of Φ_α and Φ_β.

b) In the long run, Φ_α and Φ_β are chosen to maximise

$$\mathbf{E}\Pi^*(\Phi_\alpha, \Phi_\beta) - K_\alpha \Phi_\alpha - K_\beta \Phi_\beta.$$

Let us pursue these calculations in an illuminating special case. We continue to concentrate on the competitive model, and assume that:

$$C_\alpha < C_\beta, \quad K_\alpha > K_\beta, \quad \xi = 1, \quad \text{all non - random;} \tag{3.1}$$

P is random, taking on two values $P^- < P^+$, with probabilities π^-, π^+ resp.

$$\text{Mean price is } \overline{P} = P^+\pi^+ + P^-\pi^-.$$

The most interesting case of coexistence is found to occur when the parameter values are such that all α-vessels operate annually while β-vessels operate only in good years (i.e. when $P = P^+$).

For competitive open access, long run equilibrium requires

$$\begin{aligned} \pi^+[P^+Y(\Phi) - C_\alpha] + \pi^-[P^-Y(\Phi_\alpha) - C_\alpha] &= K_\alpha, \\ \pi^+[P^+Y(\Phi) - C_\beta] \qquad\qquad\qquad &= K_\beta. \end{aligned}$$

Hence, solving for $Y(\Phi_\alpha)$ and $Y(\Phi)$,

$$Y(\Phi_\alpha) = [(K_\alpha + C_\alpha) - (K_\beta + \pi^+ C_\beta)]/\pi^- P^-; \tag{3.2}$$

$$Y(\Phi) = [K_\beta + \Pi^+ C_\beta]/\overline{P}. \tag{3.2b}$$

Genuine coexistence requires $\Phi_\alpha < \Phi$, and hence (since Y is monotone decreasing) that

$$[K_\beta + \pi^+ C_\beta]/\pi^+ P^+ < [K_\alpha + C_\alpha]/\overline{P}. \tag{3.3}$$

That is, the β-fleet, operating in good years only, has a better cost-price ratio than α-vessels, operating in all years.

On the other hand, coexistence (3.1) together with the in-season exclusion of β-vessels in poor years, [i.e., $P^- Y(\Phi_\alpha) < C_\beta$] yields

$$(K_\alpha + C_\alpha) < (K_\beta + C_\beta). \tag{3.4}$$

That is, α-vessels are more efficient in-season than β-vessels on average *if* both should attempt to operate all the time.

Finally, we are assuming that α-vessels operate at capacity during poor years (i.e. $P^- Y(\Phi_\alpha) > C_\alpha$). Together with (3.1) this implies that

$$(K_\beta + \pi^+ C_\beta) < (K_\alpha + \pi^+ C_\alpha). \tag{3.5}$$

This initially surprising result says that β-vessels are more efficient than α-vessels, when both operate only in good years. This reflects that α-vessels must operate at capacity in bad seasons too, to make up for their relatively lower efficiency in the good ones.

There are other parameter ranges that lead to coexistence in this example; some of them have been worked out in the 1988 paper.

Entirely analogous results hold for the cooperatively managed fishery. As it turns out, the domains of coexistence there are unchanged from in the competitive model (though of course the effort levels must change). This special circumstance comes about because of the extreme specialization of our model. For example, we have assumed identical production functions for the two vessel types, independent of their relative frequency in the fishery. This is a highly restrictive assumption.

4. Two prey species: competitive exclusion.

It is well-known in ecology that, in the absence of spatial and temporal effects, no more than two predators can coexist on a food niche composed of two prey species. Here we shall arrive at the analogous principle for bioeconomics.

Consider then a pool of fish, consisting of two species (denoted 1 and 2), possibly interacting ecologically, with both being harvested. Their population dynamics are given conventionally by

$$dx_1/dt = f_1(x_1, x_2) - H_1(t),$$
$$dx_2/dt = f_2(x_1, x_2) - H_2(t),$$

Here $x_i(t)$ is the population size of population i and $H_i(t)$ the rate of its harvest ($i = 1$ or 2). The biological interactions, expressed in the density-dependent growth rates $f_i(x_1, x_2)$, need not be specified at this time: the fish may be competitors, mutualists, predator and prey, or even biologically independent.

Though the species usually form an ecological community, the fish are assumed to be physically separated at times, so that it is feasible for fishing boats to target either species at will.

We shall assume that there are three types of fishing vessel in existence, which potentially may be part of the fleet. For simplicity, two are taken to be specialists, harvesting only a single fish species, and the third is a generalist, harvesting both. (Other cases are somewhat more complicated, but are handled similarly.) Thus we have:

a) α - vessels, which target fish of species 1 only;

b) β - vessels, which target fish of species 2 only; and

c) γ - vessels, which can target fish of both species,

in whatever proportion they may desire. We shall see that normally only *two* of the three vessel types can coexist in the active fleet.

In general though, if $N_\nu(t)$ is the active fleet size of ν-vessels ($\nu = \alpha, \beta$ or γ), and $h_{\nu i}$ is the individual ν-vessel's harvest function for species i, then overall harvest rates are

$$H_1 = N_\alpha \cdot h_{\alpha 1} + N_\gamma \cdot h_{\gamma 1}, \text{ and}$$
$$H_2 = N_\beta \cdot h_{\beta 2} + N_\gamma \cdot h_{\gamma 2}, \text{ respectively.}$$

We shall take the individual vessel's harvest function for species i to depend both on the vessel's targeted effort level $E_{\nu i}$ and on the current stock level of the fish. For simplicity:

$$h_{\nu i} = G_{\nu i}(x_i) \cdot S_{\nu i}(E_{\nu i}),$$

with both the catchability function G and the specific harvest function S being positive monotone-increasing functions, G convex, S concave, and $S(0) = 0$. (In this continuous model, in which a "season" is represented by a single point in time, the non-linearities are intended to capture something of the in-season changes in effort effectiveness.)

A vessel's instantaneous rate of income, $\pi_\nu(t)$ is the landings value of its harvest less the cost of its effort. For an α-vessel this is

$$\pi_\alpha = p_1 h_{\alpha 1} - C_\alpha(E_\alpha)$$

$$= p_1 G_\alpha(x_1) \cdot S_\alpha(E_\alpha) - C_\alpha(E_\alpha),$$

where p_1 is the price per unit harvest, $C_\alpha(E_\alpha)$ is the cost of effort, and we have suppressed the superfluous second subscript 1 on S_α, G_α, and E_α. Similarly,

$$\pi_\beta = p_2 h_{\beta 2} - C_\beta(E_\beta)$$

$$= p_2 G_\beta(x_2) \cdot S_\beta(E_\beta) - C_\beta(E_\beta).$$

For the generalist γ-vessel,

$$\pi_\gamma = p_1 h_{\gamma 1} + p_2 h_{\gamma 2} - C_\gamma(E_\gamma)$$

$$= p_1 G_{\gamma 1}(x_1) \cdot S_{\gamma 1}(E_{\gamma 1}) + p_2 G_{\gamma 2}(x_2) \cdot S_{\gamma 2}(E_{\gamma 2}) - C_\gamma(E_\gamma),$$

where $E_\gamma = E_{\gamma 1} + E_{\gamma 2}$.

For each vessel, long term profit is the discounted sum of net income flow, less initial entry cost:

$$\Pi_\nu = \int_0^\infty e^{-\delta t} \cdot \pi_\nu(t) \cdot dt - K_\nu.$$

We shall restrict our analysis to examining the fleet structure and effort levels that one can expect when the vessels are operating competitively, with open access to the fishery. We further restrict analysis to the case of a "rental fleet", where there are no entry costs: i.e. $K_\nu = 0$. In these circumstances each vessel, in making its current harvest effort decision, can be assumed to take into account only the current state of the fishery, ignoring the effect its own (small) share of the harvest will have on future fish stocks. Thus each vessel chooses effort, $E_\nu \geq 0$, to maximise π_ν, taking as given the current stock levels x_1 and x_2.

We begin with the easier case, that of the specialist vessels in the fleet. Thus, for an α-vessel,

$$0 \leq \pi_\alpha = p_1 G_\alpha(x_1) \cdot S_\alpha(E_\alpha) - C_\alpha(E_\alpha),$$

with the first order necessary condition for an (interior) extremum of π_α being

$$p_1 G_\alpha(x_1) = C'_\alpha(E_\alpha)/S'_\alpha(E_\alpha). \tag{4.1}$$

Combined, these two relations imply that

$$S'_\alpha/S_\alpha \leq C'_\alpha/C_\alpha. \tag{4.2}$$

It is customary in microeconomics to assume that average cost per yield C/S is a U-shaped function of yield S. Her we impose conditions that make C/S a U-shaped function of *effort* E. We introduce the effort elasticities of cost and yield:

$$\kappa_\alpha(E_\alpha) = \frac{dE_\alpha/E_\alpha}{dC_\alpha/C_\alpha} = \frac{C_\alpha(E_\alpha)/E_\alpha}{C'_\alpha(E_\alpha)},$$

$$\sigma_\alpha(E_\alpha) = \frac{dE_\alpha/E_\alpha}{dS_\alpha/S_\alpha} = \frac{S_\alpha(E_\alpha)/E_\alpha}{S'_\alpha(E_\alpha)},$$

respectively. In terms of these elasticities, (4.2) is equivalent to

$$\varepsilon_\alpha(E_\alpha) \triangleq \kappa_\alpha(E_\alpha)/\sigma_\alpha(E_\alpha) = \frac{dS_\alpha/S_\alpha}{dC_\alpha/C_\alpha} = \frac{C_\alpha/S_\alpha}{dC_\alpha/dS_\alpha} \leq 1. \tag{4.2'}$$

That is, optimal effort $E_\alpha > 0$ must occur at a level of constant or increasing costs to yield.

We now *assume* that $\kappa_\alpha(E_\alpha)$ is monotone decreasing, $\sigma_\alpha(E_\alpha)$ is monotone non-decreasing, and that the two are equal at a certain $E^O_\alpha > 0$. It's not hard to show that this implies the average-cost/yield curve C_α/S_α is a U-shaped function of E_α, with minimum at E^O_α and with the marginal-cost-to-yield curve C'_α/S'_α lying below it for $E_\alpha < E^O_\alpha$ and above it for $E_\alpha > E^O_\alpha$. The region $E_\alpha < E^O_\alpha$ is a region of decreasing costs to yield, and the region $E_\alpha \geq E^O_\alpha$ is a region of constant or increasing costs to yield.

Now let $E^\#_\alpha(x_1)$ satisfy the marginal rule (4.1), i.e.

$$p_1 G_\alpha(x_1) = C'_\alpha(E^\#_\alpha)/S'_\alpha(E^\#_\alpha).$$

Then the optimal effort level for an α-vessel at a given fish stock level x_1 is

$$E_\alpha^*(x_1) = \begin{cases} E_\alpha^\# & \text{if } E_\alpha^\# \geq E_\alpha^O \\ 0 & \text{if } E_\alpha^\# < E_\alpha^O \end{cases}$$

Entirely analogous remarks apply to β-vessels. We assume that $\kappa_\beta(E_\beta)$ is monotone decreasing, $\sigma_\beta(E_\beta)$ is monotone non-decreasing, and that their graphs cross at E_β^O. The marginal effort $E_\beta^\#$ and the optimal effort E_β^* are defined analogously to the above.

We now turn to an analysis of the behavior of the generalist γ-vessels. The condition of profitability is

$$0 \leqslant \pi_\gamma = p_1 G_{\gamma 1}(x_1) \cdot S_{\gamma 1}(E_{\gamma 1}) + p_2 G_{\gamma 2}(x_2) \cdot S_{\gamma 2}(E_{\gamma 2}) - C_\gamma(E_\gamma),$$

with $E_\gamma = E_{\gamma 1} + E_{\gamma 2}$. The (interior) first order necessary conditions, for $\underline{E}_\gamma = (E_{\gamma 1}, E_{\gamma 2})$ to be an extremum of π_γ, are

$$p_i G_{\gamma i}(x_i) = C_\gamma'(E_\gamma)/S_{\gamma i}'(E_{\gamma i}), \quad \text{for } i = 1 \text{ and } 2. \tag{4.3}$$

Combining these with the profitability inequality, one obtains as a necessary condition that

$$S_{\gamma 1}(E_{\gamma 1})/S_{\gamma 1}'(E_{\gamma 1}) + S_{\gamma 2})/S_{\gamma 2}'(E_{\gamma 2}) \geq C_\gamma(E_\gamma)/C_\gamma'(E_\gamma). \tag{4.4}$$

Alternatively, this may be written

$$\left[\frac{dS_{\gamma 1}}{S_{\gamma 1}}\right]^{-1} \frac{dE_{\gamma 1}}{dE_\gamma} + \left[\frac{dS_{\gamma 2}}{S_{\gamma 2}}\right]^{-1} \frac{dE_{\gamma 2}}{dE_\gamma} \geq \left[\frac{dC_\gamma}{C_\gamma}\right]^{-1},$$

which may be interpreted as expressing constant or increasing costs to yield.

Once again introduce the effort elasticity of cost $\kappa_\gamma(E_\gamma)$ and the effort elasticities of yield $\sigma_{\gamma i}(E_{\gamma i})$ in *each* fishery, $i = 1$ and 2. Then the condition (4.4), of constant or increasing costs to yield, becomes

$$\sigma_{\gamma 1}(E_{\gamma 1}) \cdot E_{\gamma 1}/E_\gamma + \sigma_{\gamma 2}(E_{\gamma 2}) \cdot E_{\gamma 2}/E_\gamma \geq \kappa_\gamma(E_\gamma)$$

or, writing $E_{\gamma 1} = \theta_1 E_\gamma$, $E_{\gamma 2} = \theta_2 E_\gamma$, $\underline{\theta} = (\theta_1, \theta_2)$,

$$\sigma_\gamma(E_\gamma, \underline{\theta}) \triangleq \theta_1 \cdot \sigma_{\gamma 1}(\theta_1 E_\gamma) + \theta_2 \cdot \sigma_{\gamma 2}(\theta_2 E_\gamma) \geq \kappa_\gamma(E_\gamma). \tag{4.5}$$

That is,

$$\varepsilon(E_\gamma, \underline{\theta}) \triangleq \kappa_\gamma(E_\gamma)/\sigma_\gamma(E_\gamma, \underline{\theta}) \leq 1. \tag{4.6}$$

As in the single species case, assume that $\kappa_\gamma(E_\gamma)$ is positive and decreasing, while $\sigma_{\gamma i}(E_{\gamma i})$ is positive and non-decreasing. Then $\sigma_\gamma(E_\gamma, \underline{\theta})$ is positive and non-decreasing on each fixed ray $\underline{\theta}$ = constant. Assuming that the curves $\sigma_\gamma(E_\gamma, \underline{\theta})$ and $\kappa_\gamma(E_\gamma)$ cross at $E_\gamma^O(\underline{\theta}) > 0$, then the points $E_\gamma^O(\underline{\theta})$, of constant cost to yield $\varepsilon_\gamma[E_\gamma^O(\underline{\theta}), \underline{\theta}] = 1$ form a curve Γ_γ in the positive quadrant of $\underline{E}_\gamma$ space. (If κ_γ and $\sigma_{\gamma i}$ are linear functions, Γ_γ will be an ellipse.)

As noted, Γ_γ is the curve of constant yield-to-cost. Points beyond Γ_γ (i.e. further from the origin, along a ray) are points of decreasing yield-to-cost. Furthermore, positive harvest effort can be profitable *only* at these points of constant or increasing cost-to-yield.

We now invoke a dynamic version of Scott Gordon's bionomic analysis of open access competition. We postulate that harvest will proceed at fleet capacity until fish stocks are drawn down to levels (x_1^O, x_2^O) at which active vessels in the harvest just break even: $\pi_\gamma = 0$. It may be that these levels are consistent with a harvesting-maintained steady-state, i.e. are levels of excess production:

$$f_1(x_1^O, x_2^O) \geq 0, \qquad f_2(x_1^O, x_2^O) \geq 0. \tag{4.7}$$

If so, open access competition acts to adjust the fleet size and harvest so as to maintain these levels permanently in "bionomic equilibrium".

Since break-even for an α-vessel occurs at the bottom E_α^O of its U-shaped curve, it follows that for an α-vessel to participate in the bionomic equilibrium, x_1^O and E_α^O together must satisfy the first order condition (4.1):

$$p_1 G_\alpha(x_1^O) = C'_\alpha(E_\alpha^O)/S'_\alpha(E_\alpha^O). \tag{4.8a}$$

Likewise, participation of a β-vessel at the bionomic equilibrium requires

$$p_2 G_\beta(x_2^O) = C'_\beta(E_\beta^O)/S'_\beta(E_\beta^O). \tag{4.8b}$$

Finally, for participation of a generalist γ-vessel at bionomic equilibrium, both marginal rules must hold, that

$$\begin{aligned} &p_1 G_{\gamma 1}(x_1^O) = C'_\gamma(E_\gamma^O)/S'_{\gamma 1}(E_{\gamma 1}^O) \quad \text{and} \\ &p_2 G_{\gamma 2}(x_2^O) = C'_\gamma(E_\gamma^O)/S'_{\gamma 2}(E_{\gamma 2}^O), \end{aligned} \tag{4.8c}$$

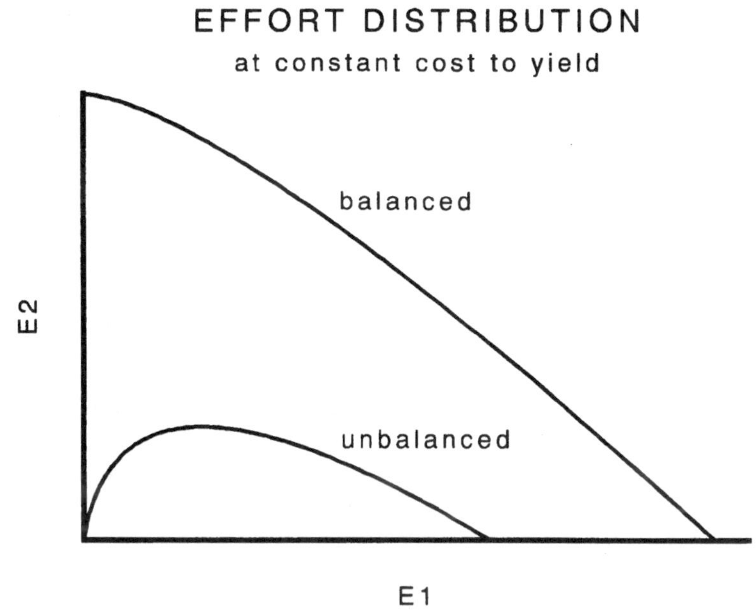

Figure 3. Equilibrium effort partitioning. (See section 4.) Feasible equilibrium distributions of effort for a generalist vessel harvesting two fish stocks. Constant cost - to - yield prevails along the curve, with increasing cost - to - yield beyond it, away from the origin. Two cases are illustrated. With sufficiently unbalanced yields - to - effort, profitable operation may not be possible through targeting solely on stock 2.

for some $\underline{E}^O_\gamma$ on Γ_γ. Through these equations (4.8c), the curve Γ_γ is mapped into a curve χ_γ of points $\underline{x}^O_\gamma = (x^O_{\gamma 1}, x^O_{\gamma 1})$ of potential bionomic production, for which dynamic equilibrium is possible.

Let us follow through the implications of this analysis. First of all it is clear that, except by accident, no more than two of the three sets of conditions in (4.8) can hold simultaneously. Thus coexistence at equilibrium is limited to two vessel types. We shall examine several possibilities.

Let us consider, as a first case, the possibility of a mixed fleet of specialist vessels coexisting in a steady state. For α-vessels, harvesting only stock 1, the constant cost to yield condition $\varepsilon_\alpha = 1$ determines the individual vessel's effort level E_α, and the marginal rule $p_1 G_\alpha = C'_\alpha / S'_\alpha$ then determines the steady-state stock level x_1. In like fashion E_β and x_β are determined for β-vessels harvesting stock 2. The equilibrium harvest levels then require fleet sizes N_α and N_β such that

$$N_\alpha G_\alpha(x_1) \cdot S_\alpha(E_\alpha) = f_1(x_1, x_2),$$
$$N_\beta G_\beta(x_2) \cdot S_\beta(E_\beta) = f_2(x_1, x_2),$$

and these equations can be solved for non-negative N_α and N_β, *provided* that $f_i(x_1, x_2) \geq 0$ for $i = 1, 2$. This last expresses the purely *biological* condition that (x_1, x_2) be a state of excess production.

Next, for α-specialist and γ-generalist vessels to coexist, then once again $\varepsilon_\alpha = 1$ determines E_α and the marginal rule (4.8a) determines the required equilibrium stock level x_1. With this known, $\varepsilon_\gamma = 1$ picks out a point $\underline{x} = (x_1, x_2)$ on χ_γ, and equations (4.8c) determine both components of $\underline{E}_\gamma$. Finally, the vessel numbers must satisfy the equilibrium harvest equations:

$$H_1 = N_\alpha h_\alpha + N_\gamma h_{\gamma 1} = f_1(x_1, x_2) \geq 0,$$
$$H_2 = \qquad\qquad N_\gamma h_{\gamma 2} = f_2(x_1, x_2) \geq 0.$$

Here however a solution with *positive* N_α and N_γ may not exist! This is because, with the effort partition of γ-vessels having been set, a draw-down to stock level (x_1, x_2) may require directing additional effort against stock 2 rather than against 1. Should this be the case, it remains possible that coexistence can occur between β and γ vessels. However, in general the coexistence of *any* vessel types cannot be guaranteed. Nor for that matter can the continuing existence of the ecosystem be guaranteed: one or both species might be driven to extinction.

As a final case, consider the possible coexistence of two distinctive types of generalist vessels, say γ and γ'. The first requirement is that their equilibrium χ curves must intersect, thereby determining (x_1, x_2). From this and the marginal rule (4.8c), the effort vectors $\underline{E}_\gamma$ and $\underline{E}_{\gamma'}$ for both types are determined. Finally, the stock equilibrium equations must hold

$$N_\gamma h_{\gamma 1} + N_{\gamma'} h_{\gamma' 1} = f(x_1, x_2) \geq 0,$$
$$N_\gamma h_{\gamma 2} + N_{\gamma'} h_{\gamma' 2} = f(x_1, x_2) \geq 0$$

for *non-negative* fleet sizes $N_\gamma, N_{\gamma'}$.

Related to this case of coexisting generalists is the ecological phenomenon of niche compression (MacArthur and Levins, 1967).

Let us consider for a moment a situation in which a single generalist vessel type has exclusive rights in a two-species fishery. The condition $\varepsilon_\gamma = 1$ restricts each vessel to operate at some point on Γ_γ, and each such point corresponds to an equilibrium stock (x_1, x_2) on χ_γ. However to achieve such an equilibrium with a pure γ-fleet, the fleet size N_γ must satisfy *both* stock-balance equations:

$$N_\gamma \cdot G_{\gamma i}(x_i) \cdot S_{\gamma i}(E_{\gamma i}) = f_i(x_1, x_2) \geq 0 \quad \text{for } i = 1 \text{ and } 2.$$

These two equations impose a compatibility condition

$$\frac{p_1 f_1(x_1, x_2)}{p_2 f_2(x_1, x_2)} = \frac{\theta_1 \sigma_{\gamma 1}(\theta_1 E_\gamma)}{\theta_2 \sigma_{\gamma 2}(\theta_2 E_\gamma)}$$

that in general picks out a single point on Γ_γ, thereby determining the γ-vessel's mode of operation in this circumstance of exclusive rights.

But of course, this effort-distribution mode will not in general be the same as that resulting when these vessels are part of a mixed fleet including also γ'-vessels, as described above. Thus γ-vessels (and γ'-vessels as well) displace their operating characteristics to accommodate the presence of competing vessel types.

All of the foregoing has dealt with vessels in open-access competition. The cooperative model, of a centrally managed fishery, may be analysed through the methods of dynamic optimal control theory. As is frequently true in this kind of analysis, the cooperative model may be reduced to the competitive one by replacing landings prices p_i by $p_i - \lambda_i$, where λ_i denotes the "shadow price" of the resource stock.

A more thorough analysis of this model will be taken up elsewhere, including a treatment of the dynamics of both competitive and cooperative economic regimes.

5. Coexistence through spatial partitioning.

The model of the previous section can be reinterpreted to describe the harvesting of a *single* species, which is geographically dispersed into biologically isolated pools. In this interpretation, x_1 and x_2 are the fish population levels in pool 1 and 2 respectively, with growth functions $f_i = f_i(x_i)$, for $i = 1$ and 2. These are to be dome-shaped curves, which are negative beyond a positive carrying capacity $x_i = \overline{x}_i$.

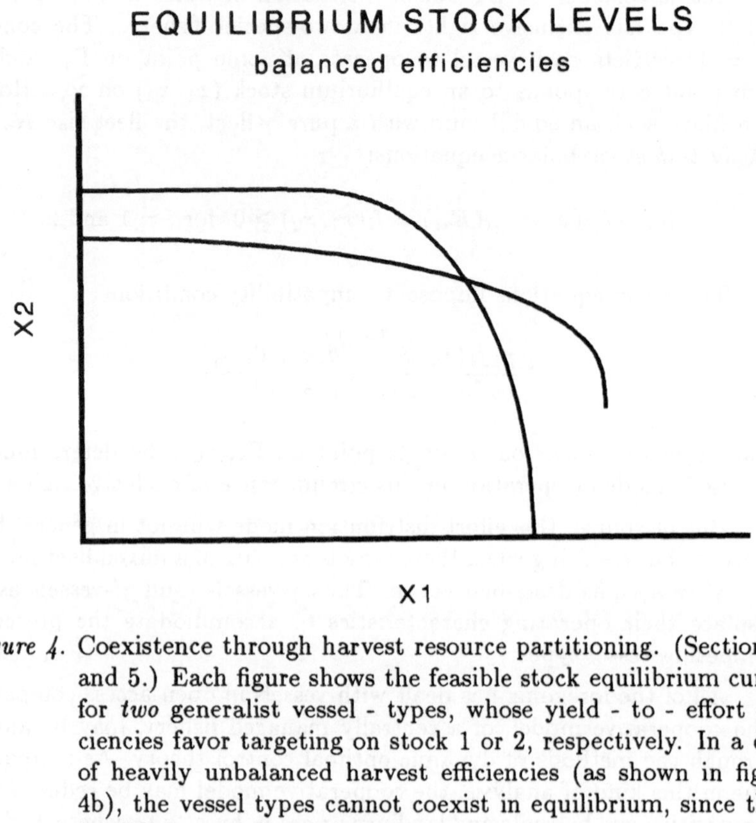

Figure 4. Coexistence through harvest resource partitioning. (Sections 4 and 5.) Each figure shows the feasible stock equilibrium curves for *two* generalist vessel - types, whose yield - to - effort efficiencies favor targeting on stock 1 or 2, respectively. In a case of heavily unbalanced harvest efficiencies (as shown in figure 4b), the vessel types cannot coexist in equilibrium, since their feasible curves do not intersect.

We assume the existence of two generalist types of fishing vessel, γ and γ', each capable of dividing harvest effort between the two pools. However γ-vessels have a comparative advantage in fishing pool 1 while γ'-vessels have an advantage in pool 2. These advantages can be expressed in the vessels' specific yield functions $S_{\nu i}(E_{\nu i})$ or alternatively in their specific harvest catchability functions $G_{\nu i}(x_i)$.

Not surprisingly, it turns out that, if the comparative advantages are sufficiently strong, then each vessel will confine its harvesting activities to its own "local' pool. With less decisive advantages, the vessels may coexist on both pools. When the relative advantages are expressed along

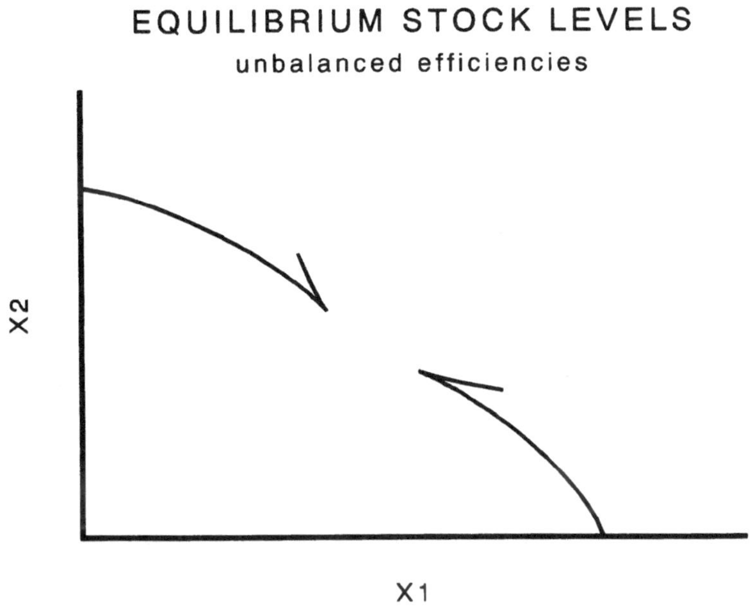

Figure 4b.

the gradient of a characteristic parameter, there will occur a threshold value, marking the boundary of the coexistence regime. This is very similar to the kind of threshold, for coexistence along an environmental gradient, that occurs in certain ecological models (Pielou 1975, Chapt. 6).

For example, consider the case where, for γ-vessels, $\sigma_{\gamma 1}(E_\gamma)$ crosses $\kappa_\gamma(E_\gamma)$ at a positive effort level, but $\sigma_{\gamma 2}(E_\gamma)$ lies entirely above $\kappa_\gamma(E_\gamma)$. This implies that γ-vessels cannot operate profitably exclusively on stock 2, though they can profitably partition effort between pools over a range of distributions $0 \leq \theta_2 \leq \theta_2^{\max} < 1$. In this circumstance, the corresponding stock-equilibrium curve χ_γ will intersect the x_1-axis (for $\theta_1 = 1$), but will bend away from and not intersect the x_2-axis.

Assuming parallel circumstances for γ'-vessels, though with the roles of the two stock-pools reversed, it transpires that the two equilibrium stock curves χ and χ' may not intersect - and this becomes more likely as the comparative advantage of each of the two fleets on its home pool becomes more extreme. Hence there is a threshold in comparative advantage beyond which the two vessel types do not coexist on either pool.

A complete analysis of this situation will be included in the study promised in section 4.

6. Summary and Conclusions.

The purpose of all of this modeling is to show how ideas borrowed from ecology can help in understanding the heterogeneous structure of a competitive fishing fleet. Obviously these concepts can apply just as well to instances of harvesting competition other than the fishery - indeed to any open-access natural resource industry.

The mechanism invoked here to drive competition is profit-maximizing behavior on the part of open-access fishing firms, with differing technological possibilities and costs. One difference is expressed in fixed costs versus variable costs, resulting in capital-intensive versus effort-intensive operations. This notion is not unfamiliar to ecologists, as a characterization of alternative "optimal life history strategies". A second technological distinction depends on relative efficiency and diminishing returns to effort in targeting alternative fish stocks. It is precisely a question of determining the "optimal foraging strategies" for the fishermen.

An important point is that what will be optimal for a particular vessel-type will depend on its ecological environment - broadly construed - including the economic motivations and technological characteristics of its competition, as well as the community ecology of the fish. The structure of the fishing fleet thereby is molded in recognizable ways, according to principles enunciated in the ecological literature: there is imposed a discrete structure on "niche space", leading to "competitive exclusion" and "niche compression" among the competitors.

Evolutionary change, a theme common to ecology and economics, has not been touched on here. However technological innovation is a powerful determinant of changing industrial structure, and is itself molded by the competitive forces within the industry. Both microevolution and macroevolution must be considered. By biological analogy, microevolution might be understood to refer to incremental design changes in vessel-types (such as enlarged hold-capacity or improved sonar for locating schools of fish). Macroevolution would mean the introduction of entirely new vessel-types or technologies. As in biology, so in economic theory, the sources of change - the raw materials of evolution - remain somewhat obscure; much clearer are the processes whereby competition exploits the new technologies to reshape an evolving industry.

Our overall theme has been that a classical ecological concept - the competitive exclusion principle - can clarify one's understanding of industrial structure in an open access natural resource industry. Many people before now have recognized useful parallels between economics and ecology, but usually it has been economic theory that has served to enlighten ecology. It is pleasing to reaffirm here that sometimes the intellectual debt can run the other way.

BIBLIOGRAPHY

[1] Armstrong, R.A. and McGehee, R., 1976. Coexistence of species for shared resources. *Theor. Popul. Biol.* **9**, pp. 317-328.

[2] Clark, C.W., 1985. *Bioeconomic Modeling and Fishery Management.* Wiley, New York.

[3] Diamond, J. and T.J. Cody. 1986. *Community Ecology.* Harper and Rowe, New York.

[4] Gordon, H.S., 1954. Economic theory of a common property resource: the fishery. *J. Polit. Econ.* **62**, pp. 124-142.

[5] Hardin, G., 1960. The competitive exclusion principle. *Science* **131**, pp. 1292-1298.

[6] Hutchinson, G.E., 1959. Homage to Santa Rosalia, or, why are there so many kinds of animals? *Amer. Natur.* **93**, pp. 145-159.

[7] Levin, S.A., 1970. Community equilibria and stability, and an extension of the competitive exclusion principle. *Amer. Natur.* **107**, pp. 413-423.

[8] MacArthur, R.H. and R. Levins, 1964. Competition, habitat selection, and character displacement in a patchy environment. *Proc. Natl. Acad. Sci. U.S.A.* **51**, pp. 1207-1210.

[9] MacArthur, R.H. and R. Levins, 1967. The limiting similarity, convergence, and divergence of coexisting species. *Amer. Natur.* **101**, pp. 377-385.

[10] McKelvey, R. 1983. The fishery in a fluctuating environment: coexistence of specialist and generalist vessels in a multipurpose fleet. *J. Environ. Econ. and Manag.* **10**, pp. 287-309.

[11] McKelvey, R. 1988. Specialist and generalists in the fishery: implications of an ecological - economic analogy. in *Mathematical Ecology,*

T.G. Hallam, L.J. Gross and S.A. Levin, eds., pp. 355-376. World Scientific, Singapore.

[12] McKelvey, R. 1989. Common property and the conservation of natural resources. in *Applied Mathematical Ecology*, S.A. Levin, T.G. Hallam, and L.J. Gross, eds. pp. 58-79. Springer - Verlag.

[13] Pielou, E.C., 1975. *Ecology Diversity*, Wiley - Interscience.

[14] Smith, C.L. and R. McKelvey, 1986. Specialist and generalist: Roles for coping with variability. *N. Amer. J. Fish. Manag.* **6**, pp. 88-89.

PARTICIPANT'S COMMENTS

As an evolutionary ecologist, whose trade is studying how natural selection via its effects on the behaviours of organisms promotes or inhibits biological diversity, I found this paper extremely interesting and exciting. McKelvey applies the same tools to understanding the diversity of competing human technologies. The parallels are striking. In natural predator–prey systems heritable variation is the strategy set, per–capita growth rates or fitness is the objective function, and natural selection provides the agent of optimization. In an open–access natural resource industry, available technologies represent the strategy set, net profit is the objective function, and rational behaviour is the agent of optimization. But the parallels go even farther. McKelvey asks how do "competitive systems" (those in which each individual is out for himself) differ from "cooperative systems" (those in which each individual is contributing towards a common goal) in promoting diversity. Biological systems also offer examples of diversity under competitive and cooperative systems. The diversity of ungulates on the Seringeti have independent objectives whereas the diversity of castes within an ant colony (or organelles within a cell) share a common objective.

McKelvey's first two models represent, in the parlance of ecology, habitat selection in time. Fixed and variable costs are important attributes of the harvester when the abundance of a resource varies temporally and the harvesters have the option of foregoing their variable costs by remaining "dormant". Variable costs determine the harvester's efficiency of patch exploitation and fixed costs determine how inexpensively a harvester can travel in time between periods of high resource abundances. The coexistence of harvester species or types is possible if there is a trade-off between fixed and variable costs among harvesters. Without knowledge (unfortunately) of McKelvey (1983) and his present work, I have recapitulated much of the essence of these works in a paper of my own (Brown 1989). Economics affords other examples as well. Within a utility company, the optimal mix of nuclear, oil, gas, and coal powerplants has been treated using fixed and variable costs (Anderson 1972, Rowse 1980). The utility must possess capacity to supply peak demand. Hence, some of the utility's capacity will go un-utilized some of the time. Powerplants with the lowest sum of fixed and variable costs (per kilowatt) should be used to meet base demand while those with the lowest fixed costs should be used to meet peak demand.

McKelvey's third model treats the coexistence of specialist and generalist fishing boats. The ability of the boats to allocate effort among fish species or patches of fish makes the problem similar to that of diet selection or patch use (the two classical questions of optimal foraging theory). As such, the combinations of coexisting fishing technologies have similarities to the diversity of species in models of habitat selection (Rosenzweig 1987). Also, the more selective behaviour of a generalist fishing boat in the face of a competitor closely follows the "compression hypothesis" (MacArthur and Wilson 1967).

In short, evolutionary ecology and microeconomics are kindred fields because each can have an optimization principle as its core. Under this core assumption, diversity is the product of tradeoffs and key axes of environmental heterogeneity. How useful is this approach to understanding the diversity of economic and ecological agents continues to be an ongoing and exciting question.

Anderson, D. 1972. Models for determining least-cost investments for electric power supply. *The Bell Journal of Economics and Management Science.* **3**, pp. 267–299.

Brown, J.S. 1989. Coexistence on a seasonal resource. *American Naturalist* **133**, pp. 168–182.

MacArthur, R. and E.O. Wilson. 1967. *The Theory of Island Biogeography.* Princeton University Press, Princeton, New Jersey.

McKelvey, R. 1983. The fishery in a fluctuating environment: coexistence of specialist and generalist vessels in a multipurpose fleet. *Journal of Environmental Economics and Management* **10**, pp. 287–309.

Rosenzweig, M.L. 1987. Habitat selection as a source of biological diversity. *Evolutionary Ecology* **1**, pp. 315–330.

Rowse, J. 1980. Intertemporal pricing and investment for electric power supply. *The Bell Journal of Economics and Management Science,* **11**, pp. 143–165.

Joel Brown

This paper examines the difficult fisheries problem of the behaviour of fishermen with different types of vessels and the resulting mix and distribution of vessel types under the different management regimes — (a) competitive and open access management, and (b) cooperative management. It uses optimal control type models to examine this problem under three different fishery situations: (1) two vessel types on a single species, (2) as per (1) but with uncertainty in the biotic and economic environment, and (3) three vessel types operating on two species.

Generally, fisheries models deal with the biological systems and fail to consider the behaviour of the fisherman. The problem of modelling a competitive fishing industry under different situations needs to be undertaken for the proper management of fisheries.

N. Caputi

MATHEMATICAL MODELING

Series editors:

William Lucas	Maynard Thompson
Department of Mathematics	Department of Mathematics
Claremont Graduate School	Indiana University
Claremont, CA 91711	Bloomington, IN 47405

Mathematical Modeling is a series of carefully selected books which present serious applications of mathematics for both the student and professional audience. The series aims to familiarize the user with new models and new methods and to demonstrate the art of constructing useful mathematical models of real-world phenomena.

We encourage preparation of manuscripts in LateX or AMS $\TeX$ for delivery in camera-ready copy, which leads to rapid publication, or in electronic form for interfacing with laser printers or typesetters.

Proposals should be sent directly to the editors or to: Birkhäuser Boston, 675 Massachusetts Avenue, Suite 601, Cambridge, MA 02139.

MMO1 *Probability in Social Science*, Samuel Goldberg

MMO2 *Popularizing Mathematical Methods in China: Some Personal Experiences*, Hua Loo-Keng and Wang Yuan

MMO3 *Mathematical Modeling in Ecology*, Clark Jeffries

MMO4 *Newton to Aristotle: Toward a Theory of Models for Living Systems*, John Casti and Anders Karlqvist, eds.

MMO5 *Introduction to Queueing Theory, 2nd edition*, B.V. Gnedenko and I.N. Kovalenko (translated from Russian)

MMO6 *Dynamics of Complex Interconnected Biological Systems*, Thomas L. Vincent, Alistair I. Mees, and Leslie S. Jennings, eds.